The Two Faces of Inca History

The Early Americas: History and Culture

VOLUME 3

The titles published in this series are listed at brill.com/eahc

The Two Faces of Inca History

Dualism in the Narratives and Cosmology of Ancient Cuzco

By

Isabel Yaya

BRILL

LEIDEN · BOSTON
2012

Cover illustration: Contador maior tezorero, Guaman Poma de Ayala (c. 1615): 360.

Library of Congress Cataloging-in-Publication Data

Yaya, Isabel.
The two faces of Inca history : dualism in the narratives and cosmology of ancient Cuzco / by Isabel Yaya.
p. cm. -- (The early Americas: history and culture ; vol. 3)
Includes bibliographical references and index.
ISBN 978-90-04-23385-0 (hardback : alk. paper) -- ISBN 978-90-04-23387-4 (e-book) 1. Incas--Peru--Cuzco--Historiography. 2. Incas--Peru--Cuzco--Kings and rulers. 3. Inca cosmology--Peru--Cuzco. 4. Inca calendar--Peru--Cuzco. 5. Cuzco (Peru)--History--Sources. I. Title.

F3429.1.C9Y38 2012
985'.37--dc23

2012027324

This publication has been typeset in the multilingual "Brill" typeface. With over 5,100 characters covering Latin, IPA, Greek, and Cyrillic, this typeface is especially suitable for use in the humanities. For more information, please see www.brill.nl/brill-typeface.

ISSN 1875-3264
ISBN 978 90 04 23385 0 (hardback)
ISBN 978 90 04 23387 4 (e-book)

This book is printed on acid-free paper.

To my parents

CONTENTS

LIST OF ILLUSTRATIONS

Tables

Figures

Maps

ACKNOWLEDGEMENTS

This book is the offshoot of an obsession with the Incas that lasted for several years. Over this time, a great number of people have encouraged and supported this work. Chief among those who made this enterprise possible is my Ph.D. supervisor, Prof. David Cahill, who demonstrated a tireless enthusiasm and intellectual curiosity for my research. I cannot thank him enough for the constructive criticism he has provided. I owe him a special debt of gratitude for trusting me with his personal notes of the Archivo Departamental del Cuzco, but also for letting me gradually empty the shelves of his personal library and transfer the precious loot to my own workspace. The Australian Postgraduate Award and the UNSW Faculty of Arts and Social Sciences scholarship (2004-2008) supported the writing of the thesis from which this book originates, as well as my research in Spain and Peru. The elaboration of this book took place across many countries, passing through the United States to eventually end in France. There, a postdoctoral research fellowship at the Musée du quai Branly (2009-2010) provided me the intellectual space and financial assistance to finalize the manuscript in the best possible frame of mind. In 2011, a Dumbarton Oaks summer fellowship gave me the enjoyable opportunity to further investigate the choice of royal wives and the political role of women in the perpetuation of Inca kingship. A portion of the results of this research is included in this book. I would like to express my special appreciation to Bridget Gazzo, the librarian of Pre-Columbian Studies, for her generous assistance. During my overseas research travels, I also received generous assistance from the archivists of the Real Biblioteca Española, the Archivo General de Indias, and the librarians of the Escuela de Estudios Hispano-Americanos in Seville. My research in the Archivo Departamental del Cuzco was greatly facilitated by the warm reception I received there from its chief archivists and the generous attention of Donato Amado Gonzales. I also greatly appreciate Meredith McGroarty's editing work on the manuscript.

Many people influenced my thinking during the writing of this book. I would like to extend my gratitude to Dr. César Itier, who taught me Quechua for more than two years and co-supervised my Diplôme d'Etudes Approfondies thesis. His feedback and encouragement have remained with me until today. In the last two years, the conversations I had with my colleagues in the Département de la Recherche of the Musée du quai Branly were extremely useful in reshaping my approach to literacy and kinship studies. I especially benefited from the brainstorming sessions that Laurent Gabail inflicted on me about the formalization of kinship. It is his master hand that hides behind the diagrams of this book. I am most grateful to members of my Ph.D. Committee, Prof. Susan Ramírez, Prof. Peter Gose, and Dr. Ian Farrington, for their encouragement and critical comments that impelled me to reconsider some of my earlier assumptions. Thanks to them, the resulting book addresses more soundly the main arguments presented in my doctoral dissertation. Prof. Tom R. Zuidema and an attentive book reviewer offered equally insightful critiques. Thank you also to those scholars who, having never met me, replied enthusiastically to my most incongruous emails about agriculture at high altitudes, astronomical observations with the naked eye, and kinship. The anonymous readers of *Ethnohistory* who reviewed the first draft of my first publication also provided suggestions that have significantly reshaped my thinking. My intellectual debt to these people is enormous. The errors and oversights that remain in this book are, of course, my responsibility alone.

These acknowledgements would not be complete without thanks to my longtime friends and to the wonderful people I met when working in different institutions. They should all know that, despite the nomadic life I lived, their reliable friendship through this time has made this enterprise considerably more enjoyable and provided a much-needed distraction when doubt crept over the blank page. And last but not least, I would like to close with a special thanks to my parents who raised my brothers and myself in the approximate language of their country of adoption, but also imparted their love of their respective motherlands. It is to them, and to all the irresistible confusions and neologisms that our upbringing created, that this book is dedicated because the flavours of our three countries are here reunited.

Note:
Chapter 3 includes material previously published in "The Importance of Initiatory Ordeals: Kinship and Politics in an Inca Narrative," *Ethnohistory* 55(1) (2008): 51-85.

INTRODUCTION

RETHINKING DUALITY IN THE ANDES: THE INCA CASE

At the time of the Spanish invasion, *Tawantinsuyu*, the empire of the four parts, embraced a vast mosaic of ethnic groups living in different ecological zones, from the desert coast, across the high plateaux, to the lowland Amazonian forest. These communities, which depended on various means of subsistence, had elaborated labour practices specific to their environment and forged discrete networks with neighbouring populations to compensate for labour and product deficiencies. Some of them were grouped in confederations that had gained control over large territories and competed with other important chiefdoms.[1] The material gathered from primary sources on ancient Peru as well as archaeological surveys suggest that Inca sovereignty began in a similar fashion, by developing matrimonial alliances and military pacts with neighbouring *etnías*. The small cuzqueñan confederation increasingly expanded its control over the land, supplanting the local and regional coalitions with an imperialist superstructure based on a centralized administration and unilateral tributary system (*mit'a*).[2] Fundamental to the maintenance of its authority was the ruling elite's exclusive access to and redistribution of rare and valuable goods originating from the various niches that formed the bountiful topography of the Inca territory. This strategy of supply regulation, adapted from pre-existing economic structures, ensured the allegiance of native lords who retained many of their former privileges, including polygyny, access to prestige goods, and exemption from subsistence production. It also constituted the grounds for labour obligations invested in building projects as well as agricultural and industrial products, particularly textiles and maize beer (*aqha,* also *chicha*). Within this ruling apparatus, not only goods but also people were subject to large-scale redistributions as a means of regulating alliance-making and ensuring political stability.[3] To achieve this design, the Incas established a series of institutions. The *akllas*, for example, were literally "chosen women" of both the royal descent group

---

[1] Broadly defined, chiefdoms are kin-based, stratified, societies in which leadership is based on genealogical considerations and ability to govern.

[2] The word *mit'a* evokes an event that occurs periodically; it also refers to the season.

[3] On the control of marriage alliances, see Alberti Manzanares 1985; 1986: 183-184; Silverblatt 1987: 87-91; Gose 2000: 89-92; Ramírez 2005: 28-30.

and provincial elites removed from their homeland and confined in state facilities. Depending on their abilities and rank, they served in religious activities as officiants and producers of ritual offerings, or were distributed as wives for important officials in name of the Inca king. The *yanakuna* were also another class of individuals originating from various ethnic backgrounds. Described imperfectly as retainers in Spanish chronicles, they were attached to the Inca nobility but exempted from tribute and reciprocal obligations to their native communities. Some of them were kinsmen (by blood or adoption) of the person they attended and may also have fulfilled prominent positions as local-level lords. The *mitmaqs*, likewise, were colonies permanently displaced from their place of origin and assigned lands in other parts of the empire. They served as loyal strongholds in newly conquered areas and as labour forces in agrarian, building, and maintenance projects.[4] All these institutions coexisted with the great masses of commoners liable for tax labour. These men and women, who made up the majority of the empire's population, were organized into a decimal administrative organization overseen by a hierarchy of Inca and indigenous officials in charge of enforcing the rules of the state.

This hegemonic machinery did not end with measures of socio-economic control. In the religious domain, the rulers would oblige their subjects to pay respects to their tutelary god, whom the Spanish identified as the Sun, and ordered the construction of temples dedicated to him throughout the conquered territory. This imperialist policy went hand in hand with a certain tolerance towards local ancestral beliefs, which continued to flourish and were sometimes incorporated into the ritual praxis of the ruling elite. Colonial records attest to the powerful influence some of these cults, particularly oracles such as Pachacamac or Apurimac, had on the course of action taken by the rulers of Tawantinsuyu. These leaders would regularly consult local deities on strategic issues and brought them offerings in return for their support and benevolence, but also required that their images be sent to the capital annually for the cleansing celebration of Citua (*sitwa*), when the governing elite paid them homage. These reciprocal relations were not always harmonious; in fact, they often implied elements of subordination that were expressly enacted in the course of the official festivals held in Cuzco. Furthermore, no deity, however influential, could escape retribution if the ruling power felt threatened. Atahualpa's destruction of the oracle of Catequil is a case in point. The

[4] Salomon 1986: 158-167, 171-172; Rowe 1982; Rostworowski 1999: 172-176.

Inca, infuriated by the prediction that his half-brother would defeat him and access the royal office, burnt down the shrine. Yet, despite these occasional antagonisms, the maintenance of provincial Andean cults, together with the development of elite networks supported by marriage exchanges, state-sponsored feasts, and the redistribution of prestige goods, were the strong poles of Inca sovereignty. These foundations were also its Achilles' heel, for they were contingent upon regional force dynamics, and therefore to the most profitable alliances. The multiple political turnarounds that ensued from the Spanish invasion brought to light the vulnerability of this system of interpersonal relationships.

The empire's provinces were not the only sites of tension in Inca times. Cuzco also was frequently the scene of contention between the various descent groups that shared political authority. These noble and non-royal ayllus, despite claiming a common ancestry, held asymmetric statuses and regularly competed for increased supremacy. Epitomizing their antagonistic relations was the dual organization system that divided them into two moieties of asymmetric values called Hanan (upper part) and Hurin (lower part), which corresponded to both a spatial and social partition of the city's residents. The kin groups' respective positions in this social structure determined in turn the political and administrative charges of its members, their ritual duties in the course of the year, their service to specific divinities, and further privileges. The Inca historical narratives, penned on paper in colonial times, evoke the origin of this classification system in discordant ways and often disagree on the incidence of this partition. Depending on the sources, it was either Inca Roca, Inca Yupanqui, or Inca Yupanqui's successor, Tupa Yupanqui, who commanded the descendants of early rulers to occupy the Lower part, while his own lineage and those of the lords who came to reign after him were given the Upper part. According to chroniclers Sarmiento de Gamboa (1572), Cabello de Balboa (1586), and Murúa (1590-1602), it was Inca Roca, traditionally regarded as the sixth king of the dynasty and the first of the Hanan moiety, who divided the city pacifically "in order to better rule and administer his kingdom."[5] In this version, Cuzco's partition appears almost as an unremarkable event, with no other effect than the spatial displacement of the future kings' descent. In contrast, Garcilaso de la Vega (1609) and Father Bernabé Cobo (1653) situate this event at the dawn of Inca history, after Manco Capac and his sister-wife emerged from the cave of Paqariq Tampu

[5] Murúa, *Historia general*: Bk. 1, Ch. 13, 59.

in search of fertile lands to settle. Arriving on Huanacauri hill, at the entrance of the Cuzco valley, they went their separate ways, he to the north and his spouse to the south, in order to convince the people they encountered to settle under their rulership. As the two congregated groups entered the site of the future capital, Manco Capac divided the space into two halves:

> The king wishes those he had brought to people Hanan Cuzco, therefore called the upper, and those the queen had brought to people Hurin Cuzco, which was therefore called the lower (…) He ordered that there should be only one difference and acknowledgment of superiority among them, that those of upper Cuzco be considered and respected as first-born and elder brothers, and those of lower Cuzco be as younger children. In short they were to be as the right arm and the left in any question of precedence of place and office, since those of the upper half had been brought by the male and those of the lower by the female.[6]

A similar description of asymmetric relations between Hanan and Hurin exists in Betanzos' early account of the city partition (1551). He situates this event within an elaborate narrative cycle relating how the ninth king of the dynasty, called Inca Yupanqui, acceded to power after vanquishing the armies of the Chanca confederation that had threatened to invade Cuzco. Soon after his investiture, the young ruler divided the city as part of necessary reforms of the politico-religious institutions, specifying that in rank and privileges, Hanan members should always prevail over the Hurin. The same Inca Yupanqui also figures as the architect of the dual organization system in Las Casas' account (1562-64), which describes the inherent asymmetry of the moieties' relations in terms of kinship grades. In this text, Inca Yupanqui bestows the most prestigious ranks of the dual organization system upon the descent groups of his closest relatives and assigns the lowest ones to his most remote kin. Here, it is no longer the historical anteriority of Hurin lords versus Hanan kings that defines moiety division, but rather the kin relations that connect individuals with the ruling Inca. In this way, the closer one stands genealogically to the ultimate source of political authority, the higher his rank and moiety membership.

The dual division system also was reflected in the spatial organization of the official festivals. Cristóbal de Molina (1575), in what constitutes the main source on Inca ceremonial life, observes that the ritual participants gathered on the main square "according to their ayllus and *parcialidades*

---

[6] Garcilaso: Bk. 1, Ch. 16; translation by Livermore 1989: 44. Cobo: Bk. 12, Ch. 3, 63.

(…) each one in conformity with the nobility of his/her lineage, divided as follows: the Hanan Cuzcos on one side and the Hurin Cuzcos on the other."[7] placing the mummified bodies of their ancestors in a like manner. Relational enmity, also, was a recurrent manifestation of dual partition in Inca ritual settings, principally at *purukaya* (second funeral) and on the closing days of the initiation festival.[8] For the latter occasion, the male novices assembled according to their moiety affiliation and engaged in a ritual battle against the opposite faction. It is said that this violent encounter, called "chocanaco" (*chuqanaku*),[9] led to frequent casualties and only ended by command of the king himself. Its particular aftermath, however, differed from the mock battle performed by the moieties at *purukaya*. Tradition attributes the creation of this last celebration to Inca Yupanqui, who established that the "people from Hurin Cuzco [would] act like losers and those from Hanan Cuzco like winners, thus representing the wars the lord had conducted in his lifetime."[10] This particular reference to actual conflicts between moieties is not isolated. Several chroniclers evoke the recurrent hostilities that placed Hanan and Hurin lords in opposition to one another over the wielding of power. The ethno-historian Concepción Bravo Guerreira (1992), basing her analysis on a diachronic reading of these texts, suggests that the Hanan lords' rise to power took place at the expense of the Hurin leaders and gave rise to repeated rebellions from the oppressed faction. In support of her claim, Cieza de León (1554) describes an insurrection of the Hurin Cuzcos under Viracocha Inca, which led to the assassination of the governor left in the city by the king at war. In response to this uprising, Viracocha Inca's successors were compelled to gradually restrict the insurgents' power, before Huayna Capac dramatically infringed upon the Hurin leader's authority. The final hostilities would have erupted not long before the Spaniards landed on the Peruvian coast, when Huascar learnt that his half-brother Atahualpa was determined to claim the royal fringe for himself. Huascar counter-attacked by changing moiety affiliation. He demanded to be recognized hereafter as a member of Hurin

[7] Molina: 78, 106, 108, 111. See also Cobo: T. 2, Bk. 13, Ch. 25, 211, 218. Rostworowski (1981) remarks that, in the sixteenth century, colonial sources used *parciliadad* as a polyvalent term for 'sector'.

[8] *Purukaya* was the second funeral of high-ranking individuals, mainly kings but also some of their prominent relatives including noblewomen. Betanzos (Pt. 1, Ch. 44, 189-190) mentions the organization of this ceremony for Mama Ocllo, the mother of Huayna Capac.

[9] "Choccanacuni: to throw stones or other things at one another" (González Holguín, *Vocabulario*: 116).

[10] Betanzos: Pt. 1, Ch. 31, 147.

Cuzco "because he intended to kill Atahualpa and all his kinsmen, and lineage that was of Hanan Cuzco and form a new lineage of Hurin Cuzco."[11]

Irrespective of the actual veracity of these events, about which scholars continue to debate, dual organization appears to have been a recurrent feature of Inca historical narratives. Accordingly, it not only determined aspects of social organization such as ranks, access to offices, and ceremonial responsibilities, it also informed various—and sometimes conflicting—traditions about its origins and the interlineage conflicts that plagued royal succession. Considering this ubiquity, a study of the different social expressions of moiety divisions and how they interconnected appears fundamental to understanding the principles of group affiliation in ancient Cuzco, and to shed light on the Incas' perception of the historical development of their kingship institution. In this book, I propose to revisit primary sources on Inca society through the conceptual framework of dual organization. In this endeavour, I follow in the footsteps of anthropologist Tom R. Zuidema, who instigated investigation into the articulation of dualistic and hierarchical principles in the religious and social representations of the cuzqueñan elite. Like him, I approach Inca beliefs and ritual action as comprising a social practice fostered by, and acting upon, a cultural framework based on dualism. Yet, in bringing a diachronic perspective to the fore, I also distance myself from a strictly structuralist approach. I argue that understanding the dialectical tensions whereby the Incas described the temporal process is instrumental to grasping the dynamics of their social and religious dualisms. Following this line of analysis, I approach dualism as a system within which the temporal process operates as a constitutive force, not as a static model of social classification. Such a framework infers the existence of a prescriptive code, but reckons this system to be flexible and innovative. This perspective is particularly illuminating because it recasts the substance of Inca narratives into the mould of history and approaches oral traditions as agents of social memory rather than mere instruments of social reproduction. What is more, I argue that Inca narratives consciously contrived this dialectical tension between structure and practice. Such an approach, as we shall see, harmonizes many discrepancies that affect colonial records of Inca dynastic accounts.

---

[11] Betanzos: Pt. 2, Ch. 2, 210.

Dual Organization in the Andes: A Theoretical Framework

Notwithstanding that dual organization is a pervasive principle of social classification in South America, it has also been described as "one of the main problems of South American sociology."[12] Even today, its definition remains contentious and covers a large typology of social expressions, including such principles as moiety formation, diarchy, and two-line relationship terminologies, all of which may or may not coexist within the same society.[13] Its most elemental definition, however, can be found in Claude Levi-Strauss' earliest study on dual organizations (1969), in which he describes his object as follows:

> A system in which the members of the community, whether it be a tribe or a village, are divided into two parts which maintain complex relationships varying from open hostility to very close intimacy, and with which various forms of rivalry and co-operation are usually associated.[14]

Furthermore, there is ample evidence from Central Brazilian and Andean cases that dual organizations almost invariably operate alongside asymmetric social dynamics. The same would have been true of pre-Hispanic and colonial societies. In 1567, Spanish chronicler Juan de Matienzo, in service in La Plata (Charcas), described the social structure in place in that region:

> The kuraka of hanansaya (upper moiety) was the principal lord of the whole province and was obeyed in everything that he said by the other kuraka of hurinsaya (lower moiety). The kuraka of hanansaya has the best place of residence and in everything else, which in this matter conforms to their order of precedence. Those of the moiety of hanansaya seat themselves at the right hand and those of hurinsaya at the left hand, each according to their order.[15]

Likewise, Jerry D. Moore's analysis (1995) of site-survey data of the Nepeña valley (Ancash) shows that the dual social organization of these communities during the Late Intermediate Period (c. AD 1000-1400) overlay an asymmetric political power in which authority and the distribution of labour were not equally shared between the two divisions. This feature, he

---

12 Levi-Strauss 1944: 39.

13 "Two-line terminologies" is a typology of kinship relationships where an individual is almost invariably definable according to two terminologies found in alternate generations (e.g. MB/FZH, or MBD/FZD).

14 Levi-Strauss 1969: 69.

15 Matienzo: Pt. 1, Ch. 6, 20.

observes, is particularly apparent in the hierarchical settlement pattern of the excavated sites.[16] Ethnographic evidence indicates as well that asymmetry is, in fact, inherent to the ideology of most non-centralized dual societies.[17] Dual organizations in lowland South America share common dynamics of leadership stratification, moiety hierarchy, class system, age grade, and ethnic distinction, even though most did not devise any form of urbanized or centralized power.[18]

Despite the identification of asymmetric features in dual social systems, scholarship on the subject has often maintained that dual organization was a conceptual model allowing societies to regulate or dialectically resolve antagonisms of a social and metaphysical nature. It would simultaneously enclose the most basic expression of inequality—the superiority of one social group over the other—and devise various ways of regulating this asymmetry. Building on the work of Robert Hertz and Marcel Mauss, Levi-Strauss (1969) sees reciprocity (i.e., the obligation to give and receive) as a universal form of human behaviour, and holds dual organization to be the most archaic expression of this functional principle. Within this framework, marriage is seen as the most basic form of exchange whereby women circulate between groups of individuals that are necessarily exogamous, regardless of their asymmetric relations with one another. Moiety division would represent the most elementary form of marriage regulation, and in its absence:

> There may be a second bipartition of the group, parallel or perpendicular to this earlier division, the moieties may embrace exogamous clans, subclans or lineages, or lastly, the modalities of marriage may depend upon specialized forms called marriage classes.[19]

Levi-Strauss insists that dual oppositions only exist in reference to a third, external element that encircles the two asymmetric poles, thereby transforming them into an entity that contrasts with the surrounding environment. Dualism thus invariably coexists with a triadic structure that either reorders clans into matrimonial classes or devises relationships of exchange in such a way that the fundamental principle of reciprocity can operate.[20] The structuralist outlook permanently influenced the study of

[16] For a different interpretation, see Netherly 1990.

[17] Nimuendajú & Lowie 1937; Lave 1977; Da Matta 1982; Crocker 1990.

[18] Hornborg 1988: 276-284.

[19] Levi-Strauss 1969: 69.

[20] An illustration of Levi-Strauss' influence in Andean studies is Olivia Harris' article (1986) on dual symbolism among the Laymi (Potosí). In this important work, she identifies

Inca society through the work of Zuidema, whose earlier contributions drew extensively from the ethnography of dual organizations in Central Brazil. Following this line of investigation, he suggests that Inca society operated as an asymmetric diarchy (1964), a proposition later adopted by Nathan Wachtel (1966) and Franklin G. Y. Pease (1991). In this system, the royal lineages, or *panacas,* were equally divided between each moiety and observed a strict hierarchical organization that determined access to water and land rights, as well as specific political and ritual responsibilities. Zuidema contends in particular that the panacas were divided into three subcategories, known as *qullana*, *payan*, and *qayaw*, operating as exogamous matrilineal marriage classes and the central regulatory structure for the Inca social order.[21] In this way, the Inca dual organization regulated the asymmetric relations of its subgroups through marriage. However compelling this argument may appear, it faces a significant methodological issue: no evidence in historical documents supports the postulate that the Inca moieties functioned as exogamous clans.

If kinship relations and reciprocity characterize the structuralist approach to dual organizations, early scholars of Gê-speaking peoples disputed that marriage practices are the regulating principle of dualism. Anthony Seeger (1981) and Judith R. Shapiro (1984) argue that not all moiety systems are ruled by exogamy, which do not inevitably condition matrimonial preferences.[22] At the heart of these critiques lay more fundamental objections, first enunciated by Edmund Leach (1961), Rodney Needham (1971), and David Schneider (1984), who have questioned the

various forms of tripartition, ritual but also economic, which, she argues, were generated to resolve the "inevitable asymmetry" of human relationships.

[21] Zuidema (1964) also advances that Inca society was hierarchically divided into five age groups associated with different politico-economic functions similar to the social organization of the Canela Indians (Brazil).

[22] In fact, like Levi-Strauss, many scholars of the Amazon basin argue that "dual organization need not involve two strictly exogamous moieties in order to qualify as a codification of symmetric alliance" (Hornborg 1988: p. 40). Karsten Paerregaard (1992) offers an eloquent illustration of this last point in his ethnological survey of the Tapay district (Arequipa). Interviewing its inhabitants about the ritual battle that takes place in early February every year, Paerregaard learnt that the purpose of this encounter was to conquer women from the opposing moiety. A similar form of ritual exogamy transpired during the festivity of the town's patron saint, when two central figures from each division concluded the celebration by calling each other daughter/son-in-law and father/mother-in-law. Yet Paerregaard observes that these expressions of ritual kinship were different from the actual marriage pattern of the community ruled by an endogamy of moiety. He concludes that the symbolic expression of reciprocity in ceremonies compensated the absence of direct social complementarity in Tapay.

relevance of isolating kinship as the central pillar of anthropological inquiry. Also reacting against the structuralist edifice, David Maybury-Lewis and his followers have approached kinship as but one social category among others and reckoned that dual organization is, "above all else, a system for discerning an order in the scheme of things and imposing it on the unruly complexity of events."[23] This view presupposes that dual systems operate as a counterpoised principle that overrides asymmetric relations and discords within society in order to create equilibrium and to harmonize ontological antagonisms. In this perspective, the study of dual organization requires an understanding of every system of cultural classification generated by a given society, for such constructions reflect a cosmology based on antagonistic forces. Within this order, ritual action is believed to be a predominant factor of dialectical regulation, through which antithetical ideas, categories, and institutions achieve stability. Needham (1973, 1987) further developed this view and, following the line of Mauss' *fait social total*, he described dual symbolic classifications as pervasive principles of human symbolic thought. In Needham's own words, social organization and symbolic forms, including the structuralist notions of reciprocity and exchange, are only aspects and manifestations of "one conceptual order."[24] Thus, symbolic oppositions in dual societies not only structure patterns of alliance and their kinship terminology, but also shape any social interactions, the order of the cosmos, and the spatial conception of the given group. These theories, together with John Murra's seminal essay (1972) on ecological and economic verticality, greatly influenced the study of dualism in Spanish South America. Although Levi-Strauss never described reciprocity as a moral value, his legacy and that of Maybury-Lewis in particular were influential in constructing the structural/cultural basis of Andean societies around the notion of exchange as an ideology of complementarity. Central to this idea were the concepts of fertility and sexual reciprocity, in which dual divisions evoke the union of male and female forces pairing to create social and natural regeneration.[25] Several ethnographical works follow this line and have addressed the significance of binary patterns in Andean social and symbolic expressions, such as dance, music, textile, language syntax, arithmetic, and, most recently,

[23] Maybury-Lewis & Almagor 1989: viii.

[24] Needham 1960: 108.

[25] Isbell 1978; Harris 1978; Platt 1986; Bouysse-Cassagne 1986; Sallnow 1987: 143-146; Silverblatt 1987: 20-21; Pease 1991: 41-42; Paerregaard 1992: 21; Classen 1993: 3, 12-13, 22-23; Bolin 1998: 120-123.

khipu coding.[26] Together, these works show the pervasive role of dualistic representations in Andean social phenomena.

Yet, the application of Maybury-Lewis' theory to dual dynamics demands a critical examination. Without dwelling upon his early disinterest for matrimonial practices, his argument that ritual performance is a means of channelling antagonisms in society raises an intricate problem related to the definition of ritual form. His argument follows Edward Tylor (1871) and James Frazer (1922) that ritual action is, in fact, the translation of a worldview, a cosmology that strives towards a harmonious relationship between its different elements. Accordingly, the enactment of dual antagonisms in ritual performance, such as in ritual battles, is primarily cathartic: it is designed to resolve conflicts. Such an approach, I believe, restricts the meaning of ceremonial behaviour to only one facet of its external functions. First, it disregards the distinctive qualities of ceremonial events in contrast to other modes of interaction and performance, which can be similarly seen as factors of social cohesion.[27] Secondly, although many moieties' encounters undeniably aim to perpetuate the fertility of the living world—and the equilibrium that such fertility implies—this rationale cannot explain why certain ritual battles invariably enact the asymmetric nature of the society and close with the oft-repeated victory of one particular faction. As mentioned previously, the Inca ritual battle performed at *purukaya* always ended with the defeat of the Hurin part by the Hanan one. In this specific case, it would be erroneous to presuppose that deadly fights and harmless competitions between social halves ultimately achieve the same harmonious purpose, or else why should their outcome vary? Understanding the distinctiveness of these encounters requires analyses of their ritual context, the time of year they are enacted, and the oral traditions to which they refer.[28]

Despite extensive data pointing to the various expressions of asymmetric dualism, the idea that dual oppositions are primarily enacted to

---

26 Cereceda 1986; Bradby 1987; Carpenter 1992; Seibold 1992; Silverman 1994: 71-93; Baumann 1996; Urton 1997; 2003.

27 For example, sporting events or theatrical plays. On this topic see Turner 1982; Schechner & Appel 1990; Grimes 2006.

28 These observations form part of a more general critique of the assumed role of ritual action. As Michael Houseman and Carlo Severi elaborate: "[The functionalist outlook] suffers from a circular line of reasoning which consists in identifying the cause with the purpose of the object studied. Ritual action is thus seen as serving to achieve optimum integration, either social or psychological, even when it involves deviant behaviour. Indeed, there is no reason to expect such an integration; to do so presupposes precisely that which is to be shown" (1998: 166).

generate equilibrium became very influential within a certain body of Andean scholarship. Irene Silverblatt (1987), for example, argues that, before Inca rule, "Andean people interpreted the workings of nature through an ideology of gender complementarity" that infused social life.[29] Descent was parallel and dual political authority followed the lines of gender complementarity. It was the rise of Tawantinsuyu that alternatively fed an imperialist ideology, which Silverblatt names "conquest hierarchy", holding women to be inferior. In this way, the Incas "transformed structures of gender parallelism, which shaped human and divine relations in the [traditional] ayllu, into institutions of imperial politics."[30] Other scholars have likewise emphasized the determining role of exploitation in the construction of asymmetric dual divisions. Along this path, Paul Gelles (1995) argues that dual social systems in Inca time were closely related to the hegemonic nature of Tawantinsuyu. While he gathers evidence that dualism in pre-Hispanic societies was a *cultural*, linguistic, and social system ubiquitous before Inca rule, he regards dual *political* divisions as the result of the Cuzco leaders' imperialism.[31] The governing elite, he argues, diffused a model of dual leadership inspired by the hierarchical organization of Cuzco, imposing their own tributary institutions on the nations they subjugated. In support of his interpretation, several chroniclers, including Hernando de Santillán (1563), the aforecited Matienzo, Polo Ondegardo (1560-70), and Cabello, claim that the Incas ordered the systematic division of their provinces into two sectors called *hanansaya* and *hurinsaya* to improve administrative control and tribute collection.[32] For Gelles, this policy introduced an institutionalized and extractive system distinct from other manifestations of conceptual dualism otherwise expressed in gender, kin, or environmental categories. By the time of the Spanish conquest, dual organization in the Andes had become the "expression of both an indigenous cosmology and state hegemony" that the colonial system perpetuated in the following centuries. Gelles' work, by examining the influence of supra-local circumstances on the reproduction of cultural models, resonates with Heraclio Bonilla's and César Fonseca Martel's (1967) early Marxist argument that sought to explain the origin of dual organization in Jesús de Machaca (Bolivia). From their perspective, no political or

---

[29] Silverblatt 1987: xxviii.

[30] Silverblatt 1987: 47. In square brackets is my addition, which aims to emphasize Silverblatt's conception of gender relations before the Inca conquest.

[31] Emphases are mine.

[32] Santillán: 105; Polo, *El Mundo*: 122; Cabello: Ch. 3, 294.

economic hierarchy had operated in this community before the Inca conquest. Instead, relations of power were strictly limited to the ritual sphere, which served as a social catalyst.

These different approaches clearly distinguish between "dual political division" on the one hand, and the "cultural logic of dualism" and its symbolic expressions on the other. They assume that the rationales underpinning these two principles belonged to separate and unrelated domains. It is undeniable that many societies hold dyadic symbolic conceptions in their art, cosmology, or environmental views without necessarily being organized socially according to binary criteria. However, it would be misleading to radically oppose the finality of these two dual processes, and to presuppose that sociological dualism is an asymmetric system subject to historical transformations, whereas symbolic dualism—and its ceremonial manifestations in particular—invariably feeds on an equilibrium ideology. This assumption confines ritual action as a particular product of the latter category. In this view, the traditional expressions of dual opposition would be but reflections of an "indigenous logic of dualism and production, whereby the division of activities into halves 'acts as the cultural medium through which the fertility of nature is conveyed into the human realm.'"[33] This perspective sees open hostilities and competitiveness in ceremonial contexts as "games", that is, a way to produce complementary life-sustaining resources out of existing oppositions, as in the case of the fertilizing blood impregnating the earth during ritual battles. Perceived through this prism, ritual action sublimates social conflicts by either minimizing or abrogating them.[34] As already stressed, this approach to ritual action is not sustainable because it largely assumes what needs to be demonstrated: that ritual is an instrument of social stability. In addition, separating the rationales sustaining dual political division and the cultural logic of dualism disregards ethnographic data showing that hierarchical and asymmetrical principles, such as concepts of leadership hierarchy and social stratification, are extrinsic to lowland South American dual organizations that have not developed a centralized power. Ultimately, a reexamination of dualist theories should not overlook the influence of historical factors in the development of asymmetric patterns, as Gelles rightly advocates. Rather, it should query the assumption that "gender parallelism" operated in the Andes as a symmetrical model before the

---

[33] Gelles 1995: 722; Sallnow 1987: 145.

[34] This view is consistent with Victor Turner's works (1969), which see ritual performances as cultural symbolic expressions aiming at harmonizing social realities.

expansion of Cuzco authority, and whether the basic form of social dualism always implements an equilibrium ideology.[35] As a few works have started to show, there is little evidence supporting the view that descent was parallel in the pre-Hispanic ayllu, or that a complementary of male and female authority operated at the *cacicagzo* level before the Spanish conquest.[36]

Tristan Platt's study of the Macha dual organization (1986) may provide an important clue in approaching the problem of asymmetry in dual organizations, and thus refine our understanding of the Inca case. In this decisive contribution, Platt notes that the Macha community, located in the northern Potosí region in Bolivia, comprehends its environment through several forms of symbolic dualism—ritual, social, and cosmological—that are epitomized by the enduring concept of *yanantin*. This term encompasses a variety of meanings. It refers not only to the relation "man-and-woman" but also describes actions or objects in perfect symmetry and equality, implying that gender complementarity is a fundamental principle of the Macha social order. Notwithstanding this meaning, *yanantin* also derives from *yana*, "servant", and thus expresses notions of asymmetry and inequality. In the light of this information, Platt concludes that "even in the case of the perfect model produced by the binary structure of the human body, a submerged relation of hierarchy is present", so that structural symmetry and hierarchy are interrelated principles of dual oppositions.[37] I ground my theoretical approach to dual oppositions on a similar premise. In doing so, I follow a post-structuralist perspective informed by the works of Louis Dumont (1986) and Terence Turner (1984), for whom binary oppositions, whether social or symbolic, are necessarily hierarchical and do not always tend to a symmetrical relation. As Dumont notes, dual oppositions are invariably different in value because they do not have the same relation to the totality they form. In the case of the right and the left hands, for example, the two bear asymmetrical values in relation to the body, but they are also "mirror images of each other", and thus "structural equivalents."[38] Moreover, the superior level of a contrasting pair always encompasses its antagonistic part when they are conceived as a whole (*totalité*). The superior level thus "distinguishes itself by its capacity to

[35] For a critic of Silverblatt's depiction of the Inca empire operating as a patriarchal institution, see Gose 2000.

[36] Julien 2000: 25-26; Graubart 2007: 161-167.

[37] Platt 1986: 256.

[38] Turner 1984: 337.

define the global order of the society."[39] In the case of dual organizations, the values and characteristics of the upper part come to define the identity of the whole society in contrast to a third entity or relation.[40] Social oppositions are thus bidimensional in essence, their symmetrical and asymmetrical aspects being *both* essential in relation to the social totality they constitute. Viewed in this light, identity and contrast in dual organizations constitute interrelated principles that are innate to one another. This approach has the merit of integrating a possible diachronic framework to the study of dual organization systems. Indeed, assuming that elementary structures of social hierarchy are intrinsic to dual social divisions would explain how certain dual organizations such as that of the Incas evolved from a basic moiety model to a fundamentally unequal social system.[41] Following this line of thought, I approach asymmetry as an ideological principle inherent to Inca dual organization, and not exclusively as a hegemonic model that the Cuzco nobility devised at a politico-economic level simultaneously with their expansion, nor as a contingent principle elaborated *a posteriori* to regulate asymmetrical relations within the descent group. At the same time, I question both moieties' relation to the social unity they form, especially as they are depicted in historical narratives and ceremonial settings. My line of reasoning is not to deny the impact of historical process or the influence of supra-local circumstances on the development of hierarchical institutions in dual organization. Rather, I argue that the inherent nature of binary oppositions predisposes such changes.

The issues relative to dualism in Inca society are vast and cannot be treated exhaustively in this book. The following pages therefore focus on binary structures across three case studies: the representations of moiety

---

[39] Barraud et al. 1984, in Parkin 1992: 55.

[40] To illustrate his theory, Dumont (1980) draws an example from the Indian caste system in which the Brahmins and the Kshatriyas, he argues, are two groups versed in their own specializations and therefore dependent on the other's domain of competence outside their own field. This apparent complementarity however fades as the spiritual sphere represented by the Brahmins is in fact superior to the Kshatriyas who originated from them. Thus, at the superior level, the Brahmins encompass the Kshatriyas and represent the global order of the society.

[41] Dumont's followers, among whom his own student Serge Tcherkézoff, criticized the structuralist and culturalist views in a number of publications and singled out Needham's work for representing the most delusive aspect of these theories. Needham responded to these vehement critiques. Both sides incessantly claimed that their theoretical opponents had misinterpreted their own argument. For an analysis of these events, see Parkin 1992; 1996: 79-86; 2003.

opposition in Inca historical narratives and ritual actions, the ruling elite's conception of the cosmological order, and its ceremonial calendar. In exploring these themes, this book draws on a diversity of published materials elaborated in colonial times by early conquistadors, appointed and self-proclaimed chroniclers, vice-regal administrators, and missionaries. It also heeds the voices of those men and women who were simple witnesses to events unfolding before their eyes and whose words were inscribed in lawsuit reports.[42] By putting this material into context and unravelling the motivations and biases of their various authors and contributors, I intend to reconstruct some basic traits of the Inca religion and to locate them more securely in the wider debate on dual organizations. Such an undertaking is confronted with different challenges, which this study meets via a multidisciplinary approach that draws from the fields of anthropology, history, and, to a lesser degree, philology. In sum, the contribution of this book lies not in the disclosure of new data, but rather proceeds from a systematic re-examination of available sources in order to identify the inconsistencies that bedevil depictions of Inca society. In reopening this research agenda, I would like to emphasize a particular approach to ritual action. Following a relational approach, I choose to examine ritual action as a specific mode of social production that does not merely reflect a given religious credo, ancestral cosmology, or frozen social order, but rather shapes and reconfigures the attendants' relationships to one another. Through this perspective, cultural constructs, whether narrative or ritual, are not fixed, in essence, nor do they depict a static reality. My aim is not only to understand how the Inca nobility devised ritual means of interacting with its environment, but also to explore how their festivals intended to transform the celebrants' social order. In this task, the understanding of kin terminology, rules of descent, and marriage strategies is instrumental in defining the relationships at play in historical and ritual representations. I will focus particularly on the practice of royal succession, which did not follow static rules of perpetuation, but instead constituted a dynamic system that enabled the reinvention of kin and social relations at the nomination of each ruler. This is because royal inheritance depended on political

---

42 The current state of knowledge on *khipus* precludes us from including in this list any data registered before the Spanish invasion. As we shall see in chapter 1, the future of this field appears nevertheless promising thanks to the endeavour of dedicated scholars who, in recent years, have produced several important works. See Murra [1973] 2002; Ascher & Ascher 1981; Mackey et al. 1990; Arellano 1999; Fossa 2000; Urton 1998, 2003; Quilter & Urton 2002; Salomon 2004; Charles 2007.

alliances contracted as much within the Inca descent group as outside of it, with local leaders.

Like Spanish historiographers four centuries ago, today's ethnographers of the Inca past are faced with the unavoidable requisites of naming and describing native categories in the words of their own language of communication. Even though such renderings are inevitably doomed to reflect only imperfectly the cultural objects under scrutiny, I consider that ethnography (past and present) records cultural representations, not the objective realities in social behaviour they empirically stand for. If the anthropologist's practice of translating culture still calls for a new justification, the answer may partly reside in decentering the encounter situation, wherein the values and perspectives of every participant engaged in cultural dialogue/social interaction are similarly shaped by and acting upon those of their interlocutors.[43] Through this perspective, any attempts at emic interpretation that take into account individual and interindividual mechanisms remain legitimate.[44] Arguably, Andean ethno-historians are confronted with more intricate issues than those of modern anthropologists because they manipulate data that agents or subjects of the colonial power had collected for a variety of motives that are not easily discernible. In order to refocus Inca historiography on the discourses of the people it describes, I aim to partly restore the historical quality of pre-Hispanic narratives. In this endeavour, an understanding of the ethnocentric biases of post-conquest informants, although instrumental in unlocking the narrative strategies of colonial actors, only partly harmonizes the many discrepant accounts of Inca society. Hence, a major assumption of this book is that Inca historical consciousness, however much forced into foreign categories, permeated the chronicles and can be recovered in part. To unveil this original material, I follow the path of recent scholarship that adopts a comparative literary analysis to identify in primary sources the traits proper to distinctive traditions/genres of pre-Hispanic historical narration. Studies by Catherine Julien (2000), Galen Brokaw (2003), and José Mazzotti (2008), although resulting from different avenues of thought, employ this approach to bring to the fore a series of discursive mechanisms and linguistic categories of pre-Hispanic origin that had infused the chronicles. This mode of investigation presupposes that Inca historiography was indeed formulated through colonial discursive genres but was also shaped, to varying degrees, by some of the features that characterized earlier indigenous

---

43 Ortiz 1940.
44 Sperber 1985, 1996.

narrative conventions. This book is not a textual analysis comparable to Julien's work, but it elaborates on her methodology and offers an answer to a question she left unsolved: what aims did Inca historical discourses pursue?

Précis of the Chapters

The underlying theme sustained throughout this book is the reassessment of Inca historical traditions through the lens of dual oppositions. In a spirit of coherence with its object of study, this inquiry unfolds in two parts. The first examines historical productions as instruments of legitimacy in the hands of the ruling elite and as reflexive discourses on the success of their sovereignty. The second investigates social time constructed around the festivities of the state religion. Chapter 1 introduces the reader to the core issues affecting the primary sources that inform the analysis. It discusses the different media of historical transmission used in Inca times and investigates the contexts in which recitation of these narratives occurred. It then turns to question how the colonial situation affected the transcription of these traditions into a format conforming to European literary conventions. Instead of approaching these texts as the result of the encounter between two immutable traditions, I see them as accommodating sites where discourse was fed as much from Western paradigms as native ones, allowing the emergence of new cultural norms. This chapter thus offers an alternative exploration into primary sources that harmonizes textual inconsistencies by crediting the Incas with a multi-layered historical consciousness. It argues that the pre-Hispanic ruling elite devised a plurality of historical representations comparable to those of European and other non-Western societies. It is therefore deceiving to extract from the chronicles a single storyline that appears more coherent or closer to a realistic portrayal of past events, and to consequently attribute the discrepant information to Spanish misinterpretation or political agendas. A constructive approach to primary sources requires that we discern the purposes that Inca historical traditions served in order to better perceive their discursive traits.

Chapter 2 analyses the ways in which antagonistic relations were depicted in Inca historical traditions on the one hand, and how they were enacted and recast in ritual action on the other. It starts by investigating dualistic and hierarchical principles within the structure of the ruling elite's kinship relations, and posits that the formal model of the conical clan not only clarifies the underlying structure of Inca descent, but also enlightens the triggering mechanisms of Inca succession wars and moiety conflicts. The chapter identifies two conflicting historical discourses that served to legitimize the royal pretenders' claims to the kingship office. These contrasting narrative genres were intrinsically linked to the Inca descent group's dual division and shed light on the narrative and ritual depictions of moiety conflicts.

Chapter 3 extends this analysis to the relationships between the Inca ruling elite and provincial lords, with a specific focus on the metropolitan festivals that involved foreign celebrants. It examines these interactions as an instance of dual opposition that took up and extended the characteristics of Cuzco moiety division. In doing so, it reviews the ceremonial events, notably *sitwa* and *qhapaq hucha*, in which asymmetric relations were clearly enacted. The chapter, however, makes the case for interpreting the initiation ritual, also described as the most elitist ceremony of the Inca calendar, as an explicit celebration of the ruling elite's alliance with foreign ayllus as a means to insure the viability of Inca sovereignty. This festival ritually reaffirmed marriage networks that established channels of assimilation allowing native lords to strategically position their kin inside the Inca royal ayllu.

Chapters 4 and 5 focus on unravelling the narrative threads interwoven in the foundation narratives of Inca history by identifying in this literature the cultural productions specific to each dual division. I suggest that historical traditions can be divided into two groups, each of which encloses a particular, and mutually conflicting, representation of the mythical past that reflects the moieties' respective perspective. Each corpus depicts a distinct cosmology headed by different divinities that were complementary in overseeing water regulation throughout the annual cycle. Although it would have been tempting to speculate about the miscellaneous entities and concepts that peopled the Inca cosmological world, I restrict this analysis to the study of the major ancestor-gods worshipped by the ruling elite. The mythical acts of Wiraqucha and P'unchaw in particular are closely interwoven with the foundational narratives of the royal dynasty. This connection allows for a rigorous inquiry into the mechanisms sustaining the divine legitimacy of Inca kingship and casts light on the different facets of the state solar religion. By examining the content of primary sources through a cultural framework informed by dualistic principles, these two chapters attempt to account for, and ultimately harmonize, the inconsistencies blurring Spanish depictions of Inca religion. They investigate the native paradigms that facilitated Wiraqucha's merging with the almighty God of the Christian tradition whose deeds on earth were identified with the evangelist journey of an apostle. They set out to unravel the distinct rationales sustaining the religious and social reforms attributed to a young ruler, ordinarily identified as Pachacuti Yupanqui, who devised the subordination of the Sun deity under Wiraqucha. I argue that these institutional changes were simultaneous with a change of divine authority that saw the destitution of the old Sun in favour of Wiraqucha.

Finally, chapter 6 further develops the theme of dual division in Inca religion with an inquiry into the ceremonial calendar observed by the ruling elite. It examines the month-by-month interrelations between ritual praxis and the Inca cosmological order. In doing so, it delineates the characteristics of each seasonal division in the annual cycle, and focuses on situating these celebrations within the broader context of social practices such as agrarian labour, herding, tribute payment, and migratory flows of labour force from different ecological niches. Moreover, works on archaeoastronomy are employed to draw a coherent picture of the time of year during which these festivities took place. They offer a timeframe dividing the annual cycle into two almost equivalent periods, each headed by a different aspect of the Sun god.

A Note on Nomenclature and Spelling

In the course of this argument, a choice has been made—albeit approximate—to describe, spell, and translate a number of Inca institutions and traditions. Concerning the spelling of native words, I conform to the official orthography of contemporary southern Quechua. Although this standardization presents the inconvenience of flattening the historical evolution of the language and its dialectical diversity, it allows for a more comprehensive reading. I make allowances in two particular instances. First, I use the customary spelling for well-established words in the literature, such as personal names and place names (Huayna Capac, Coricancha, Huarochirí), but I do not extend this criterion to the denomination of supernatural beings, such as Wiraqucha or P'unchaw. In this case, I consider the standardized spelling to be particularly enlightening for the analysis. Secondly, where there is no consensus on the normalization, especially concerning festival names, I adopt the spelling formulated by the colonial sources, regardless of inconsistencies. Finally, for the purpose of convenience, I always indicate the plural of native words with an *-s*, and do not use the Quechua suffix *-kuna*, which is not strictly equivalent to the English plural form.

When referring to the political and territorial unit subjected by the Incas, I use the term "empire" because I believe it best describes the dynamics of pre-Hispanic sovereignty. Although the concept and its related notion of imperialism have inevitably taken different forms throughout time and space, they emerge as key categories "constantly contested and reaffirmed in the present with reference to the past."[45] I follow Colas (2007) in outlining some basic traits to describe empires in contrast to other forms of polity. Firstly, unlike nation states that favour territorial stability and homogenized identity, empires are built upon expansion, which they largely achieve through conquest. As a result, their geographical borders are constantly redefined to incorporate a culturally, religiously, and linguistically diverse population. Secondly, within the land under their jurisdiction, empires devise a hierarchical order whereby the centre exploits the periphery. The capital centralizes the institutions of imperial dominance wherefrom the elite controls the production and distribution of wealth, and propagates an image of cultural superiority. Finally, empires implement order by providing for public infrastructures (roads, laws, official religion and language, etc.). In Tawantinsuyu, the Incas ruled over a great diversity of communities and larger chiefdoms ethnically different from them. The governing elite exercised coercion over its regional polities that were incorporated into the ever-expanding imperium. While the Incas based their supremacy upon pre-existent systems of commodity exchange and did not seek to annihilate local traditions, they also devised ideological instruments of authority in order to install cultural, religious, and economic ascendancy over their subjects. The similarities between the imperial dynamics of Tawantinsuyu and other historical empires also fostered the rhetorical purposes of early Spanish chroniclers who paralleled their infrastructures with those of Ancient Rome and other exotic polities, such as China or Japan.[46] Ultimately, these comparisons sustained the central arguments elaborated by early Modern theologians and jurists on the degree of *civilitas* they believed Inca society had attained. Likewise, for European settlers, sovereignty was intrinsically linked to the concepts of private property and territoriality. Recent studies, however, suggest that neither of these two notions would have applied to the Inca socio-political organization. In the pre-Hispanic Andes, agricultural lands were not owned but

---

45 Colás 2007: 5.

46 MacCormack 1991a, 2001, 2007; Pagden 1986: 146-197.

rather worked by labour assignments on a rotational basis. According to Susan E. Ramírez, Inca rulers "did not conquer geography so much as people or nations" because workforce was the true pillar of their imperial machinery.[47] Men and women were subjected to tribute obligations for state endeavours (e.g., infrastructure work) and cult sustenance, as well as for the personal benefit of the ruler and his lineage. The conquistadors, the chroniclers, and the colonial administration failed to recognize this system and applied concepts of private estates and territorial boundaries where they never existed, thereby facilitating the implementation of the *encomienda* system. To what extent, then, are Spanish impressions only partial and biased representations of an empire that was also formed by the unique dynamics of Andean society? Outlining the distinctiveness of Inca imperial institutions requires excavating the many layers of interpretation that have shaped the early historiography. The purpose of this book is precisely to understand one particular aspect of Tawantinsuyu as an imperial power: the forging of its elite's identity through narrative, ritual, and cosmological means. In order to penetrate further into this issue, we shall now turn to the examination of the primary sources that inform our knowledge of the Inca realm.

---

[47] Ramírez 2005: 24.

PART ONE

THE POLITICS OF INCA HISTORY

CHAPTER ONE

MAKING HISTORY: KNOTTING AND UNRAVELLING DISCURSIVE THREADS

> Kachins recount their traditions on set occasions, to justify a quarrel, to validate a social custom, to accompany a religious performance. The storytelling therefore has a purpose; it serves to validate the status of the individual who tells the story, or rather of the individual who hires a bard to tell the story.
>
> (Leach 1954: 265)

Inca studies today are the legacy of centuries-long scholarship that originated in the controversy over Spanish legal rights to dominion in the Americas. Early examples of this primary literature reveal little concern over the history of ancient Peru. The first Europeans to set foot in Tawantinsuyu recorded primarily the military and political hostilities that precipitated the fall of Inca supremacy in the Andes. As the backdrop for these events, the chroniclers evoked, sometimes with bewilderment, the empire's extensive infrastructures, with its hierarchy of state and local authorities, as well as its abundance of natural resources. This focus on socio-political control and wealth shifted starting from the early 1550s, when two major events took place that permanently shaped the nature of the discussions. First, the festivals that the ruling elite had hitherto performed in Cuzco were irrevocably banned as practices. From then on, Spanish chroniclers would have needed to rely on the oral reports of their Inca informants and first colonists to describe the splendour of the bygone festivities.[1] Most importantly, this ceremonial proscription coincided with the First Lima Council, assembled in 1551-52, whose declared intention was to set a uniform Church policy against native religions labelled as "idolatries". To fight more effectively the Gentiles' ignorance, Church officials began to produce extensive literature on Andean belief, starting with descriptions of the Inca state religion. Secondly, years of civil war between Spanish factions, concurrent with the decline of the indigenous labour force and the abuses perpetuated in the *encomiendas*, increasingly strained regal officials now looking for ways to instigate the Crown's legal rights over

[1] Duviols 1977a: 111.

Peru's wealth. Theologians and jurists were called upon to delineate the foundations of these titles, sparking decades of controversy that focused largely on defining the status of Amerindians as human beings. The famed debates that ensued placed two main schools of thought in opposition to each other. The first, led by Juan Ginés de Sepúlveda, stated that the deviant behaviours and vices observed amongst Amerindians demonstrated their irrationality, therefore impeding the articulation of just governance and the development of civil societies. Following the medieval lines of the *Requerimiento*, which deprived infidels of political jurisdiction and rights to property, Sepúlveda suggested that the Crown had legal title to launch a just war on the opponents of the Christian faith and *ergo* on the Amerindians who defied the true God with their heathen beliefs. In view of these legal provisoes, he argued, the Spanish king had the right to claim ownership of America and to subjugate its people.[2] This dogmatic thesis provoked a firm condemnation from the partisans of the new Scholasticism and pupils of Francisco de Vitoria, who had elaborated a unique interpretation of the Crown's lawful titles. Vitoria's assertions were based on a reformulation of the Thomist concept of natural law, which he saw as a pervasive principle of human nature governing the moral standards of all rational beings. For no one can lack reason and policy by virtue of his nature; all men are entitled to dominion over their material goods and actions, including the newly "discovered" Amerindians. The new scholastics argued that the existence of a form of religion in America before the Spanish conquest could not have offended God's law because previously peoples of the New World had lived in ignorance of the Gospel.[3] To argue their case, both parties referred to the official, albeit composite, chronicles of the conquest, which in turn began to be influenced by the very contentions of this theological and legal controversy.

It is in this argumentative context that Spanish chroniclers transcribed the lives and military deeds of the eleven or so Inca sovereigns known to have ruled before the European invasion. Their writings, assembled out of miscellaneous oral traditions, responded to imperatives resulting directly from the controversy. They constituted a plethora of legal arguments for the Spanish Crown's exploitation of American resources, justifications for the evangelization of native populations, or juridical supports for obtaining individual privileges. The viceroy Francisco de Toledo (1569-81), in particular, initiated as soon as taking office a zealous policy of justification of

---

[2] Pagden 1990: 15-19.

[3] Grisel 1976.

colonial legal rights that led to the proliferation of historical documents on the bygone empire. In this process, he systematized an already existing procedure known as the *visita general*. These inquiries, designed to maximize the efficiency of the colonial tribute policy and thus future revenues, intended to define the political status of pre-Inca chiefdoms, as well as the nature of their economic commitments to the former ruling elite.[4] In parallel, they sought to establish the Crown's dominion by denouncing the barbarity and illegitimacy of the Inca conquest. The stated aim was to acknowledge the prior status of the provincial chiefdoms as the first and rightful occupants of the land.

It is these controverted sources, permeated with the evangelist and political agendas of their authors, that provide the basis for present-day scholarly debates on the history, religious beliefs, and ritual practices of the Inca elite. Whether they were histories, bilingual lexicons, or administrative reports, such documents were never elaborated upon out of a spirit of disinterested ethnological inquiry (if such an enterprise exists), and necessarily reveal aspects of the dominant society in which they were produced. With time, the enforcement of the colonial system created new demands on the remnant indigenous elite, whose members were required to provide judicial proof of their noble ancestry (*probanza*) in order to benefit from social and economic privileges, or for them to participate in the prestigious festivals of the Catholic liturgy. Indigenous informants were compelled to assemble conclusive, and sometimes biased, reports of their kin group's blood ties, and to detail the past involvements members of their family had in the ritual and administrative life of the Inca empire.[5] In this context, the agendas of both colonizers and indigenous Andeans cropped up in the pre-Hispanic politics that circumscribed the original elaboration of Inca narratives. Embedded in this literature are the discordant voices of native informants—Inca nobles, provincial lords, and their scribes—who were quick to understand that words on paper bore significant weight on the resolutions of the new administration. It follows that in the process of

[4] In order to control more efficiently the indigenous working force, the Spaniards created Indian nucleated settlements (*reducciones*). These were located among larger administrative districts or *repartimientos* governed by a hierarchy of local lords and headmen who insured that each able-bodied man paid a head-tax or tribute. The caciques and members of the Inca nobility who could prove their filial relations with the ancient royalty were exempt from tribute. In such a context, it was evidently preferable and even beneficial to establish the nobility of one's ascent.

[5] Duviols 1979a; Adorno 1988: 13-32; Urton 1990: 46-70; Stern 1993: 121-128; Cahill 1998, 2000.

elaborating a history of the Inca empire, Spaniards were not only confronted with categories foreign to their own referents, which they selected, translated, and interpreted, but they also handled information that reflected the post-conquest conflicting interests of different ethnic groups and their social strata. The resulting historiographical record resembles a bundle of entangled threads that served to establish the status, rights, and privileges of individuals in the colonial system, and to determine their disposition towards the "true religion".

These records, despite their layers of embedded reflections on paganism and politics, are our richest source of information on the pre-Hispanic world. Our understanding of Inca institutions thus lies *in extenso* on the critical analysis of the information enclosed in this corpus. As scholars have long acknowledged, a major conundrum of this literature is how to make sense of the widespread divergences that affect the historical accounts of Inca expansion. Although influential chroniclers from the 1550s began spreading a relatively standardized and linear picture of the Inca dynastic succession, a comparative analysis of all primary sources reveals that they often disagree on the number and identity of these kings, attributing to different figures the same conquest, political reform, or ritual institution.[6] Obscuring the matter further, these contradictions not only affect different texts, they may also permeate a single one so that the same author would appear to contradict himself. These discrepancies, it goes without saying, complicate all attempts at reconstructing the historical evolution of Inca expansion; they likewise render problematic the identification in these texts of chronometric indications for the interpretation of the materials brought to light by archaeologists. In response to these issues, historiographical works have multiplied from the turn of the nineteenth century onwards, seeking to evaluate the objectivity of colonial sources by investigating the chroniclers' agendas and their empirical knowledge of the culture they so vividly describe. Despite shedding considerable light on the politics of transatlantic colonial literature, these studies have not yet proposed conclusive interpretations of the chronicles' pervasive discordances. This chapter intends to lay the foundations for the literary and anthropological analysis of the material enclosed in primary sources, however discrepant they may appear. It reviews the conflicting accounts that characterize this literature, before discussing the critical readings that modern scholars have developed around these issues. In

[6] On this topic, see Covey 2006b and Ramírez 2006a.

doing so, I wish to highlight the values as well as the oversights that affect these studies. I suggest that a careful examination of the political issues and cultural practices surrounding the production of historical narratives among the Inca ruling elite in pre-Hispanic times opens a promising avenue of research to unravel the tangled threads of their history.

Inconsistencies in Content

Dissonances in primary sources on Inca history are varied and infused with descriptions that, it should be noted forthwith, do not necessarily touch aspects related to colonial policies. In this way, post-conquest politics and the corollaries of cultural encounter may explain part of the chronicles' incongruities, but these falsifications coexist with bewildering discrepancies that require specific commentaries. Thus, to approach this material is to be confronted with contradictory lists of kings that inconsistently jumble the ruling sequences, conquests, deeds, and reforms of different lords. In addition to these heterogeneous data, a standardized picture of the Inca royal genealogy made its appearance approximately twenty years after the conquest among Spanish historians looking to establish an authoritative account of the past. This kingship list, imparted by a handful of "reputable" chroniclers, records the reigns of eleven to twelve rulers before the fratricidal war between Atahualpa and Huascar broke out. This list then rapidly achieved authority up to the present day, as modern studies refer to it customarily as the most reliable dynastic picture for the reconstruction of Inca expansion.

Juan de Betanzos is the earliest chronicler known to us to have recorded this standardized genealogy (Table 1). The list appears in the opening pages of his *Suma y narración de los Incas*, which he wrote during Antonio de Mendoza's tenure as viceroy of Peru. The traceability of Betanzos' sources and his personal acquaintance with a segment of the Inca nobility—he married doña Angelina (also Cuxirimay Ocllo), member of Inca Pachacuti's lineage—explain in great part the aura of credibility that surrounds his text today. Most noticeably, he had a long career in the service of the viceregal administration. Prior to becoming one of the earliest interpreters of the Crown, he had availed himself of various native testimonies about the Andean past and supervised the interviews of the *khipukamayuqs* who provided the information for another important, albeit controverted, primary source: the *Discurso sobre la descendencia y gobierno de los Incas* (1543-1608). The royal genealogy in both this last document and the *Suma*

Table 1. List of Inca rulers according to four chroniclers.

Betanzos (1551)	*Cieza de León* (1554)	*Sarmiento de Gamboa* (1575)	*Garcilaso de la Vega* (1609)
Manco Capac	Manco Capac	Manco Capac	Manco Capac
Sinchi Roca	Sinchi Roca	Sinchi Roca	Sinchi Roca
Lluque Yupanqui	Lluque Yupanqui	Lluque Yupanqui	Lluque Yupanqui
Capac Yupanqui	Mayta Capac	Mayta Yupanqui	Mayta Capac
Mayta Capac	Capac Yupanqui	Capac Yupanqui	Capac Yupanqui
Inca Roca Inca	Inca Roca	Inca Roca Inca	Inca Roca
Yahuar Huacac Inca Yupanqui	Inca Yupanqui	Yahuar Huacac	Yahuar Huacac
Viracocha Inca	Viracocha Inca	Viracocha Inca	Viracocha Inca
Inca Yupanqui Pachacuti Inca	Inca Urco	Inca Yupanqui Inca (*also* Pachacuti Inca)	Pachacuti Inca
Yamque Yupanqui	Inca Yupanqui		Inca Yupanqui
Tupa Inca Yupanqui	Tupa Yupanqui	Tupa Inca Yupanqui	Tupa Inca Yupanqui
Huayna Capac	Huayna Capac	Huayna Capac	Huayna Capac

y narración follows a similar template of linear succession, but it remains unclear which text anticipated the other.[7] A contemporary of Betanzos, Pedro Cieza de León finalized in the early 1550s his monumental *Crónica del Perú*, in which he too proposes a genealogy of twelve Inca kings. Yet, his sources are different from Betanzos'. Thanks to his position as secretary to governor Pedro de la Gasca (1547-50), Cieza gained access to the archives of the early conquest but also interviewed the descendants of two lords, Viracocha Inca and Huayna Capac. Cieza made use of these records, together with the information accumulated in the course of his extensive travels, to compose the three parts of his chronicle. Of these different volumes, the second one deals with the foundation and rise of the Inca empire. Arguably, the *Suma* and the second part of the *Crónica del Perú* enjoyed no or little readership during the colonial period.[8] Therefore, much of the material they enclose and their original wording remain largely unreproduced in other chronicles. Twenty years after their completion, however, Sarmiento collected independently the main lines of a similar

[7] For a critical historiographical overview of the on-going debate concerning the authorship and dating of the *Discurso*, see Domínguez Faura 2008, 2010 and Julien 2009.

[8] Presumably Father Gregorio García is the only author to have consulted the *Suma* before Jiménez de la Espada published it in 1880 (Hamilton, in Betanzos 1996: xii).

history for his *Historia General Yndica*. This book, commissioned by Francisco de Toledo, is a well-known example of material gathered to conform to the colonial enterprise of the newly appointed viceroy. It is also a significant ethno-historical document, for it echoes the voices of selected members of the Inca nobility and descendants of the provincial elite who were called upon to confirm the authenticity of its contents. As we shall see, Sarmiento, Cieza, and Betanzos' accounts of royal deeds differ on various occasions, but their genealogical models of Inca kingship are similar. This dynastic picture henceforth prevailed in the colonial literature on ancient Peru, with widely read chroniclers such as Garcilaso de la Vega reproducing it in their work.

In addition to providing a royal genealogy, some of these chroniclers inform us about the royal lineages that descended from late Inca kings. Cieza and Betanzos, while acknowledging the existence of distinctive *nación* or *linaje* within the ruling elite, do not explicitly describe the mode of affiliation that characterized these groups. Sarmiento is the most prolix author on this subject and his commentaries held a long-lasting imprint on modern perception of the Inca dynasty. Each panaca, he claims, was founded at the death of a ruler and belonged to either of two moieties that spatially divided Cuzco into residential quarters. The lineages proceeding from the early kings of the dynasty were affiliated to and populated Hurin Cuzco, the Lower part. As for the descendants of the latter sovereigns, they belonged to and occupied Hanan Cuzco, the Upper part (Table 2).

A glance at Table 1 summarizing the Inca genealogy in these respective sources, however, reveals several discrepancies. Cieza, for example, supplants the reign of Yahuar Huacac, traditionally believed to be the seventh king of the dynasty, with that of an unknown figure, Inca Yupanqui, who would have been brutally murdered in Cuzco and, for that reason, "did not receive the mortuary honours of his predecessors, nor was he mummified."[9] The name of Inca Yupanqui appears to be conflated with that of Yahuar Huacac in Betanzos' and Sarmiento's texts but, contrary to Cieza's comments, both authors certify that his reign was prosperous and closed with the honours generally rendered to each Inca sovereign. Another peculiarity distinguishes Cieza for, unlike other chroniclers, he argues that the older son of Viracocha Inca, called Inca Urco, was invested with the royal fringe before his cadet overthrew him. As for Betanzos, he inverts the rules of Mayta Capac and Capac Yupanqui, but also lists Yamque Yupanqui, a relative of his noble Inca wife, to Pachacuti's succession. Finally, unlike the

---

9 Cieza: Pt. 2, Ch. 37, 110-111.

Table 2. The Inca royal lineages (panacas) in primary sources. Original spelling has been preserved regardless of inconsistencies.

	Sarmiento (*1575*)	*Acosta* (*1590*)	*Discurso* (*1543-1608*)	*Garcilaso* (*1609*)	*Cobo* (*1653*)
Manco Capac	Chima panaca ayllu		Chima panaca	Chima panaca	Chima panaca
Sinchi Roca	Raura		Raorao panaca	Rauraua panaca	Raurahua panaca
Lluque Yupanqui	Auayni panaca ayllu		Chiguayunin ayllu	Hahuanina ayllu	Ahucani ayllu
Capac Yupanqui	Apu Mayta		Apumayta ayllu	Apu Mayta	Apumayta ayllu
Mayta Capac	Usca mayta panaca ayllu		Uscamayta ayllu	Usca mayta	Uscamayta ayllu
Inca Roca	Vicaquirau panaca ayllu	Uizaquirao	Vicaquirao ayllu	Uncaquirau	Ayllu Vicaquirao
Yahuar Huacac	Aucaylli panaca	Aocailli panaca	Aucaylli panaca	Aylli panaca	Aucayllu panaca
Viracocha Inca	Sucsu panaca ayllu	Coccopanaca	Sucsu panaca	Sucsu panaca	Sucsuc panaca
Pachacuti Yupanqui	Hatun ayllu Inaca panaca ayllu	Inaca panaca	Inaca panaca	Inca panaca	Yñaca panaca ayllu
Tupa Yupanqui	Capac ayllu	Capac ayllu	Capac ayllu	Capac ayllu	Capac ayllu
Huayna Capac	Tumipampa ayllu	Temebamba		Tumi panaca	Tumibamaba panaca

aforementioned authors, Garcilaso inserts the reign of a king named Inca Yupanqui after that of Pachacuti Inca. In doing so, the Mestizo writer conforms to the genealogy of early chroniclers, including Damián de la Bandera (1557), Santillán, and the scribes of the *Relación de Señores Indios* (c. 1580), whose three texts display close similarities throughout.[10] Later sources, such as the *Compendio y descripción de las Indias Occidentales* by Antonio Vázquez de Espinoza (1629), record as well the life and deeds of this lord sometimes called Capac Yupanqui. *Memorias* from the late

[10] The genealogy recorded in Bandera and the *Relación de Señores* is identical, but Santillán's list of kings appears more original. He is known to have exploited various sources to write his *Relación*, including Bandera's description of the Huamanga province (1557), Castro and Ortega Morejón's *Relación*, as well as Polo's manuscripts. Yet, his chronicle, which answered a royal questionnaire on Inca customs and tribute system, records data that none of these documents mention. See Lohmann Villena 1966; Andersson 1968; Porras Barrenechea 1986: 324-327; Sánchez-Concha Barrios 1996.

Table 3. The Inca kingship sequence in three early chronicles.

Bartolomé de Segovia (*1553*)	*Damián de la Bandera* (*1557*)	*Castro & Ortega Morejón* (*1558*)
Inca Viracocha	Pachacuti Yupanqui	Capac Yupanqui
Tupa Inca Yupanqui	Viracocha Yupanqui	Tupa Inca Yupanqui
Huayna Capac	Inca Yupanqui	
Huascar/Atahualpa	Tupa Inca Yupanqui	Huayna Capac
	Huayna Capac	
	Atahualpa/Huascar Inca	

Table 4. The Inca kingship sequence in three early chronicles (suite).

Hernando de Santillán (*1563*)	*Pedro Pizarro* (*1571*)	*Relación de Señores* (*c. 1580*)
Pachacuti	Inca Viracocha	Pachacuti Yupanqui
Viracocha	Tupa Inca/Inca Pachacuti	Viracocha Yupanqui
Yupanqui or Capac Yupanqui	Huayna Capac/Amaru Inca	Inca Yupanqui
Inca Yupanqui	Two unnamed successors	Tupa Inca Yupanqui
Tupa Inca Yupanqui		
Huayna Capac	Huayna Capac	Huayna Capac
Huascar Inca/Atahualpa		Atahualpa

seventeenth century included members of his descent for the office of *alférez real,* and the Spanish administration eventually recognized his panaca alongside eleven other royal lineages from 1721 onwards.[11]

Equally discrepant are the reports of early conquistadors and provincial informants that differ greatly from the ruling list repeatedly found in official sources of the 1550s. Pierre Duviols (1979c) and Bravo Guerreira (1992) already noted that Cristóbal de Castro and Diego de Ortega Morejón (1558), who conducted interviews among the residents of the Chincha valley, as well as the group of Bandera, Santillán, and the *Relación de Señores Indios*, together with Pedro Pizarro (1571) and various informants of the inquiries headed by viceroy Toledo (1571-72), all commence the Inca kingship genealogy with late rulers of the Hanan moiety, either Inca Viracocha or Pachacuti Yupanqui. When Manco Capac is listed first, Pachacuti follows after him (Tables 3, 4, and 5).

[11] Vázquez de Espinosa: Ch. 97, 384; Dean 1999: 241, n. 2; Amado Gonzales 2002.

Table 5. The Inca kingship sequence according to three informants of the Toledan inquiries (1572).

Alonso Pomaguaca	*Antonio Guamán Chuco*	*Baltasar Guamán Llamaca*
Manco Capac	Viracocha Inca	Manco Capac
Pachacuti Inca Yupanqui	Pachacuti/Capac Inca	Pachacuti/Capac Yupanqui
Tupa Inca Yupanqui/ Capac Yupanqui	Tupa Inca Yupanqui	Tupa Yupanqui Inca
Huayna Capac	Huayna Capac	Huayna Capac
Huascar Inca	Huascar/Atahualpa	

These testimonies recall Polo's unconventional ruling list, later adopted by José de Acosta in his *Historia natural y moral de las Indias* (1590). A jurist and twice *Corregidor* of Cuzco, Polo recorded detailed information on the ceremonial practices of the former elite. One of his major achievements, completed thanks to his personal contacts with the local nobility, was the discovery and subsequent disposal of several mummy bundles of late Inca rulers. In 1559, Jéronimo de Loaysa, then archbishop of Cuzco, charged him with an enquiry into indigenous customs and beliefs. The data Polo gathered during this project, notably among *khipu* specialists, informed the manuscript that the Third Council of Lima proceeded to publish after his death, *Los errores y supersticiones de los indios* (1585). This text is unique in starting the kingship sequence with the Incas of the Upper moiety, followed by those of the Lower moiety, hence shattering the antiquity of Hurin versus Hanan lords.[12]

The foundational narrative of the Inca ancestors' migration to the Cuzco valley also comes in a variety of versions, even though a stabilized transcription of this event spread from the 1580s onwards. Cieza, Betanzos, and Sarmiento provide the earliest, most detailed account of this myth, which is described similarly in a handful of other chronicles, including those of Cabello and Murúa. Although the latter two authors would have reproduced a narrative from a lost manuscript attributed to Father Molina,[13] the testimony of Cieza, Betanzos, and Sarmiento is most likely a transcript of first-hand materials. They all claim that in times immemorial a man called Ayar Manco emerged from a cave southeast of Cuzco together with two or

[12] Another originality of Polo's list is the inclusion of two rulers after Mayta Capac, both named Tarco Huaman. The issue has been thoroughly discussed by Duviols (1979c).

[13] Like Sarmiento, Molina participated in the Toledan inquiries. For an overview of the sources used by Murúa, see Ávarez-Calderón 2007; Julien 2000: 166-185.

three brothers and their sister-wives. The location of this birthplace is a common point of divergence between chroniclers: some situate it in the Lake Titicaca Basin, and others in the Cuzco region, a little more than one hundred kilometres from the city that the Incas built thereafter. For those who set the scene in the high altitudes of the Collao province, it was the divinity Wiraqucha who commanded Manco and his brothers to set forth on a journey in search of arable lands;[14] for the other authors it was, in fact, the Sun who ordered the migration. In his *Comentarios*, Garcilaso intertwined these two versions in an elaborate narrative that he claims to have heard as a child from his maternal uncle.[15] In both variants, the god addresses Manco as his son and presages that he will give birth to a dynasty that would subjugate many nations. Following this apparition, Manco explored the land for several years and made short-term settlements accompanied by his brothers who, one by one, were transformed into stone shrines. Eventually, he and his sister-wife established residence in the Cuzco valley among other ayllus that already lived there. Cohabitation with these first occupants was initially contentious, but Manco's descendants soon imposed their authority upon the region and founded a dynasty of rulers made known through the Spaniards as the Incas.

The early expansion of Inca sovereignty generated another series of divergent stories. According to most sources, including the substantial narratives of Betanzos, Sarmiento, and Cieza, it would have taken approximately eight generations for the lords of Cuzco to quell their neighbours' hegemonic pretentions before the rise of Pachacuti Yupanqui marked a historical turn. The youngest son of Viracocha Inca, Pachacuti acceded to the kingship office after repelling the Chanca invasion that his older brother and favourite for royal succession failed to confront. It is said that Viracocha Inca, too old to fight on the battlefield, fled Cuzco with his court when he heard that the neighbouring Chanca confederation had gathered a powerful army to seize the city. Pachacuti, disapproving of his father's conduct, immediately embarked on assembling support to confront the enemy. On the eve of the decisive battle, the young Inca had a vision of a radiant deity claiming to be his father. By all accounts, the tutelary figure assured Pachacuti Inca of his divine support and imminent victory, yet the

[14] Among the chroniclers to have reported this tradition, albeit with different details, are Gómara, Zárate, who used the former extensively, Castro and Ortega Morejón, Pizarro, Molina and Gutiérrez.

[15] Although it was claimed that Garcilaso's reference to his maternal uncle was a figment to establish the authority of his text, it remains that his version of the origin myth is not unique.

name of this deity remains uncertain. Betanzos identifies him with Wiraqucha, while others, such as Sarmiento, Molina, or Cobo, depict him as the Sun. Still, the prediction occurred as specified and Pachacuti effectively subjugated the Chancas, thus ensuring his election as *Sapan Inca* (Only king). His reign was marked by a series of institutional and religious reforms intended to strengthen the authority of the noble houses that assisted him in seizing power. He also placed the god that had appeared to him in a vision as head of the official religion, and ordered that a statue of him be cast in gold and sheltered in the Sun temple. This redistribution of divine and worldly powers anticipated Pachacuti's offensive campaigns that led to the extension of Inca jurisdiction over a multi-ethnic land. Besides this shift from local politics to imperialism, Betanzos also credits the reformer king with the spatial and social division of Cuzco into moieties. Sarmiento, however, contests this last attribution and, like Cabello and Murúa, he imputes the dual partition of the city to Inca Roca, the ninth king of the dynasty. Gutiérrez de Santa Clara associates this episode with Tupa Yupanqui in his *Quinquenarios o Historia de las guerras civiles del Peru* (c. 1590), while Cobo argues that it was Manco Capac who initially divided Cuzco.[16] Finally, all the more divergent is Garcilaso's text, which attributes Pachacuti's every feat to Viracocha Inca.

Historiographical Responses to the Chronicles' Discrepancies

Inca researchers have striven to identify the origins of the dissonances pervading primary sources since the earliest modernist studies of the nineteenth century. In this process, they have incriminated two major agents that could have simultaneously made an impression on the texts: one was endogenous to the Inca narratives and related to their arguable mythical nature, and the other was exogenous to them and referred to the colonial actors. The first argument suggests that the dynastic accounts of the ruling elite did not aim to transmit a factual past, nor did they even aspire to interpret real events by incorporating the supernatural or the divine into the plot, but instead offered an exegesis of Inca institutions. In doing so, they employed a mythical language whose structure was typically multivalent and aberrant, but which metaphorically imparted a social model. The apparent linearity articulating the representation of Inca royal succession in

[16] Cobo: T. 2, Bk. 13, Ch. 10-11, 73-77; Gutiérrez: 213-215; Cabello: Ch. 3, 25, Murúa, *Historia general*: Bk. 1, Ch. 13: 58-59; Garcilaso: Bk. 1, Ch. 16, 41.

primary sources would have been a colonial adaptation of a native hierarchical principle of social organization based on temporal depth. Exploring this theoretical ground, Zuidema has authored works that significantly renewed our views of pre-Hispanic societies. Inca history, he claims, "integrated religious, calendrical, ritual and remembered facts into one ideological system, which was hierarchical in terms of space and time. This incaic hierarchical ideology should not be confused with the Western linear conception of history imposed by the Spanish."[17] Zuidema suggests more precisely that the kings whose memory had been preserved were the apical ancestors of the ten panacas that had coexisted in the Cuzco valley from remote times; any living ruler would have originated from either of these lineages. He adds that the reigning lists recorded in colonial documents did not reflect the kings' chronological order of succession, but instead the hierarchical position of each descent group in its respective moiety. The group members' status determined in turn their functions within the religious and administrative system, since each panaca was in charge of a distinct month of the ritual calendar and attended its own tutelary divinity. In this way, the chroniclers would have manipulated the temporal framework of the original dynastic accounts by affixing onto them a factual structure conformed to European literary conventions and historical narration. The temporal markers enclosed in the Inca corpus, including the sequential record of kings or the conquest of new subjects, formed a "mythical and political structure" alien to the objective course of events. These episodes, therefore, could not be exploited for chronological reconstruction. A similar perspective on Inca narratives shapes Pease's works. The Peruvian historian argues that the structure of Andean memory did not comply with Western historical categories but instead followed a ritualized scenography. For Pease, this fundamental aspect of native traditions deceived the Spaniards into documenting the Inca expansion as a series of theatricalized scenes of military campaigns. These narrations, he argues, did not keep an account of successive conquests in time. They were a collection of myths and ceremonial descriptions that "did not conform to temporal, spatial or individual patterns of recognized history; the chroniclers ordered—reordered—this information in a chronological fashion, harmonized with the sequential presence of kings."[18] Following his and Zuidema's line of thought, the discrepancies occurring repeatedly throughout colonial texts resulted from the mythical/ritualized nature of

---

[17] Zuidema 1982c: 173-174.
[18] Pease 1994: 17. See also Pease 1995: 96-105.

the Inca discourse on the past and its transcription according to Western formats.

Contrasting with this perspective on pre-Hispanic accounts, the approach described as "historicist" proposes to elaborate a normalized chronology of Inca expansion based on the dynastic list of reliable chronicles. Today, a general consensus agrees on the sequences of this standardized genealogy and associates with it a chronological scale elaborated in the first half of the twentieth century. John H. Rowe, responding to Philip A. Means' works (1931) on the periodization of Andean history, established in 1945 an absolute chronology of the last phase of Inca dominance. He starts his reconstruction in 1438, the alleged year of investiture of Pachacuti Yupanqui, the ruler who would have initiated the expansion of Inca sovereignty beyond the Cuzco region. For Rowe, however, prior to the advent of this king, the course of history as narrated in primary sources is highly speculative. He claims that the confusion and lack of plausibility permeating the accounts of the early Inca period made them akin to legends and thus improper for chronometric use. Notwithstanding these considerations, the archaeologist assigns an undeniable historicity to the material relative to late Incas and, more specifically, to the data contained in Cabello's *Miscelánea Antártica*.[19] Although he never lived in Cuzco, Cabello is one of the rare chroniclers to date with precision the official tenure of each ruler in accordance with the information he claims to have extracted from *khipu* registers. According to Rowe, Cabello would have interviewed non-Inca native specialists to produce a series of reigning dates that "are the most plausible set, ancient or modern, which has yet been proposed."[20] These data, he continues, are "not only reasonable, they are some of the most conservative figures ever suggested"[21] so that their historical verisimilitude appears a crucial argument attesting to their objectivity. On the basis of these criteria of validity, Rowe discredits a number of other sources also mentioning the kings' age at death or length of reign but offering unrealistic lifespans (Table 6).

Yet, appraising the reliability of primary sources on the premise of their historical plausibility raises several methodological issues. Although it is manifestly more credible that Inca lords ruled an average of fifty years

---

[19] There are two manuscripts of the *Miscelánea Antártica* that present various dissimilitudes. One is held in the New York Public Library, and the second is in the collections of the University of Texas, Austin. See the "Introduction" by Luis E. Valcárcel: ix-xl, in Cabello.

[20] Rowe 1946: 203.

[21] Rowe 1945: 277.

Table 6. Estimated length of the Inca dynasty according to five chroniclers.

	Discurso (*1542*)	*Sarmiento* (*1572*)	*Cabello* (*1586*)	*Gutiérrez* (*c. 1590*)	*Guaman Poma* (*c. 1615*)
Manco Capac	?	† 144 yrs rules 100 yrs	† in 1006 rules 60 yrs	† 120 yrs	† 160 yrs
Sinchi Roca	† > 70 yrs	† 127 yrs rules 19 yrs	† in 1083 at 77 yrs	† 50 yrs	† 155 yrs
Lluque Yupanqui	rules > 50 yrs	† 132 yrs rules 111 yrs	† in 1161 rules 78 yrs	† 120 yrs rules 60 yrs	† 130 yrs
Mayta Capac	rules 50 yrs	† 112 yrs	† in 1226 rules 65 yrs	rules 65 yrs	† 120 yrs
Capac Yupanqui	rules > 60 yrs	† 104 yrs rules 89 yrs	† in 1306 at 80 yrs	† 114 yrs rules 65 yrs	† 140 yrs
Inca Roca	† > 80 yrs	† 123 yrs rules 103 yrs	† in 1356 rules 50 yrs	† 100 yrs rules 40 yrs	† 154 yrs
Yahuar Huacac	rules > 40 yrs	† 115 yrs rules 96 yrs	† in 1386 rules 30 yrs	rules 80 yrs	† 139 yrs
Virarocha Inca	rules > 70 yrs	† 119 yrs rules 101 yrs	† in 1438 rules > 50 yrs	† 120 yrs rules 85 yrs	† 124 yrs
Inca Yupanqui	† > 80 yrs	† 125 yrs rules 103 yrs	† in 1473 rules 36 yrs	† 80 yrs rules 55 yrs	† 88 yrs
Tupa Inca	† > 80 yrs	† 85 yrs rules 67 yrs	† in 1493 rules > 30 yrs	† 80 yrs rules 50 yrs	† 200 yrs
Huayna Capac	rules > 50 yrs	† 80 yrs rules 60 yrs	† in 1525 rules 33 yrs	† 65 yrs rules 40 yrs	† 86 yrs
Length of dynasty	± 650 yrs	± 950 yrs	± 520 yrs	± 620 yrs	1496 yrs

rather than a century, most of the dynastic narratives recorded in the chronicles do not contain precise intimations of plausibility upon which we may estimate their factual value. Lacking such clues, which objective criteria should we consider in order to identify the authentic account of an event amidst many conflicting versions? Even when such intimations exist, how do we determine that they are of pre-Hispanic origin and not introduced by chroniclers to give a historical substance to their discourse? Finally, are we truly in a position to be the judges of the historical objectivity of Inca narratives when we do not possess independent cross-checking material? Archaeological research is only beginning to provide self-sufficient chronometric data. Incidentally, recent excavation reports appear to indicate that Inca imperial expansion would have started as early as the fourteenth century, more than a hundred years before the alleged year of investiture of the conqueror king Pachacuti Yupanqui.[22]

Whichever interpretational grid scholars adopted in their reading of Inca narratives, they all endeavoured to evaluate the authenticity of the material recorded in primary sources by investigating the chroniclers' reliability. These studies led to significant advances as they brought to light conclusive evidence that colonial actors instigated a great number of textual discrepancies. These dissonances resulted either from the chroniclers' miscomprehension of the language and cultural framework of their indigenous informants, or through strategy. In this way, several authors altered intentionally parts of the Inca past to legally establish their personal pretensions and thereby benefit from the prerogatives that the colonial system offered. Betanzos, for example, is the only historian to refer to Yamque Yupanqui, his wife's grandfather, as Pachacuti's favourite heir to the royal office. As Nowack (2002) reveals, the intrusion of this figure into the dynastic list of the *Suma y narración* is not a coincidence. With this embellishment, the chronicler imputed his family-in-law with a noble ascendance that would have legitimized his intention to claim before the Crown the privileges reserved to the indigenous nobility. Duviols (1979a) also unveils that the second part of the *Discurso* served as a *probanza de méritos* to establish Don Melchior Carlos Inca's noble status and his grandfather's services to the Crown.

These works of historiography were fundamental in modifying perceptions of age-long sanctioned chronicles. The authority of Garcilaso's *Comentarios* suffered particularly from a similar critical scrutiny. The son of a Spanish captain and an Inca noblewoman,[23] Garcilaso spent his youth in Cuzco before embarking for the Iberian Peninsula in his late teens. At the end of his life, more than thirty years after leaving Peru, he turned to writing influenced by humanist culture, Neo-Platonist ideals, and the writings of the Jesuit Father Blas Valera. In his *Comentarios*, Garcilaso employs Renaissance rhetorical strategies to describe Inca rule favourably. Like Cieza and Valera before him, the Mestizo chronicler praises the good governance implemented by the former elite and evokes its edifying mission to civilize the empire's subjects. Inca kings, he claims, were magnanimous rulers who controlled the redistribution of wealth and goods fairly, and

---

[22] Covey 2006.

[23] Garcilaso's mother was Isabel Suárez Chimpu Ocllo, daughter of Tupac Huallpa, the first "puppet" ruler to have been given the *maska paycha* from Spanish hands after the execution of Atahualpa.

aimed to eradicate their subjects' superstitious beliefs and replace them with the laws that the "natural light" had taught them. By the late nineteenth century, the increasing disclosure of his partisan involvement in the politico-juridical debates of his time and his poor empirical knowledge of the field blemished Garcilaso's earlier position as a privileged beholder of native culture. Prominent scholars, such as Marcos Jiménez de la Espada (1879), Heinrich Cunow (1896), Rowe (1945), and Roberto Levillier (1956), claimed that his account of Inca history and religion was "entirely fanciful", and rejected the use of his "utopian novel" for the study of Tawantinsuyu.[24] The Mestizo historian was not alone in being severely appraised in this way. Means (1928), who actually disagreed with the views of his contemporaries on Garcilaso, declared the "utter worthlessness" of the documents assembled on the initiative of viceroy Toledo. He argued that the political enterprise underlying the collection of these works necessarily affected the veracity of their information. For him, "history based upon materials such as these is worse than useless; it is positively noxious to anyone who seeks to know the truth."[25] Most critiques today tend not to be as Manichean and instead focus on the historical validity of specific themes or episodes of the Inca past repeatedly found in primary sources. Among these, researchers place particular scrutiny upon reports of native beliefs because "gentile religion" was a major issue at stake in the definition of the Crown's lawful titles to Peru. Several chroniclers, for example, advance the idea that the Incas had devised a form of monotheism in the worship of a god called Wiraqucha, whom they regarded as their creator and whose cult Manco Capac and his successors spread throughout the Andes for the benefit and salvation of their subjects.[26] The parish priest Cristóbal de Molina even suggests that the ruling elite believed in a Heaven where the righteous rested next to the Maker, while the evil ones tormented by the Devil suffered from starvation and thirst in hell.[27] Luis Jerónimo de Oré (1598), Cabello, and Murúa also relate how Pachacuti Yupanqui would have speculated on the real nature of this god and understood that the Sun divinity was subordinate to him. All three authors report this event in terms reminiscent of the Aristotelian argument of the First Cause.[28] References

[24] On this issue, see Villarías Robles (1998: 286-291); Fernández (2004: 135-137).

[25] Means 1928: 497.

[26] For Murúa and Garcilaso, this "All Merciful" god was Pachacamac, a deity of the central coast.

[27] Molina: 111-112.

[28] For a critical study of this episode see MacCormack 1991a: 258-260; Duviols, in Pachacuti Yamqui: 61-64; Taylor 2003.

to Ancient Rome also became a rhetorical implement to articulate the providential nature of Tawantinsuyu. By comparing Cuzco to Rome, the defenders of the Incas' cause connected the former with the civilization that had become the crucible of Christianity, thereby implying the Incas' predisposition both to receive and to spread the true religion.[29] Other chroniclers claim that the elaboration of religious concepts was directly related to the Incas' ability to conceive a rational tongue.[30] Dominican fray Domingo de Santo Tomás best articulates this idea in the introduction of his *Lexicón* (1560), in which he praises the "Cuzco language" for its syntactic and grammatical complexity and exalts its function as a vehicular tongue. For him, the Incas' ability to articulate rational concepts and their ideation of a supreme divinity demonstrated the work of natural reason and showed the natives' propensity to receive the Word of the Gospel. For others, finally, these elements constituted the evidence that an apostle had trodden ground in Peru before the Spanish arrival. Supporting this view, Murúa, Guaman Poma de Ayala (c. 1615) and Colla chronicler Pachacuti Yamqui (c. 1615) narrate in evangelistic terms the journey on earth of either god Wiraqucha or Tunupa.[31]

As researchers gradually shed light on the many issues related to the collection and transcription of indigenous discourses, they attributed the data discrepancies contained in primary sources to the imprint of European literary traditions, individual biases, and political agendas. For historian María Rostworowski, the reasons for the records' apparent confusion "must be sought in the interpretation of Andean sociopolitical structures by sixteenth-century Spaniards whose mentality prevented them from understanding a society with organizational and recording traditions radically different from their own."[32] Hence, in the course of the twentieth century, scholars appraised the chronicles' historical accuracy on the basis of their author's affinities with the Spanish or indigenous elite, their presence on the site, or the general coherence/structure of their prose. It was assumed that the more factual coherence a text displayed, the more likely it followed a historical tradition upon which one could establish the foundations of a sound account of the past. Ethno-historians gradually disregarded the works of chroniclers who had been overtly sympathetic to the

[29] MacCormack 1991a, 2001, 2007.

[30] Mannheim 1991: 68-71.

[31] Duviols 1977a: 56-70; MacCormack 1991a: 312; Pease 1995: 44; Bouysse-Cassagne 1997: 159-160. Las Casas and Cieza make a reference to this tradition, but do not support it.

[32] Rostworowski 1999: viii.

arguments of the Controversy inasmuch as the ideological stands diffused through their lines would have dramatically distorted the pre-Hispanic material they enclosed. Likewise, they regarded with increased suspicion the testimonies of those who had never set foot in Peru because they had not witnessed the social practices they described or had not collected first-hand information. Among these unfortunate authors, Gutiérrez de Santa Clara is scathed with an enduring ill reputation despite the richness of the ethno-historical data he recorded. To the chronicler's discredit, historian Marcel Bataillon revealed that Gutiérrez extensively copied material from his predecessors, namely, Cieza, Francisco López de Gómara (1552), Augustín de Zárate (1555), and Diego Fernández (1571). His monumental *Quinquenarios* also contains a variety of information on Inca society that no other sources ostensibly support, prompting Bataillon to dub him the "writer-novelist". Adding to his unpopularity, Gutiérrez sets the beginning of the Incas' primordial migration in the Lake Titicaca Basin, which is a version of the origin myth that some scholars regard as a colonial construction. Among them, Martti Pärssinen suggests that the chroniclers who locate the emergence of the Inca ancestors in the Collao area may have mingled the original narrative starting near Cuzco with the epic journey of the god Wiraqucha, born from the waters of Lake Titicaca. Although he points out that many Spaniards situate the Incas' birthplace in the altiplano, Pärssinen notes that "Cieza de León, Betanzos and many others firmly argued against this view in the 1550s (yet) it did not prevent Gutiérrez from making his claim again with new modifications based on his own imagination."[33] Even today, Gutiérrez's chronicle remains marginalized.

Polyphony of the Inca Dynastic Accounts

Attributing the origin of all textual discrepancies to colonial actors presupposes that the Inca historical tradition formed a corpus whose content was fixed as well as coherent, and which we would need to refine from European contaminations to extract its original narratives. Yet, primary sources explicitly contradict this alleged homogeneity of the Inca accounts on two distinct points. First, they provide evidence that the ruling elite of Cuzco transmitted their oral history through a variety of formats, using modes of recollection dependent on the contexts and means of recitation. At least two different types of devices served as media of historical transmission,

---

33 Pärssinen 1992: 61.

including pictorial representations on tablets (*tablones pintados*), and ritual chants with which the *khipu* or knotted-cord recording system was associated.[34] Unfortunately, we know little about the series of tablets adorned with gold on which the Incas would have painted "the most noteworthy events of their ancient history."[35] The only mentions of these objects come from Cristóbal de Molina, who collected information among the Cuzco nobility in the late 1560s, and from Sarmiento. The former claims that the tablets featured the myth of Manco Capac's emergence from the Paqariq Tampu cave, as well as the life and conquests of every ruler "painted in accordance with their mode of figuration."[36] According to Sarmiento, the elaboration of these pictorial accounts dated from Pachacuti Yupanqui. This king would have consulted first with the Inca holders of tradition and listened to them recounting the "antiquities, origin and most noteworthy events of their past."[37] He then had this material painted on tablets and placed in the secluded room of a temple dedicated to the Sun called Puqin kancha. This site was a pilgrimage station southwest of the capital where access to the pictorial accounts was restricted to the ruler and the few learned men who had acquired the knowledge of retelling the stories they represented.[38] Some isolated documents also suggest that the Incas used painted boards to register the kings' reigning dates, as well as matters regarding legislation.[39] Upon taking office, viceroy Toledo ordered the reproduction on cloths of the last known example of pictorial records of the Inca past. The task was entrusted to indigenous experts who created four *lienzos*, as they became known in Spanish inventories. The first depicted the

[34] Although I cannot expand on the subject here, it should be noted that Inca studies are increasingly bringing our attention to the role of textile geometric motifs (*tukapu*) and ceramic iconography in the transmission of cultural norms (Cereceda 1986; Silverman 1994; Cummins 2002a, 2002b; Frame 2007); others point out to the instrumentality of the visual world and the spatial perception of landscape in enacting social memory (Hill 1988; Rappaport 1990; Howard-Malverde 1990; Arnold, Jiménez & Yapita 1992; Gow 1991; Boone & Mignolo 1994; Abercrombie 1998). To this day, analyses of these forms of pictorial and representational communication remain subject to a wide variety of approaches. While some posit that the *tukapu* patterns adorning Inca garments could be read as logographs, others suggest more persuasively that these motives were not linked to oral speech but registered forms of social knowledge corresponding to people's activity within the empire's structured organization. Viewed in this way, iconography on vessels and textiles possessed a cognitive significance that operated as an alternative to orality.

[35] Sarmiento: Ch. 9, 49; Ch. 30, 95.

[36] Molina: 50.

[37] Sarmiento: Ch. 30, 95.

[38] Sarmiento: Ch. 9, 48-49; Molina: 49-50; Acosta: Bk. 6, Ch. 8, 290.

[39] Porras Barrenechea 1999: 122-162.

Incas' origin myth and the epic on earth of their god Wiraqucha; two others featured the portraits of each ruler and his wives, together with representations of their deeds. The last one, which was remarkably longer in size, contained the genealogy of the Inca dynastic group. On January 1, 1572, members of the former ruling elite authenticated the content of the completed cloths, which were then sent to King Philip II with Sarmiento's *Historia Indica*. The *lienzos* remained some sixty years at the repository of the royal collection in Madrid before meeting a tragic fate in 1734, when a fire consumed the building and part of its collection.[40] After this event, Toledo's *lienzos* disappeared from the royal inventories, and with them the latest pictorial accounts of Inca history vanished. A few references in primary sources suggest, however, that the colonial Inca nobility carried on the practice of painted accounts until the early seventeenth century. Cobo, for example, alludes to similar representations on *cumbe* textiles, perhaps copies of the *lienzos*, which Polo would have made use of to document his writings. The Jesuit father claims to have seen one of these "tapestry" cloths, which he describes as "no less delicate and carefully represented than if it were on fine royal fabric."[41] Garcilaso also attests to have received in 1603 a petition from several Inca noblemen, enclosed with a royal genealogy on fine European cloth that detailed Manco Capac's descent down to Paullu Inca.[42] Yet, none of these objects have survived.

The Inca ritual chant was another medium of historical transmission, anchored in a secular tradition observed throughout the Andes. The original purpose of these chants was to remember and honour local chiefs who had served their community with dedication and temerity.[43] For Cieza, they were comparable to the *romances*, or "historical poetic narration", of the Iberian Peninsula that centred on the deeds of single individuals. He explains that in pre-Hispanic Cuzco no oral account of a king's deeds was allowed to circulate during his lifetime. Only once a ruler had passed away would a group of elders, originating from his kin group, gather to determine whether the dead had been sufficiently magnanimous to consider preserving his memory. If the assembly agreed that he deserved reminiscence, it was ordered that a chant be composed in which he should be praised in such ways as that "all the people should be astonished to hear of deeds so

40 Julien 1999.
41 Cobo: Bk. 12, Ch. 2, 59.
42 Garcilaso: Bk. 9, Ch. 40, 295-296.
43 Ramírez 2005: 150.

mighty."[44] The song was recorded on *khipus*, the device composed of knotted dyed strings of wool or cotton. It was intoned, accompanied with music and dance, at every royal investiture, wedding, and funeral heretofore, when the life stories of exemplary Inca rulers were recited in turn. Only specialists could interpret them in the mandatory presence of the deceased whose life was being retold. Murúa suggests they were responsorial song-performances that lasted three to four hours each: a lead reciter opened the narration and was echoed regularly by a chorus of both men and women.[45] If anyone dared to evoke the content of the chants outside the prescribed ritual circumstances, he or she was severely punished. However, if a ruler "had been negligent, cowardly, or vicious, or preferred pleasure to the labour of extending the bounds of the empire, it was ordered that such a king should receive little or no mention."[46] In this way, ritual chants did not constitute an exhaustive record of Inca history, but rather offered a selective appraisal of the achievements accomplished by one ancestral figure.

The author of the *Noticia del Perú*, identified as Miguel de Estete (1535), claims that these chants were the principal means for the Incas to remember "the events and wars that occurred in the past; and even though [they] lack writing, we have memory of those who accomplished great feats by way of these *cantares*."[47] The same chronicler had the opportunity of hearing these songs performed at the investiture festival of Manco II (1533). The gathering took place in Cuzco after the combined Spanish-Inca expedition had defeated Atahualpa's remaining allies. In the midst of this chaos, Pizarro's endorsement of the young ruler did not suffice, and the native nobility expected Manco to organize the accustomed celebration. The event lasted over thirty days, during which people assembled on the main plaza to drink their plenty and dance. For his part, Manco began his ritual accession by retreating for three days and visiting the mummy bundles of every royal lineage's ancestor in their separate estates. The corpses were then transported to the main plaza and positioned on their grandiose litters next to one another. As the festival unfolded, the celebrants chanted the life and conquests of every mummified lord, praising their might and expert command. With the songs performed, the main minister of the Sun

---

[44] Cieza: Pt. 2, Ch. 11, 28.

[45] Murúa, *Historia del origen:* Ch. 6, 176. Betanzos (Pt. 1, Ch. 17, 86) indicates that the chosen women were the ones who sang these chants.

[46] Cieza: Pt. 2, Ch. 11, 28. See also Santillán: 103; *Discurso*: 3.

[47] Estete: fol. 11r.

admonished Manco publicly, advising him to follow in the footsteps of these ancestors and to apply a comparable audacity and valour to his governance. A powerful spirit of emulation would have emanated from the recitation of these narratives, reinforced by the physical presence of past rulers witnessing the investiture. Interestingly, a similar sentiment would have proceeded from the exposition of the mummy bundles to the male novices on the final days of their initiation (*warachiku*). This presentation occurred after the probationers had competed in several trials to demonstrate their physical strength and endurance. For this occasion, the celebrants intoned "several chants in praise of their false gods and past Inca kings", after which the young boys asked their forebears to make them as "fortunate and valiant as [they] had been in their lifetime."[48] In both events, the historical chants conveyed exemplary models of successful leaders, prompting the ritual participants to act accordingly.

The exclusive attachment of these epics to a particular individual was such that their transmission remained within the kin sphere of the ruler whose life was being remembered. In this way, every panaca was in charge of keeping alive the memory and vital energy of its mummified ancestor. They cultivated his lands with the help of several *yanakuna*, but also clothed, fed, and entertained the deceased daily in the palace he had occupied in his lifetime. Designated officiants would occasionally transport his corpse to the estates of deceased peers with whom the mummy was believed to engage in conversation.[49] The perpetuation of his past feats participated in this continuous actualization of the ancestor's vital strength. His renewed vigour, in turn, ensured the prosperity and health of the descent group. The subjection of historical chants to individual lineages meant as well that they were contingent upon the interests of panaca members seeking to assert the legacy of their own ancestor. In Cuzco, where the various descent groups maintained tense relations with each other and competed for influential offices, differences in the perception of historical events would have been particularly exacerbated. Violent wars and bitter reprisals often plagued royal successions, which left the weakened panacas out of the political reconfiguration in progress. The most widely documented conflict that placed two pretenders to the royal office in opposition is certainly the war in which Atahualpa and Huascar were

---

48 Molina: 107-108. Se also Gutiérrez: Ch. 64, 254.

49 Pizarro: 42; Betanzos: Ch. 17, 85-86; Ch. 41, 181-183; Cieza: Pt. 2, Ch. 11, 27-29; Santillán: 112; Cobo: T. 2, Bk. 13, Ch. 10, 163-165. See also Alonso Sagaseta 1989: 114-115; MacCormack 1991a: 68-71; Gose 1996b: 4-5, 8-9; Bauer 2004: 162-177.

engaged when Pizarro set foot in Tawantinsuyu. Historians already noted that the panacas' factional views permeate the primary sources on this event, resulting in competing reports of the conflict. However, a particular episode of this war reveals the strategic importance of "owning" a history among the Inca ruling elite. In the course of this struggle for power, several of Huascar's wives and children, as well as his most loyal allies in Cuzco, were slaughtered. In addition, Atahualpa had the leading members of Tupa Yupanqui's panaca killed and ordered that the mummified body of this late Inca be burnt because his descent had supported Huascar's election to the royal office.[50] He also sent his most loyal captains to kill his rivals' *khipukamayuq* and destroy their records because "a new era had to start."[51] With the physical annihilation of this ancestral figure and the massacre of those who perpetuated his deeds, Atahualpa was striking a strategic blow to weaken the ideological apparatus of a rival panaca. He knew that condemning the memory of a forebear to oblivion was denying his descent any legitimacy in the religio-political game that was being redefined at each royal succession.

In this highly stratified society where kin ties and exceptional merit defined access to authority, historical narratives had been instruments of political legitimation and thus objects of contestation well before the implementation of the Spanish administration. By means of the *khipu* records, the panacas established the eminence of their ancestry and kept account of their forebears' remarkable actions, thereupon asserting their members' rights to prestigious offices. Their recordkeeping of past events perpetuated exemplary models that were concerned exclusively with illustrious individuals of the ruling caste. They were designed to create awe and flatter the membership of specific royal households. In keeping with this line, they scarcely mention the non-royal ayllus that made up Cuzco's population, even though some of these groups claimed a common origin with the Inca nobility and participated actively in the official festivals. Finally, a lineage lacking a historical tradition was not only deprived of political influence, it was also devoid of the aura that the recitation of these accounts conferred in ceremonial contexts. The performers rivalled in narrative talent to exalt the deeds of their lineage's head and ensure their audience's engrossment. Touching on the subject of speech virtuosity, John M. Schechter (1979) argues that the Incas' appreciation for rhetorical skills is reflected in González Holguín's early bilingual dictionary. This source

[50] Sarmiento: Ch. 67, 162-164.
[51] *Discurso*: 3-4.

lists a great variety of entries appraising vocal and speech quality. Some indicate respect and praise for those who speak with authority and spirit, while other entries assess negatively inappropriate or mistaken language and singing. Historical song-performances provided an ideal setting for expressing forcefully contrasting views of the past, which necessarily resulted in a dissonant corpus of historical accounts. As Frank Salomon observes:

> In societies where plural kinship corporations associate at the apex of power, historical truths are held to be plural, proprietary, and almost incommensurable. The attempt to reconcile or merge overlapping testimonies from different corporate sectors (e.g., lineages, castes) of a given society into a synthetic image of "what really happened" is misguided because in many cases historical truth is felt to be relative to, and the property of, the group "owning" a tradition.[52]

The Inca historical narratives were no different from similar productions worldwide and across time. An extensive literature exists on the political dimensions that inform discourses on the past, whether they are mythologies, genealogies, or life stories of historical individuals. Edmund R. Leach famously observed that the Kachins of highland Burma produced multiple versions of traditional tales that varied according to contrasting vested interests. "There is no 'authentic version' of Kachin tradition to which all Kachins would agree, there are merely a number of stories which concern more or less the same set of mythological characters and which make use of the same kind of structural symbolism (…), but which differ from one another in crucial details according to who is telling the tale."[53] As it were, the rituals afferent to these mythical constructions did not tend to replicate any organic order of social structure, but rather constituted a language of argument among factions. Following up on this contention, Arjun Appadurai (1981) qualified the discourses on the past as "an aspect of politics, involving competition, opposition and debate". His analysis of corporate interests in the control of a temple in Madras City shows that the past plays a fundamental role in the expression and resolution of conflicts in the present day. The norms that govern antagonistic versions of the past are sufficiently flexible to create a space of *debatability*, which can allow new forms of actions and regulate social change. In both cases, historical truth hinges less on the objective course of events than on the socio-political setting in which the narratives are enunciated. It is thus subject to multiple

---

52 Salomon 1999: 84.
53 Leach 1954: 266.

variations through time and cannot be reduced to an immutable, authoritative picture of the most credible occurrences. In fact, verisimilitude in historical accounts can be misleading when one tries to evaluate the factual validity of the data they enclose. Eric Jolly (2002) illustrates this point with clarity in his study of the variability of Dogon epics (West Africa). Among these, the *baji kan* is a public harangue that itinerant storytellers, versed in the art of dramatic performance, improvise by taking inspiration from local traditions. In doing so, they construe foundational figures as stereotypical heroes whose craftiness and tenacity ensured their access to power after they overcame excruciating ordeals. With its rhetorical apparatus and the appearance of a linear account, the *baji kan* forges a "trans-regional, consensual and empowering past" aimed at enthusing a wide audience. In contrast, the *Abirè nu* is a collective and repetitive chant that excludes improvisations. Chanted in prophetic tones at funerals, its purpose is to illustrate the vanity of power by retelling the past of plains communities under Peul dominance in the eighteenth and nineteenth centuries. Although this narrative format has remained relatively stable through time and evokes real events, Jolly notes that ethnographers have overlooked the *Abirè nu* for evidence of past events. Instead, they comply with the native view that the *baji kan* is a recipient of abiding and more "historical" accounts. This is because the latter conveys forcefully a credible picture of illustrious forebears, which serves the elites' current politics and is shared by individuals regardless of their ethnic and religious background. On the other hand, the prophetic tonality and repetitive scansions of the *Abirè nu* override its historical thrust. In this case, the enunciative modalities wrongly determine judgement of the narratives' objective content.

These studies urge us to question the relevance of investigating the historical veracity of the Inca dynastic accounts on the basis of their seeming credibility. They likewise support the view that narrative discrepancies in primary sources may be not only the results of colonial contaminations, but also indications of conflictive, yet coexisting, views of the past that pre-dated the Spanish conquest. The exclusive attachment of Inca historical elaborations to a specific lineage, their political dimension, and the variety of narrative media in which they were registered were conducive to alternative perspectives on the past. Moreover, the Inca historical records penned in primary sources intertwined different types of discourses, genres, and voices. These texts integrate and often coalesce a variety of statements, from descriptive data relative to social norms and prescriptions (e.g., marriage preference; royal succession rule) to reflexive discourses in which individuals experience these norms in a variety of fashions (e.g., an account of the circumstances in which one chooses to marry an unconventional partner; the epic story of a young upstart who accesses power through unorthodox means). These different types of statements may refer to realities that are not necessarily congruent, notwithstanding their veracity within their own category of speech. In addition, reflexive discourses yield different literary categories that define their genre. The products of these genres may convey contrasting but equally valid perspectives on the past that sometimes inform each other. In ancient Greece, *mythoi* and *historia* were not mutually exclusive elaborations; their differences lay *de facto* in their rhetorical treatment of the past. Myths, amongst other things, answered questions on the origin of institutions that reckoned historical effect as the doings of gods among men. In the fifth century BC, Herodotus introduced the notion of inquiry (*historia*) into the past with a focus on human affairs to explain the succession of causal events that led to the Persian Wars. Although the course of his history is the result of men's social action, Herodotus punctuates his discourse with myths to instil necessary dramatization. In all of that, the question of truth or plausibility in the historical or mythical inquiry was rarely a matter dealt with in his time. As Peter G. Bietenholz explains: "both distinctions, history as against myth and truth as against falsehood, were basically separate, and both were optional, rendered inconsequential by the sovereign right of each *logos* (tale) to be told."[54] Historical consciousness in ancient Greece did not result from the systematic appearance of a dialectical objectivity that condemned mythical constructs to oblivion. These two modes of representation coexisted, and what explained their discrepancies were their contexts of transmission and aims. The concept of time is likewise relative to the format and purpose of the discourse. The Maya of the Classic Period (c. AD 250-900) devised two modes of dating events and measuring time. One tradition presented historical process in a linear fashion from a fixed starting point. It described the genealogy of noble households and their migrations as well as conquests, and defined the type of authority that enabled particular individuals to rise to power. These narratives coexisted with another type of historical elaboration that reckoned the same occurrences within repeating cycles linked to cosmic times. Their discursive qualities were more sophisticated and the events they described involved the direct agency of supernatural forces. Nancy M. Farriss (1995) suggests

54 Bietenholz 1994: 24.

that these latter constructs supported a political model that regulated succession on a rotational basis, alternating every four years. In contrast, the linear record established which person(s) could assume power at any given time. These considerations bring attention to two interconnected avenues of research into a renewed analysis of Inca historical accounts. One is the study of the socio-cultural circumstances that informed the production and public utterance of these narratives; the other is the comparative literary analysis of this corpus with regard to unravelling the narrative threads and the different forms of discourse entangled in the chronicles. This twofold approach, far from rejecting the pertinence of previous historiographical works, should be complementary to them. There is no doubt that the political issues associated with colonization and the literary conventions of Modern Europe framed in part the chronicles' elaboration. These factors, however, do not elucidate the pervasiveness of narrative discrepancies in the Inca foundational sources, nor would they have silenced altogether the pre-Hispanic voices that had elaborated them originally. As Gary Urton already noted, colonial documents contain prefigured elements shared by both native informants and the representatives of the dominant power.[55] Their hybrid discourses resulted from the choices individuals made to appropriate and combine certain elements of traditions, thus internalising an Andean substratum whose characteristics allowed for a rationale of assimilation. Ultimately, as Peter Burke writes, "what the historian needs to investigate is the logic underlying these appropriations and combinations, the local reasons for these choices."[56] Moiety relations and dual cultural productions offer circumscribed objects for the analysis of the socio-political and cultural circumstances in which the Inca historical narratives were elaborated. They took over various manifestations associated with asymmetric power, ritual violence, political tensions within the royal descent group, and the historical development of the dynasty, all of which provide a rich corpus for a comparative analysis of its descriptions in Inca narratives. The resolution of Inca history's seeming contradictions requires that we identify the different types of statement entangled in the chronicles, as well as the modalities proper to pre-colonial discourses. By isolating for a particular narrative cycle the contrasting sociological data associated with each version, a clearer picture of Inca historical elaborations emerges.

---

[55] Urton 1990: 62.

[56] Burke 1997: 209.

CHAPTER TWO

FILIATION IS HISTORY: THE INCA DIALECTICAL VIEW OF THE PAST

With the development of a new approach to Inca historiography, by the mid-1960s some radical reassessments of the empire's chronological expansion and its dynastic politics emerged. Zuidema (1964) first proposed an implied asymmetric diarchy in the Hanan and Hurin division of the conventional kingship list. He claimed that this genealogical record provided a model of political order, not a register of factual events. In his view, the Inca narratives perpetuated a body of myths relative to the royal lineages' ancestral founders, who could not be considered historical figures. Duviols (1979c), despite his disagreement with Zuidema on this last issue, furthered the diarchy thesis. He suggested that the chroniclers had indeed misinterpreted native testimonies in such a way that they had shaped into a linear account what had initially been a simultaneous coregency of two lineages. The *Sapa Inca* would have proceeded from the Upper part, while his *segunda persona* originated from the Lower one. To resolve the imbalance in the number of rulers traditionally documented in each moiety, Duviols suggested that the Hanan sequence of nine generations corresponded to the most accurate dynastic span. Thus, not every Hurin lord would have been remembered. As Peter Gose (1996a) argued conclusively, the primary evidence supporting this model is extremely scant and problematic. Even though there are widespread precedents of Andean dual socio-political structures at the provincial level, colonial sources concur in describing a unilateral wielding of power at the apex of the Inca hierarchy. Whether or not the sovereign delegated the implementation of his decisions to a second-in-command, entrusted the conduct of his army to an able relative, sought advice from a high-ranking counsellor, or appointed a governor in Cuzco while at war does not dismiss the autocratic nature of the Inca ruling system. Most monarchies enforce similar governing instruments that do not call into question the king's absolute authority. Likewise, the fact that the local interests of each moiety were entrusted to its own leader does not necessarily entail an institutionalized diarchy at state level in which power is effectively divided in separate realms. Far from this view, the chronicles describe an Inca ruler who cumulated supreme authority over all domains, be they ritual, military, or political. Of course, other members of the elite held important offices in these spheres, including the lord identified by the Spanish as the high priest of the Sun. None of them, however, appear to have held independent or exclusive power. The king and the king alone was head of Tawantinsuyu. The Inca ceremony of royal investiture, which was centred on the divine election of a single individual regarded as being the unique son of the Sun, reinforces this picture.

Ultimately, the diarchy model does not explain why the Inca dynastic accounts report two modes of moiety opposition: one based on kin relations and the other on temporality. To grasp the cultural foundations of this asymmetrical partition, a sound understanding of the nobility's social structure is required. Since the moiety system and its relational antagonisms are primarily linked to kingship, my analysis of Cuzco's dual social division focuses on determining the features of Inca dynastic rule. Some traits of social classification, including stratification by kinship, bilateral transmission of ranks, siblingship, distinction between senior and junior lines, and marriage preference, support my reading of the royal descent group as a conical clan. I argue that this formal model, which Kirchhoff (1949) was the first to identify in the Andes, clarifies the dynamics of the Hanan/Hurin antagonisms and unveils some of the mechanisms that shaped historical consciousness in ancient Cuzco. As opposed to previous interpretations of the kingship succession rule, I propose that Inca narratives reported two competing strategies to access the imperial office: one was based on the primogeniture rule, and the other on merit. For any aspirant to the *maska paycha* (royal headdress), both his position in the dynastic line and his personal aptitudes—his ability to lead, command, and conquer—conditioned his accession. These two qualities gave rise to two contrasting narrative genres that depict individuals through the idiom of dual classification. Moiety oppositions in Inca historical narratives and rituals not only reveal aspects of the underlying structure of the nobility's kinship relations, they also contrast these social patterns with the violent realities of royal succession. Taken together, these narrative genres reconcile structure with history by allowing the contingencies of the succession wars to play an active role in legitimizing kingship. As such, expressions of antagonism and affinity in Cuzco cannot be read as literal translations of actual marriage networks, as they can in many dual organizations consisting of exogamous moieties. Rather, they translated two opposing political strategies to access kingship that preserved the architecture of the Inca prestige hierarchy.

The Inca Conical Clan

Royal Inheritance Rule

A stirring debate in Inca studies concerns the identification of the royal succession rule on which colonial sources are divided. Chroniclers such as Betanzos, Fernández, and the authors of the *Discurso* identify categorically patrilineal primogeniture as the principle of kingship inheritance and emphasize the hierarchical predominance of the firstborn child over younger siblings. They maintain that the royal headdress went to the eldest son of the king's legitimate descendants by his principal spouse, whose progeny were called *piwi churi*, literally "firstborn sons."[1] Supporting this claim, the life stories of individual rulers written down by Father Cobo present the dynastic line as an unbroken succession of senior sons. However, several chroniclers also record variants of the same historical narrations, where the youngest child actually succeeded his father's office. In fact, the very same colonial texts that explicitly identify primogeniture as the royal succession rule provide examples contradicting this prerogative. Hence, Murúa's astonishment that Manco Capac became king "although some people say he was the youngest and last of the Inca brothers" to have emerged from their birthplace, the cave of Paqariq Tampu.[2] Betanzos records the same information, as do Cieza and Las Casas, who name Manco Capac the last of the Ayar siblings.[3] Other narratives, including the stories of ascent to the imperial office of Inca Yupanqui, Tupa Yupanqui, and Huayna Capac, tell how these lords took over supreme authority after outshining, and sometimes killing, their eldest brother.[4] Santillán defines the rule of royal inheritance more precisely in his *Relación*, directed at Crown officials: "the order regarding succession among the Incas was that the son succeeded his father, except when the Inca had several sons, succession

[1] Betanzos: Pt. 1, Ch. 16, 78; *Discurso*: 12; Fernández: Pt. 2, Bk. 3, Ch. 5, 80. *Piwi* designated "the first-born daughter or son" and *churi* is employed by a man to refer to his sons, see González Holguín, *Vocabulario*: 545. Cieza also suggests on several occasions that the eldest son was the legitimate royal heir, see Pt. 2, Ch. 8, 22; Ch. 31, 95; Ch. 33, 101; Ch. 35, 107.

[2] Murúa, *Historia general*: Bk. 1, Ch. 2, 39.

[3] Betanzos: Pt. 1, Ch. 3, 17; Cieza: Pt. 2, Ch. 6, 13; Las Casas: T. 2, Ch. 250, 393.

[4] Betanzos: Pt. 1, Ch. 6-9, 21-41; Sarmiento: Ch. 26-33, 84-99; Ch. 42, 115-116; Ch. 55-56, 138-139.

was not necessarily given to the eldest, but to the one the father preferred."[5] For Fernández as well, the primogeniture rule could be infringed upon and preference given to a brother of the legitimate heir if he demonstrated more valour.[6] Perhaps it is Murúa who best summarizes the qualities required of an Inca monarch:

> If the eldest son of the *Coya* (principal wife) was incompetent and not apt to govern, he was never allowed to succeed; they searched among the other sons she had with the Inca, a man to rule wisely over them and who would pursue the conquests; such a one the Inca appointed as his heir. And if he could not be found, they picked the bravest, most astute and valiant man at war among the sons of the Inca's other wives, the most trustworthy one who would expand his ancestors' kingdom, who would govern with most justice, and who proved spirited and sagacious in campaign, this one the Inca elected as successor.[7]

Rostworowski (1960) was the earliest scholar to describe this situation as "succession by the most able" whereby any son of the deceased lord could inherit the royal fringe if he proved able in battle and apt to govern. She thus finds little truth in the existence of primogeniture among the Incas and believes instead that references to this rule in primary sources reflected the Spaniards' own concept of inheritance. Most Andeanists, including Zuidema, Duviols, and Pease, have since adhered to this view.[8] However compelling this interpretation might be in accounting for the multiple Inca succession wars, it nevertheless dismisses too promptly the various and detailed information indicating that primogeniture operated in the royal descent group. Betanzos in particular is extremely specific in describing this rule. In doing so, he stresses the queen's descent in determining succession:

> This lady had to be a direct relative of the Inca on both the paternal and maternal sides without the least trace of *wakcha q'uncha* [Hurin Cuzcos]. The Inca received this lady as his principal wife the day he received the borla fringe of office and royal insignia. The sons this lady had were called *piwi churi*, which means « legitimate children », and the eldest of these sons was lord of the empire and the legitimate heir (...) After receiving this woman as his wife or before this one, he had fifty other women who were his sisters and relatives, but the children he had with these would never

[5] Santillán: 108.

[6] Fernández: 80.

[7] Murúa, *Historia general*: Bk. 2, Ch. 14, 368.

[8] Pease 1991: 88, 107; Zuidema 1964: 122-128; Duviols 1979c.

> succeed him in office unless he was the son of the legitimate *piwi* wife, as they say.[9]

Bewildering as it may appear, Betanzos is the same author documenting the life stories of several kings who had infringed upon the inheritance rule he outlines. The chronicler reports at length how Inca Yupanqui, the youngest son of Viracocha Inca, ascended to power after his eldest brother, who was also their father's preferred heir, fled from Cuzco on the eve of the Chanca invasion. He likewise retells the royal election of Tupa Inca, whom he presents as a cadet son of Inca Yupanqui. Despite the apparent discrepancies between the subject theme of these narratives and the succession rule he outlines, it is unlikely that Betanzos intentionally deformed or misunderstood the material made available by his informants. Other chronicles present similar peculiarities but, most importantly, the degree of precision Betanzos unveils in his description of primogeniture, including the mention of a native word to describe the legitimate children, should preclude us from drawing hasty conclusions. To harmonize these conflicting data, an alternative position would be to consider primogeniture as the *ideal* rule of Inca royal succession. Betanzos' *Suma y narración*, like most chronicles, combines different types of enunciation, that is, it intermingles sociological and descriptive materials together with narrative elaborations in the form of dynastic accounts. In a single text thus coexist the articulation of prescriptions and social norms, as well as reflexive discourses on these norms. Here, the specification of the ideal criteria of inheritance—its normative description—coexists with the narration of events that prompted royal pretenders to infringe upon the established order. It would be counterproductive, even misguided, to elect one of these versions as the true reflection of past realities. The information they enclose belongs to different categories of speech and evokes realities that are not necessarily congruent. Even more importantly, many aspects of kingship narratives, including the effects they aimed to have on their audience or their purposes compared with other types of discourses on the past, would escape us if we were to overlook the ideal principles of royal succession. Moreover, European and modern Western societies did not and do not have a monopoly on primogeniture, which was largely observed in various monarchies throughout the world. In the Inca realm, as in other forms of polities, the reality of the political arena meant that this prerogative did not always prevail. The *de facto* viability of an expansionist state like Tawantinsuyu

9 Betanzos: Pt. 1, Ch. 16, 75.

calls for rules to be transgressed, especially when they may impose the nomination of an unreliable candidate for the most demanding role. It follows that birth position among siblings or within a given lineage did not inevitably endow individuals with the corresponding title or office. Status and rank in ancient Cuzco, as in most societies with social stratification, can be seen as "a latent quality, a potential that must be 'activated' through experience and deed,"[10] all the more so when the perpetuity of kingship was at stake. For this reason, every Inca pretender to the royal office, whatever his genealogical position relative to the former ruler, had to evince his warlike abilities, including his aptitude to coalesce powerful allies under his command. Was the imperative to demonstrate personal merit mutually exclusive from a prescribed rule of inheritance? Not necessarily. Any formal model of social organization represents an ideal, a structure of reference that aims to ensure stability and the perpetuation of successful institutions. As such, it is not concerned with variations in the political context, although on the ground historical actors regularly violate rules and rework their institutions. This clash between the necessary established order and historical contingencies may prompt a society to produce a polymorphous discourse on past events in which individuals legitimize their position through different means. One tradition may emphasize a person's genealogical status and the other his personal aptitudes, although both aim to establish eligibility for a coveted charge. These two sources of legitimacy, however contradictory, can also be cumulated to confer ultimate authority. This was precisely the strategy adopted by Hawaiian kings in the eighteenth and nineteenth centuries when Europeans recorded two conflicting discourses on the origin and deeds of past rulers. In this archipelago, genealogical seniority was the prescribed rule of royal inheritance. Following this principle, Hawaiian genealogical chants depicted the dynastic line as "an uninterrupted descent from firstborn to firstborn."[11] Still, in Hawaii, another type of discourse, the historical narrative, contradicts the uniform content of these chants and lists, as there were, several instances of secondary wives' sons succeeding to kingship after triumphing over their older brother.[12] None of these representations of past events was more valid than the other; it was their context of transmission and aims that explains their discrepancies.

---

[10] Carr (2005: 240) about Havana Hopewell societies of North America.
[11] Valeri 1990a: 169.
[12] Sahlins 1985; Valeri 1990b.

Primary material on Inca kingship presents similar discordances that offer a window into native historical constructions. It appears that some pre-Hispanic narratives emphasized the perpetuation through time of the structural features of royal descent, such as inheritance rule, marriage preference, or affiliation patterns. These traditions accounted for the success of the dynastic line by highlighting its structural stability and duration: firstborns succeeded firstborns, kings married their full sisters, and Hanan lords succeeded chronologically to Hurin leaders without conflicts. These traditions constituted only one facet of Inca history. Other narratives detailed the intrigues to seize power realized by outsiders to the senior line. They exalted the individual qualities of kings, their ability to govern and conquer new territories having shattered the prescribed rule of royal inheritance. These traits concur with Julien's (2000) seminal analysis of Inca historiography in which she establishes through textual analysis that colonial sources indifferently enclose native material from two distinct discursive genres: the genealogical accounts and the life stories of individual rulers. They also reveal that the Incas, like many other societies, did not conceive of affiliation principles as mere rules of social reproduction. The disruption of the established order, and thus historical contingencies, ensured that imperial governance fell to the hands of the most able commander. To understand how these two narrative genres opposed each other, we must now turn to the analysis of royal descent relations. A survey of the constitutive aspects of the Inca elite's kinship system reveals a strong adequacy with the formal model of the conical clan, a basic form of many Polynesian social organizations where it is known as "status lineage" or "ramage."[13] Kirchhoff (1949) drew the main lines of this argument, which Jenkins (2001) has since furthered. Here, I offer an analysis of the royal descent group distinct from theirs by interpreting data discrepancies in primary sources as dialectical patterns inherent to Inca historical constructions.

[13] Largely recognized in Polynesia and Micronesia, this form of social organization is called 'conical clan' by Kirchhoff 1955, Hage & Harary 1996; 'ramage' by Firth 1957, Sahlins 1958; and 'status lineage' by Goldman 1970. It has been argued that the Maya were also organized into status lineages (Hendon 1991). I chose to use the designation 'conical clan' throughout this study because it is consistent with the previous references of this model in the field of Inca study and because the term 'clan' best conveys the extended nature of Inca descent.

The Qhapaq Ayllu

Since Kirchhoff's earliest description of the conical clan, anthropologists of all persuasions have revisited some of its aspects. In the light of the work of these critics, it is possible to delineate a number of characteristics attached to this particular model of social organization. First and foremost, in its anthropological definition, a clan is a unilineal descent group that may encompass and unite a series of ramifications or lineages that recognize a theoretical common ancestor. The particularity of the conical clan is that descent is not necessarily traced through a single line (male or female) but may privilege cognatic principles of social stratification. It is also a system of stratification by kinship whereby every member of the social group is ranked according to closeness to the apical ancestor. To borrow a formulation by anthropologist Valerio Valeri, "temporal proximity—genealogically reckoned—to a source-event (the founding of the lineage inscribed in an ancestor's procreation) determines status."[14] A logical implication of this structure is the hierarchy of senior and junior lines and its correlate: the relational asymmetry between elder and younger brothers, so that primogeniture, and sometimes ultimogeniture, is the ideal criterion for determining rank.[15] Following this rule, within the nuclear family, a firstborn son theoretically possesses a rank superior to his cadets because he stands nearer to the deified ancestor in the descent line. His descendants likewise outrank his younger siblings' descendants. The same rationale governs the composition of the entire descent group, which can be subdivided into several lineages originating from the children of the original forebear. These ramifications are thus organized hierarchically according to the degree of proximity that their respective founder maintains with the common apical ancestor. A representation of this normative model shows a main line descending from the primogenitor, and from which proceed junior divisions of declining ranks, so that A is superior to B, which is itself superior to C, etc. (Figure 1).

Similar principles applied to the ayllu, the Andean basic organizational structure.[16] Historical material reveals that it originally functioned as a

---

[14] Valeri 1990a: 159.

[15] Leach (1954) observes that the Kachin of Burma rank lineages through a rule of ultimogeniture.

[16] Sixteenth-century sources loosely describe the ayllu as *parcialidad*, genealogy, family, lineage, caste, nation, segment of inalienable community lands, and barrio of Toledan *reducciones*. Comparing contemporary ethnography with historical material on Andean polities reveals that the configuration of the ayllu varied across time and space. Centuries

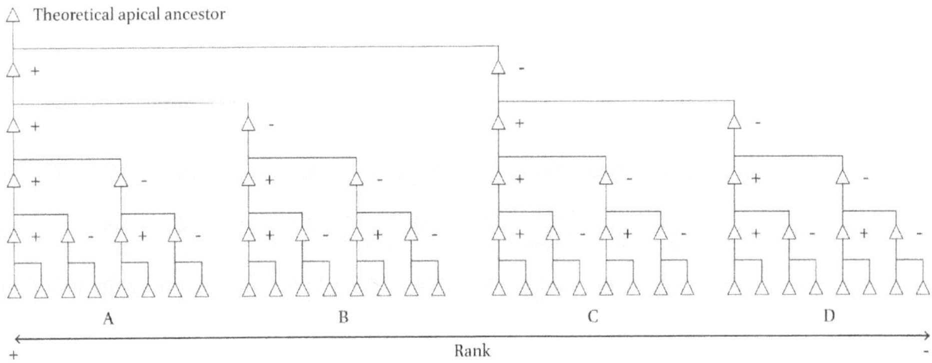

Figure 1. Model of the conical clan following Sahlins' representation (1958).

non-localized segmentary organization of preferred unilineal descent, most commonly in its patrilineal variety, which may consist of "multiple patrilineages (or, in principle, matrilineages) insofar as any given member can trace descent from the 'founder' or apex via a given child of the 'founder'."[17] Its members collectively shared resources borne by a common forebear and were ranked according to their genealogical distance from this ancestral figure.[18] In Cuzco, the founder of the royal ayllu was Manco Capac, who held this position by virtue of his supernatural ancestry. According to oral traditions, he was the son of the Sun, a divinity who stood as the utmost ancestor of the dynastic descent group. Several sources suggest that the royal ayllu was composed of a main senior line from which proceeded a variety of ramifications called panacas, which formed the sublineages of Cuzco's asymmetric moieties. Each of these ramifications traced descent from one of the bygone rulers whose life stories had been preserved and who were regarded as direct descendants of the apical ancestor. Garcilaso is particularly explicit on this point:

> Although they all descended from King Manco Capac, each claimed descent from one king or another, saying these descend from this Inca, those from that Inca, and so on for all the rest. This is what the Spanish historians refer

of Spanish rule followed by the postcolonial era of state formation altered its political and economic orientations.

[17] Salomon, in Avila 1991: 22.

[18] For an overview of the question, see Cunow 1929; Tschopik 1946: 539-540; Mishkin 1946: 441; Bastien 1978: xxiii-xxiv; Webster 1980: 184-185 especially n. 2; Rostworowski 1981; Godoy 1986: 723-728; Allen 1988: 33-34, 108-109; Salomon & Urioste, in Avila 1991: 19-23; Urton 1990: 76-77; Isbell 1997: 98-100; Rostworowski & Morris 1999: 780; Salomon 2004: 62-64; Ramírez 2005: 245, n. 35.

> to when they say confusedly that such and such Inca founded one line, and another a different one, suggesting that they were of different stock. But the lineages were in fact all the same, as the Indians show when they apply the common name Capac Ayllu, "august lineage of the royal blood", to them all.[19]

This text supports Seler's (1894) early hypothesis that the Qhapaq Ayllu was the Incas' main dynastic line, a hypothesis that Zuidema (1986) later refined. Many primary sources suggest likewise. For Betanzos, Las Casas, Santo Tomás, Román y Zamora (1575), and Gutiérrez, the name of this ayllu meant "royal lineage" or "lineage of the king(s)."[20] The lexical entry for *qhapaq* in most colonial dictionaries lists "rich", "powerful", and "renowned". The Incas also employed this word "when they wanted to bestow a more important title on the king,"[21] and used it to describe foreign rulers who had founded their own royal dynasty, such as Chimu Capac, Colla Capac, or Tocay Capac.[22] Finally, the Qhapaq Ayllu is the only lineage that chroniclers designate consistently as an ayllu, the other descent groups being called panacas. This latter term likely derives from *pana*, the Quechua for sister/cousin from a male Ego's perspective,[23] so that it may thereof denominate "sister-lines". Cobo also indicates that the main dynastic line created by Manco Capac did not belong to any moiety division: "Manco Capac, as the head and trunk of both *parcialidades* of Hanan Cuzco and Hurin Cuzco, was not counted in their divisions; the other kings were either from Hanan Cuzco or Hurin Cuzco."[24] In accordance, the rulers whose memory and corpses had been preserved, except for the apex of the senior line, can be seen as heads of hierarchically ranked collateral lines. Garcilaso offers a last enlightening illustration of this segmentary organization. In the final chapter of his *Comentarios*, the chronicler mentions a 1603 *probanza* that several Incas of royal blood would have dispatched to him, Don Melchior Carlos Inca, and Don Alonso de Mesa, asking for an exemption from the colonial tribute. The letter listed each litigant's noble descent and was enclosed with a painting on textile of approximately 1 metre 25 in

---

19 Garcilaso: Bk. 7, Ch. 9, 107.

20 Betanzos: Bk. 1, Ch. 32, 150; Las Casas: Ch. 251, 581; Santo Tomás, *Lexicón*: 248; Gutiérrez: Ch. 50, 214; Román y Zamora: Bk. 2, Ch. 12, 25-26.

21 Betanzos: Bk. 1, Ch. 27, 132.

22 Julien 2000: 27.

23 "Pana o pani: hermana como quiera, prima hija de hermano o hermana de padres" (Santo Tomás, *Lexicón*: 336); "Pana: hermana del varon, o prima hermana, o segunda, o de sus tierra, o linaje o conocida" (González Holguín, *Vocabulario*: 277). See also Zuidema 1964, 1980: 77-78; Regalado de Hurtado 1996: 46; Rostworowski 1999: 15-16.

24 Cobo: T. 2, Bk. 12, Ch. 8, 72. See also Acosta: Pt. 6, Ch. 20, 306.

length. It represented "the royal genealogical tree from Manco Capac to Huayna Capac and his son Paullu". Garcilaso reveals that each Inca king figured with the name of his particular lineage in addition to the designation *Qhapaq Ayllu* because "this title belongs to all in common and implies that all descend from the first Inca Manco Capac. Then they give another specific title to the descent of each king, with different names, to indicate those which originate from such and such a king."[25] Altogether, these data strongly support the coexistence of all panacas since the dawn of the Inca dynasty, a picture coherent with the lineages' homogenous demographics in the early colonial period (Julien 1998).

Descent Stratification

These panacas and a certain number of non-royal ayllus formed the subgroups of Cuzco's asymmetric moieties. In this structure, the kin groups of the Upper part were more highly esteemed and ranked than those residing in the Lower part of the city. This dual dimension of social disparity operated as well within the organization of each moiety. Both Hanan and Hurin Cuzco were internally divided into an upper and lower *parcialidad*, generating four quarters of asymmetric value identifiable with the *suyus*.[26] The upper part included the demographic divisions of *chinchaysuyu* and *antisuyu*, whereas the lower part covered the less prestigious quarters of *qullasuyu* and *kuntisuyu*. In turn, each of these four sections obeyed a principal lord so that each moiety was administered by two leaders of unequal power. The respective positions of these dignitaries were comparable to that of the *cacique principal* and his *segunda persona* in colonial times, which meant that the power of decision was ultimately given to the headman of the Hanan subdivision when the interest of the whole moiety was at stake.[27] This stratified model of socio-political organization, which Rostworowski calls quadripartition, is abundantly documented for the southern Andes in the colonial era.[28]

---

25 Garcilaso: Bk. 9, Ch. 40, 281.

26 Zuidema 1964: 42-43; Zuidema 1986b: 188; Pease 1978: 109-110; 1986: 229-230; Pärssinen 1992: 171-173; Ziólkowski 1996: 113-118.

27 In Spanish chronicles and other colonial documents (lawsuits, *visitas*), the expression *segunda persona* refers to the second-in-command who assisted the leader of a community, or *cacique principal*, in a variety of duties. This function reflects the division of the Andean ayllu into sub-groups (minimal ayllus, moieties), which were headed by different individuals. In this way, the *segunda persona* was the headman of the community's lower-ranking division.

28 Rostworowski 1999: 177-181; Ramírez 2005: 25. Some traditions emphasize this quadripartite model more. For example, the narratives of the Inca ancestors' emergence out of

Associated with these principles of hierarchical division was a pattern of contrasting generations. The panacas descending from Hurin leaders, traditionally believed to have ruled in the early period of Inca dominion, inhabited the Lower part, whereas the lineages proceeding from Hanan lords, whose epics celebrated the expansion of Inca territory outside the valley, peopled the Upper, more prestigious half. Similar features equally divided several Aymara-speaking communities of the southern Andes and other settlements of the Peruvian central highlands. In this latter area, the Upper division was associated with the Llacuaz, the bellicose newcomers who had subjugated the long-established population called Huari, who went to live in the Lower part. Colonial lexicons also associate *hanan* with the term *hawa*, which refers to objects and people located outside or above, whereas *hurin*, whose stem *ruri-* probably derived from a non-Cuzqueñan dialect, means "inside."[29] *Uri*, for instance, relates to the yields, tubers, or roots that grow underground.[30] These etymologies, like the historical narratives, associated Hanan with the outside populations and the world above, while Hurin referred to the first settlers and insiders' category.

The noble lineages and non-royal ayllus also performed ritual duties corresponding to their moiety affiliation, but whereas most kingship narratives mention the existence of eleven panacas in total, colonial records of Inca celebrations allude to the participation of only ten of them. Five were affiliated with the Lower part, and the other five with the Upper part. According to the descriptions of the latter, Huayna Capac's descent, otherwise listed as the Tumibamba Panaca, appears to have had no ritual commitment. Hence, during the cleansing festival of *sitwa*, there were ten panacas and ten common ayllus that assembled in the main plaza according to their moiety and *suyu* divisions in order to symbolically expulse illnesses. The Hanan side consisted of three royal lineages associated with *chinchaysuyu*—the Qhapaq Ayllu, Hatun Ayllu (also Iñaka Panaca), and Wikakiraw Panaca—and two more associated with *antisuyu*—the Suksu Panaca and Awqaylli Panaca. As for Hurin, it was composed of three lines

---

Paqariq Tampu mention the Ayars' formation into four couples of common ancestry. Sarmiento (Ch. 13: 61) indicates as well that Manco Capac originally divided Cuzco into four parts. Garcilaso (Bk. 1, Ch. 18: 43-44) also reports the work of a supreme being from Tiahuanaco who divided the world into four quarters and distributed them among four chieftains. Finally, Fernández (Pt. 2, Bk. 3, Ch. 6, 83) and Gutiérrez (Ch. 64, 254) stated that the Inca initiation ritual was restricted to members of the elite, composed of "four *parcialidades* or ayllus, Hanan Cuzco, Hurin Cuzco, Tampu (Apu) and Masca (Payta). The Incas regarded these four ayllus as their own as these were the true Incas".

29 Cerrón-Palomino 2002.

30 González Holguín, *Vocabulario*: 350, 357.

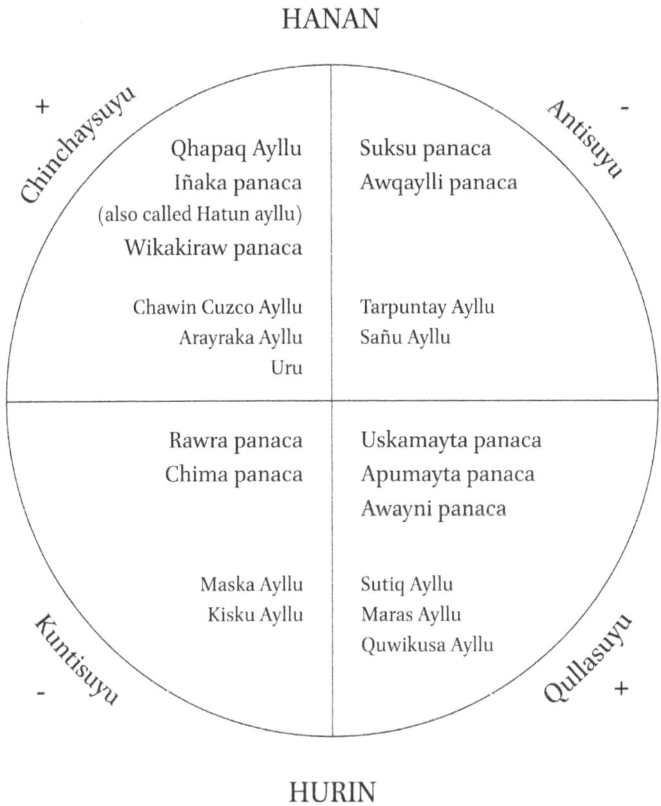

Figure 2. Ritual distribution of the panacas and non-royal ayllus.

affiliated with *qullasuyu*—the Uskamayta Panaca, Apumayta Panaca, and Awayni Panaca—and two others affiliated with *kuntisuyu*—the Rawra Panaca and Chima Panaca (Figure 2).

The association of each descent group with a specific *suyu* also determined its ritual responsibilities for the *wakas* (shrines) covered by the *seqe* system. This sacred and geopolitical system was a complex web of imaginary lines (*seqes*) that radiated from a central point in Cuzco to cross hundreds of shrines in its surroundings. It was instrumental in mapping out water rights and determined the lineages' cooperation over the management of the valley's irrigation system.[31] As during *sitwa*, the ritual maintenance of this system involved ten panacas and ten non-royal ayllus with

[31] Sherbondy 1979, Zuidema 1986b, Bauer 1998.

the exclusion of the Tumibamba Panaca. Each group was allocated a *seqe* corresponding to its *suyu* division and attended the needs of the *wakas* it traversed. The *Relación de los adoratorios de los indios* (c. 1560), attributed to Polo, reveals that the spatial organization of the *seqe* system followed complex hierarchical rules. Each *suyu* was crossed by nine *seqes* regrouped in triads (except for *kuntisuyu,* which included fourteen of them). Within these subdivisions, the *seqes* were known as *qullana*, *payan*, and *qayaw*, which were three categories that also described the internal tripartition of the Inca ayllu. They served as hierarchical principles of classification in which the *qullana* division held the most elevated position, followed by *payan* and *qayaw*.[32] This overview of Inca social stratification points to a remarkable hierarchical system in which each segment of the royal descent group was successively divided into basic kin units of asymmetrical status. At the head of this prestige hierarchy stood the dynastic senior line, the Qhapaq Ayllu, which encompassed a series of ramifications ranked according to their genealogical distance from the founder of the descent group.

Patterns of Affiliation

Some colonial information may seem to contradict the identification of the Qhapaq Ayllu with the main dynastic line of the royal descent group, particularly the sources describing this ayllu as Tupa Yupanqui's kin group. These data, however, can be harmonized with the picture of royal descent outlined so far by virtue of two principles that underlay the Inca political system: social stratification based on kinship, and the living king's status as son of the Sun (*Intip churin*). This title, which corresponded to Manco Capac's metonymic position at the apex of the royal ayllu, was bestowed upon a new Inca as soon as he achieved lordship. Since closeness to the apical forebear determined ranking, every panaca figured its position in the prestige hierarchy according to its genealogical distance from the

---

[32] The magistrate Ulloa Mogollón indicates that communities in the Collagua province "governed themselves according to what the Inca had ordered, which was, that each ayllu and *parcialidad* nominated a cacique. There were three ayllus called collona, pasana and cayao; each one of these ayllus had three hundred Indians and one principal whom they obeyed" (*Relaciones geográficas de Indias*, T.1: 330). In seventeenth-century Cuzco, the Ayarmaca-Pumamarca Ayllu followed the same tripartition. It was divided into two moieties, each one composed of the *qullana*, *payan* and *qayaw* divisions where the qullana division held the most elevated position. A 1689 official inquiry conducted in the Cuzco diocese reveals that several villages at that time included maximal and minimal *qullana* ayllus, invariably listed as the primary ayllu of the community (Villanueva Urteaga 1982). The same record indicates that, in the parish of Capacmarca (Chumbivilcas), the headmen who administrated the three ayllus of the town invariably came from the *qullana* division.

newly appointed king, whose lineage stood thereafter as the senior line. In a theoretical world where firstborns only inherited the *maska paycha*, the distribution of ranks would have remained stable over time. However, on the ground, the most able contender could access power and elevate his lineage to the most prestigious rank of the descent group. In this way, royal successions often prompted a disruption of the pre-existing social order, which modified the position of the Qhapaq Ayllu, the senior line. When the conquistadors first set foot in Tawantinsuyu, the last king to have assumed office in conformity with native rules of inheritance was a member of Tupa Yupanqui's panaca called Huayna Capac. He died sometime between 1526 and 1528, leaving two of his sons to compete ferociously for the royal title. Like his predecessors, once invested with power Huayna Capac became the living head of the Inca royal ayllu. His panaca and its founding figure were assimilated thereupon into the senior line of the prestige hierarchy. He was also the last king to receive the traditional mortuary honours during which the dead monarch's descendants assembled to record his deeds, including the early achievements that proved his ability to rule, his later conquests and socio-political reforms, etc. This material was the last elaboration of the life story genre that was entirely realized following pre-Hispanic rules of historical composition. So when the Spaniards began collecting information on the Inca past, they recorded the structure of the royal descent group as described at the time of Huayna Capac's death, when Tupa Yupanqui was regarded the ancestral head of the Qhapaq Ayllu. The writing process of colonial politics then stabilized this picture of the prestige hierarchy as a permanent institution. This interpretation also accounts for the absence of Huayna Capac's lineage in colonial records of Inca festivities. Otherwise known as the Tumipampa Panaca, Huayna Capac's descent group would have endorsed the title of "royal lineage" as long as it produced kings and its members attended the metropolitan celebrations as delegates of the Qhapaq Ayllu.

The practice of reordering the prestige hierarchy each time a different panaca overthrew the royal house in power would have ceased with the Spanish conquest. The interference of foreign authority in electing a new ruler, the rapid disappearance of the kingship ritual apparatus, and the new administrative imperatives of the colonial system soon put an end to this practice. Arguably, even the outcome of the fratricidal war between Huascar and Atahualpa would not have changed the order of things. Huascar belonged to Tupa Yupanqui's panaca, like his father, so that his accession to the royal office could not have called into question the pre-existing assimilation of his lineage to the Qhapaq Ayllu. As for Atahualpa, he claimed descent from Pachacuti Yupanqui. Although his faction murdered Huascar, which allowed him to assume kingship, not all Inca *principales* considered his investiture in Carangue legitimate. Unlike every other royal heir, Atahualpa had not fasted or retreated to the Sun temple of Cuzco before receiving the *maska paycha*, nor had he been recognized thereafter as the son of the royal dynasty's godly ancestor. Finally, upon his execution in the hands of the conquistadors, no customary funeral commemorated Atahualpa's brief achievements, which had been too insignificant to shatter the former political structure. Before these last events, however, Atahualpa attempted to enforce the legitimacy of his nomination by killing the living members of Tupa Yupanqui's descent group. His main motivation was not so much to eliminate Huascar's supporters as to bring his own panaca's genealogical position closer to the apical head of the dynasty. Garcilaso reveals this scheme in an enlightening passage of his *Comentarios*: "Because the lines of Huayna Capac and his father Tupa Inca Yupanqui were *the closest ones to the royal tree*, Atahualpa strove to eradicate them above all the others, and thus very few escaped his cruelty."[33] Indeed, the annihilation of Tupa Yupanqui's panaca would have elevated Pachacuti Yupanqui's lineage, to which Atahualpa belonged, to the highest possible rank in the prestige hierarchy because it was next in status.

An interesting account, recorded in almost similar terms by Las Casas, Román, and Gutiérrez, describes the royal ayllu as effectively composed of ten noble lineages, of which the Qhapaq Ayllu was the most prestigious line. It places kinship relations at the core of Cuzco's dual organization and emphasizes the asymmetric positions of every ramification with regard to the senior line. Significantly, it attributes the panacas' distribution into unequal moieties to a reformer king, Inca Yupanqui or Tupa Inca, immediately after his accession to power. In this way, it suggests that the nomination of the new king induced a disruption of the noble lineages' positions in the authority structure. We do not know with certainty the reasons behind the similarity of these three versions. The practice of copying being pervasive in the colonial era, it is conceivable that these chroniclers transcribed the same source. Gutiérrez's case, however, remains enigmatic, and it is likely that he collected this account independently from the other two.

> [The king] divided Hanan Cuzco, which was the most prominent district, into five parts: the first and most important he called Capac Ayllu, which

[33] Garcilaso: Bk. 9, Ch. 40, 296. Emphasis is mine.

> means "the king's lineage"; he gathered a great multitude of people from this faction and part of the city. He named the second Iñaca Panaca; the third Zucsu Panaca, the fourth Aucaylli Panaca, the fifth Vicaquirau Panaca; he assigned a great number of people to each of these districts; thereby dividing up the city into factions. He made *capitán* of the first district his eldest son who was to inherit the kingdom; he appointed his father and his descendants in a transversal line to the second and third [districts]; the fourth to his grandfather and his descendants in a transversal line; the fifth to his great grandfather in the same line (…) He divided Hurin Cuzco, which was the Lower half of the city, into five parts or *parcialidades*: he called the first Uscamayta, and appointed as its *capitanes* the descendants of the second son of the first Inca; the second he named Apumayta, where he instituted as *capitán* and *capitanes* the second son of the second Inca and his descendants; he named the third *parcialidad* or faction Haguaini, for which he designated as *capitán* and *capitanes* the second son of the third Inca and his descendants; he named the fourth quarter Raurau Panaca and entrusted its *capitanía* to the second son of the fourth Inca and his descendants; he named the fifth Chima Panaca and the second son of the fifth Inca and his descendants became its *capitán* and *capitanes*.[34]

Several observations may be drawn from this account. First, it indicates that the leadership of the Qhapaq Ayllu was entrusted to the Inca's firstborn destined to inherit the kingship title, which would confirm that royal succession ideally followed the rule of male primogeniture. Secondly, it combines both kinship and temporal features to explain dual opposition, but unlike other versions of moiety formation in which Hanan lords succeed Hurin lords in a continuum, it operates a reversal of the temporal succession of Hurin/Hanan by associating the kings of the Upper part with the ascendant lineal kin of the reformer king. Here, Hanan Cuzco is connected with past generations, albeit ones not remote in time. This picture contrasts with Murúa and Cobo's account of the same event, which depicts royal descent as an undisrupted genealogy from father to firstborn proceeding from Manco Capac. In this variant, Inca Roca divided Cuzco pacifically by allocating the Lower moiety to the descent of former kings, and the Upper moiety to the lineages of those who were to rule after him. Here, in contrast, the heads of the Hanan panacas are related to each other through direct descent and are ranked by function of their genealogical position from the living Inca. It is the degree of proximity to the reformer king that determines their hierarchical position. The same principle orders the hierarchical organization of the Hurin panacas, except that the first

34 Las Casas: Ch. 251, 581. See also Román y Zamora: Bk. 2, Ch. 12, 25-26; Gutiérrez: Ch. 50, 214.

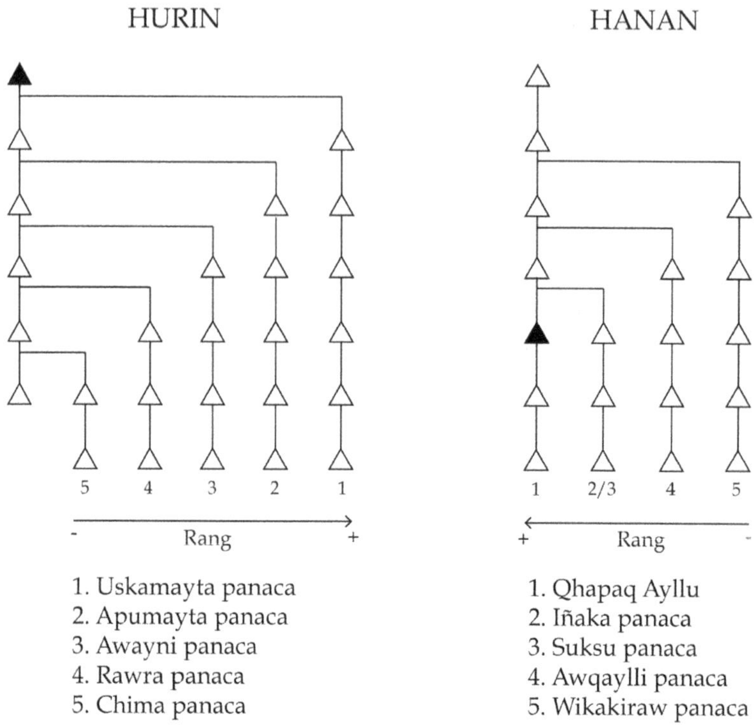

Figure 3. Genealogical links connecting the panacas' founding ancestors according to Las Casas, Román, and Gutiérrez.

king of the dynasty is now the genealogical referent. Clearly, the narrative refers to a system of social stratification based on kinship. On the one hand, the lineage configuration of the Lower moiety hinges on a fixed referent that was the apical forefather of all Inca lords, Manco Capac, and is inscribed in an ancestral past. On the other hand, the panaca organization of the Upper moiety is structured around the living king, within a temporality covering a close past and a near future. The representation of the kinship links that bond the lineages of each division reveals that the same layout principle governed the panacas' hierarchical organization (Figure 3). The account, however, employs two distinctive narrative formulations in describing this rule in order to differentiate between the two moieties. Concerning the Upper part, it emphasizes the direct descent relationship between the ruling Inca and the other heads of lineages that were his son, father, grandfather, and great-grandfather. When turning to the Lower part, the narrative draws attention to the collateral lines through the repetitive mention of the secondary sons of past rulers as panaca heads, so that collaterality (Hurin) is here in opposition to direct descent (Hanan). This discursive subtlety takes on its full significance in light of the sibling relationships described in other sources.

Siblingship

In his narrative on the origin of the moiety division, Garcilaso alludes to birth order as a hierarchical ordering principle. He indicates that members of the Upper half "prevailed in office" over the residents of the Lower part, much like firstborn and elder brothers command more respect than their younger siblings.[35] Although this particular passage has often been read as a colonial elaboration, it is not isolated. The 1569 *memoria* of the Qhapaq Ayllu referring to land inheritance also describes the internal divisions of this lineage in similar terms. It reveals that the asymmetric categories segmenting every ayllu (*qullana, payan, qayaw*) referred to degrees of relational distance that followed principles of siblingship.[36] The manuscript records evidence before the Crown of the noble status of the three heads of the Qhapaq Ayllu's divisions. The first man to testify was a lord of the Qullana ayllu division who regarded himself and his kin as "grandchildren" (*hawa*) of Tupa Inca Yupanqui. As for the members of the other two sub-ayllus, called *payan* and *qayaw*, they traced their ancestry back to Amaru Tupa Inca and Tupa Yupanqui, respectively, who were both brothers of the great king.[37] In sum, the 1569 *memoria* indicates that the highest line of the maximal ayllu traced descent through lineal kin, whereas its minor sublineages did so through collateral kin. This pattern reproduces the dual opposition of the moiety system described by Las Casas, Román, and Gutiérrez in their complex narrative. This analogy is made all the more revealing when considering that colonial lexicons associate the word *hanan*, which refers to the Upper part, with *hawa*, which designated both the outside world and the grandchildren of either sex.[38]

Emphasis on birth order is also reflected in the kinship terminology, which included specific terms for elder (*kuraq*) and younger (*sullka*) siblings.[39] Furthermore, these words were associated with appreciative and

---

35 Garcilaso: Bk. 1, Ch. 16, 41.

36 Published in Rowe 1985.

37 Pachacuti Yamqui (230) confirms that Amaru Tupa's chakra belonged to the Qhapaq Ayllu.

38 Arellano Hoffmann 1998: 475-476.

39 Other evidence also suggests that age order determined social position among the Incas. Bandera, Castro and Ortega Morejón, Santillán and Guaman Poma all indicate that

depreciative judgement, respectively. *Kuraq* implied notions of superiority, esteem, and preference. In its verbalized form, it referred to "rewarding in excess more than deserved or someone more than others". In contrast, *sullkacha-* described "giving less than others or less than deserved, to attach less importance to someone in regards to others, or to put someone in an inferior place."[40] Betanzos adds a further dimension to seniority as a principle of social differentiation in assimilating younger brothers with the sons of kings with their secondary wives. He records that men associated with Hurin Cuzco were called *wakcha q'uncha*, literally "the poor and orphaned sister's son."[41] This expression not only puts the focus on collateral relatives, it also infers an absence of ancestry and lack of riches thereof. Here, the *wakcha* of the descent group's collateral/junior branches (orphaned, poor) are in opposition to the members of the *qhapaq* senior line (royal, rich, powerful, renowned). Yet, this contrast of antonyms only superficially refers to a distinction in material wealth or in the mode of affiliation. Above all, it relates to an enduring aspect of the Andean redistributive system whereby individuals have to mobilize a significant network of kin to participate in the reciprocal aid giving them access to resources and services. Even today, extended kinship ties ensure more opportunities for wealth and power, while *wakcha* continues to designate people with limited relatives who suffer from indigence.[42]

Betanzos also places patterns of filiation at the heart of Hanan Cuzco membership, while associating Hurin Cuzco with kin by alliance. In his account of the moieties' origin, the chronicler reports that Pachacuti Yupanqui distributed the Upper part to the "relatives and descendants of his lineage in a direct line" and ordered the "illegitimate sons of lords, born of women foreign to this nation" to settle in the Lower part.[43] Examined

both men and women were hierarchically organized according to broad categories, or age groups, as it was customary in other Andean dual organizations. These chroniclers provide detailed lists of this system assigning family and communal tasks appropriate to each age set. The *aklla*, the chosen women who served in religious activities or as wives of the elites, were likewise organized according to age-rank categories. See Bandera: 62; Santillán: 106-107; Guaman Poma: 195 [197] to 234 [236], 170-209. Studies on this issue include Rowe 1946: 156, Zuidema 1986a: 86, Classen 1993: 63, Jenkins 1993; Rostworowski 1999: 167-170, Salomon & Urioste, in Avila 1991: 19-21.

[40] González Holguín, *Vocabulario*: 56, 331.

[41] Betanzos: Pt. 1, Ch. 16, 78; González Holguín, *Vocabulario*: 167; Santo Tomás, *Lexicón*: 280.

[42] Spalding 1974: 68; Isbell 1978: 76-78.

[43] Betanzos: Pt. 1, Ch. 16, 78. This statement has severe implications for the distribution of panaca lands. Concerning the nature and allotment of these lands, see Moore 1973: 33-38; Rostworowski 1993a: 105-110; Ziółkowski 1996: 257-285; Sherbondy 1996; Rowe 1997.

Table 7. Structural oppositions of the Inca dual organization.

HANAN	*HURIN*
– Highest social status	– Lowest social status
– Outside/above	– Inside
– *Hawa*: grandchildren, lineal descent	– *Q'uncha*: sister's son, collaterality
– *Qhapaq*: rich, royal, with extended kin	– *Wakcha*: poor, orphan, with no extended kin
– Firstborn sons	– Cadets/secondary sons
– Children of principal wife born inside the royal senior line	– Children of secondary wives born outside the royal senior line

together with his terminology for moieties, Betanzos' description presents the Hurin residents as the offspring of kings with secondary wives. These women were addressed as *pana* (sister/cousin) owing to their marriage, although they did not belong to the main dynastic line. In addition, this description sheds light on the royal ayllu's underlying structure. It associates firstborns with high-ranking insiders to the main dynastic line, whereas their younger brothers' position is equated to that of outsiders to the Qhapaq Ayllu. Here, Hanan Cuzco stands for the apical ancestor's direct line of descent whose living head was the ruling king. Hurin Cuzco, in contrast, regroups the junior branches, or panacas, proceeding from the cadets or "illegitimate" sons born from the union of kings and their secondary/foreign wives. Altogether, these elements articulate the Inca dual organization around relational dichotomy (cognates/affines) and gender opposition (Table 7).

Sarmiento, Murúa, and the authors of the *Discurso* actually support this dual division of the royal ayllu but, contrary to Betanzos and the corpus of Las Casas, Román, and Gutiérrez, they trace its origin back to ancestral times. They report that Manco Capac decreed before his death that leadership of the panacas should be entrusted to the second son of the late king, whereas his eldest son by his legitimate wife and sister was to succeed to the state.[44] In an isolated chapter of the Wellington manuscript, Murúa associates this particular distribution of authority with the moiety divisions.[45] Although the chronicler later credits Inca Roca with the foundation of the Hanan and Hurin districts, this excerpt clearly refers to a

44 Sarmiento: Ch. 14, 63-64; *Discurso*: 12-15, 19-21.

45 The first known version of Murúa's chronicle is the Loyola manuscript, first published in 1911, which was a copy of a manuscript held in a private collection in Ireland. The

principle of dual organization similar to that described in the reforms of Pachacuti Yupanqui and Tupa Inca:

> The first thing Manco Capac did was to put order to the city, by establishing a village where there were fields and farmhouses before, and by dividing it into two ayllus or districts (...) He named his son Sinchi Roca head of one ayllu who was to inherit the crown after his death. He distributed the rest among his descendants in a transversal line, decreeing that the Inca kings' second sons should rule the ayllus and *parcialidades* as their head and principal leader.[46]

Moreover, age hierarchy, and more specifically the asymmetry between firstborns and cadets, implied rivalry when associated with patterns of dual opposition. Pachacuti Yupanqui, Tupa Yupanqui, and Huayna Capac all seized power at the expense of their older brothers. The best-documented example of this type of antagonism is the fratricidal war that placed Huascar and Atahualpa in opposition. The former was an insider to the main dynastic line owing to his affiliation with the Qhapaq Ayllu, whereas the second belonged to a collateral line, Pachacuti Yupanqui's panaca, and was therefore an outsider. The two brothers' positions in the royal descent group replicated patterns of moiety asymmetry. Yet, rivalry between siblings was a constitutive pattern of Inca royal succession precisely because hierarchy and competitiveness were paramount in the relations brothers held with each other. In fact, siblingship provided the idiom through which reciprocity and competitiveness were expressed. A tangential analysis of the ethnographic data from the Cuzco region assists in further clarifying this aspect. Among the three ayllu divisions, *qullana* is a word still in use today to describe a specific agrarian function. The *qullana* conducts labour in the fields; he is the first to plough the *chakra* and the first to carry the yields to the storehouse at harvest time.[47] Garcilaso and Guaman Poma's depictions of Inca life offer a vibrant illustration of communal work in the fields, where men joined forces in small groups to perform agricultural tasks.[48] In his study of agricultural cooperation (*ayni*) in the Cuzco region today, Mario E. Tapia describes the same practice and makes reference not only to the *qullana*, but also to his attendant, called *qayawa*, whose name

Wellington manuscript was discovered in 1946 and is now in the Getty collection. It was the original text that Murúa presented to Philip III in 1616.

[46] Murúa, *Historia general*: Bk. 1, 42-43.

[47] Beyersdorff 1984: 83; Cerrón Palomino, Ballón Aguirre and Chambi 1992: 71.

[48] Garcilaso: Bk. 5, Ch. 2, 228; Guaman Poma: 250 [252], 224.

brings to mind the third division of the Inca ayllu tripartition, the *qayaw* category:[49]

> At dawn the field owner (*chakrachikuy*) calls those who will work with him (...) They form groups of three persons called *masa* that are supervised by a *qullana* or *capitán* whose role it is to distribute the work fairly, according to the age and strength of every member of the *masa* (which, among other things, allows group competition) and he sets the pace of work. The *capitán* is usually a kinsman, son-in-law or brother of the field owner. Another role is that of *qayawa* or second-in-command, who inspects the completion of the work.[50]

Competitiveness and prestige are the two leading aspects that characterize the function of *qullanas* who routinely compete against other *masa* leaders with the assistance of their second-in-command, the *qayawas*. Among the Ch'eqec today, the *qullana* is also a high-ranking *cargo* due to its heavy financial obligations. For a year, he provides food and drinks for all community cooperation projects, which he has the authority to begin and close ritually.[51] In other parts of the southern Andes, funerals and the cleaning of irrigation canals are overseen by a *qullana* and a *qayaw* whose function is to maintain discipline among the celebrants.[52] In these different examples, they are heads of a social hierarchy operating in a context of collective work, for the benefit of their ayllu or entire community. They thus exercise authority over an organized group of individuals who perform complementary roles in the completion of labour and ceremonial rites.[53] The tripartite division of Inca social organization was based on similar principles of complementary and communal obligations that entailed hierarchy and competitiveness. This conclusion is all the more consistent in the context of the *seqe* system, where the internal organization of the *suyus* operated

---

49 Zuidema notes that "callahua" was another written form of "cayao". It also probably became "cayhua" (*qaywa*) by metathesis in certain dialects. I believe that all of these early Spanish spellings refer to the *qayawa* function. However, this word cannot mean "beginning or origin" as Zuidema suggests following Jorge A. Lira's disputable spelling of "kkálla" (*qaylla*). Nor should it be associated with the Spanish "callao" that came to designate certain port cities of South America (Zuidema 1964: 165 n. 159). If *qayawa* has any linguistic link with *qaylla*, which remains to be demonstrated, then its meaning is one of "edge", "border" or "close by".

50 Tapia, in Morlon 1992: 57-58. *Masa* refers to in-laws.

51 Webster 1981: 621.

52 Cáceres Chalco 2001.

53 There exist other examples of tripartite work distribution throughout the Andes. During the cleaning of Saywite's irrigation canals (Abancay), for instance, the work force was divided out into three groups of a hundred men, each division working respectively in the first, middle and last sections of the canal (Sherbondy 1987: 136).

according to the three categories *qullana, payan*, and *qayaw*. In this regard, it should be noted that although the *seqe* came to denote a line, a ray, or a limit in colonial time, it also—and perhaps originally—described arable land divided into several cultivation parcels.[54] The *seqe* not only complemented other modes of (asymmetric) demographic organization, like the *suyus* or the moieties, its meaning also suggests a connection between the ritual organization of the Cuzco region and the system of dividing work in the fields. Each triad of *seqe* was attended by a panaca and two ayllus of lesser status who were individually in charge of a section of the ritual landscape. Despite their hierarchical difference, each group was tied by its ritual duties to a certain number of *wakas* and, similarly to agricultural labour, it was the mutual completion of every sacrifice and offering to these shrines that made the ritual efficacious.

Marriage Preferences

Relational tensions between siblings often induced rivalry to the kingship office because in practice, rules of royal inheritance were open to more than one pretender. Guaman Poma summarizes this situation as follows:

> To be *qhapaq apu Inca* (perfect king), one needs to be the legitimate son of the Queen mother, *qhapaq apu quya*, and married to his sister or mother. And he must be called by his father the Sun to his temple to be named king. They did not consider whether he was the eldest or youngest [son], but whether he was the one elected by the Sun as legitimate.[55]

Among the conditions that determined access to the royal office, Guaman Poma claims that marriage with the full sister, or theoretically with the mother, was required; and he was not alone in asserting this. Several chroniclers note that marriage with the biological sister reportedly occurred between members of the nobility, even though it was strictly prohibited among commoners, incurring a death penalty.[56] For Betanzos, the king had to marry a woman "from his family and lineage, one of his sisters or first cousins", which were both referred to as *pana*.[57] With words similar to Guaman Poma's, Cieza explains: "it was ordered that he who was to become king should take his sister, legitimate daughter of his father and mother, as wife in order that succession to the kingdom remains in the royal

[54] Beyersdorff 1984: 102. Bouysse-Cassagne (1986: 201) drew a similar conclusion concerning the *suyu*, which she sees as a 'subdivision' or 'distribution'.

[55] Guaman Poma: 118 [118], 96. See also Santillán: 108.

[56] Fernández: Pt. 2, Bk. 3, Ch. 9, 85-86.

[57] Betanzos: Pt. 1, Ch. 16, 78.

household."[58] To set the royal example, the founder of the dynastic line, Manco Capac, married his sister, Mama Ocllo. Cieza also reports Sinchi Roca scrupling to marry his son to the daughter of a foreign lord because "it was contrary to the rule established and ordained by his father."[59] As for Murúa, he attributes this form of incestuous union systematically to every king. There is also a consensus among chroniclers to say that Tupa Yupanqui married his full sister, Mama Ocllo, on the day he was invested with the *maska paycha*.[60] Yet, it appears from the life stories of Inca rulers that few of them resorted to this type of union. Royal incest in Inca Cuzco, as in pharaonic Egypt or eighteenth-century Hawaii, remained an ideal rather than the actual convention.[61] Its practice among Inca kings is nevertheless illuminating because marriage with a full sister is the most elementary combination to ensure high-ranking offspring in stratified societies where status is bilaterally transmitted, that is, when both the male and female lines determine a person's rank.[62] Betanzos identifies this practice when he insists that the Inca's principal wife (*paqsha*) had to be "a noblewoman from both the paternal and maternal line."[63] These conditions made her a *quya*, which was a status held by "any woman whose birth qualified her as a possible spouse for the Inca."[64] Bilateral principles of affiliation are also a common feature of conical clans.[65] This is because in such a system, the collateral lines necessarily decline in status with each

---

58 Cieza: Pt. 2, Ch. 10, 25. See also Murúa, *Historia general*: Bk. 1, Ch. 4, 48.

59 Cieza: Pt. 2, Ch. 31, 95.

60 Betanzos: Pt. 1, Ch. 26, 127; Sarmiento: Ch. 43, 118; Cobo: T. 2, Bk. 12, Ch. 14, 83. Huayna Capac, also, would have married his principal wife on the day of his ascent to the imperial throne (Pachacuti Yamqui: f. 31, 243), but the practice seems to have varied. See Hernández Astete 2002a: 118-119.

61 Rostworowski 1960; Bixler 1982a, 1982b; Davenport 1994.

62 Julien 2000: 259-260; Jenkins 2001: 169.

63 Betanzos: Pt. 1, Ch. 16, 78.

64 Julien 2000: 35.

65 Here, bilaterality refers only to patterns of transmission of title and rank. The existence of this principle does not necessarily mean that the Incas reckoned descent as bilineal, as Silverblatt (1987) suggests. In most societies, rights and obligations may be transmitted according to different modes of affiliation. The classification of these rights and obligations, and the identification of their transmission rules ultimately allow their association with a particular type of descent system (Needham 1971). If there is a cognatic tendency in the Inca kinship system, it also displays strong patrilineal features that are not merely ideological. All noblewomen claimed membership to one of the royal lineages founded by an ancestor-king. Manco Capac, the founder of the descent group was also male. Although some queens were mummified and received a cult, none was head of a panaca, nor was the Moon deity regarded as a divine ancestress, at least among the Incas. Finally, the queen traced descent from the Sun god alone, which entitled her to the status of "daughter of the Sun".

generation, for their members grow farther from the founding ancestor. This characteristic, which Geertz calls the "sinking status pattern", implies that individuals have to increase their parents' ranks in order to reproduce the dynastic continuity.[66] This particular increment is obtained through marriage. The model is therefore open to bilateral principles of affiliation so that the combined ranks of parents produce a higher rank in their children, and thus (metonymically) increase their genealogical closeness to the apical ancestor. Importantly, bilateral reckoning of ranks has another implication on marriage preferences, which the Hawaiian case best illustrates:

> Since the son's rank depends as much on his mother as his father, a dynasty can automatically reproduce its rank only if its male members consistently marry their sisters. However, brother-sister marriage is allowed to happen only exceptionally; otherwise the collateral lines would steadily lose rank for lack of access to the rank of the senior line (Valeri 1972). The dynasty must reproduce, then, by way of a trade-off between it and its collaterals: in exchange for political support, it gives the collaterals part of its rank through its women. But unless the ruling line periodically shuts itself off by making few brother-sister (or uncle-niece) marriages, its rank becomes equal to that of some of its collaterals, who may then wrest the rule from the patrilineal successors.[67]

Colonial sources reveal that a similar situation applied to the Inca royal descent group. While primogeniture was the ideal rule of inheritance whereby firstborns succeeded their fathers to kingship, their high-ranking collateral kin, who were structurally associated with second-born sons, could usurp the office that their older brother should have acquired by virtue of birth. This model of affiliation can reproduce itself through two types of marriage preferences that Spanish chronicles detail for us. Firstly, they mention the royal union with a full sister, in which both parties held a direct genealogical relationship to a common apical ancestor. Such a practice, correlated with primogeniture, ensured dynastic continuity. Secondly, primary sources indicate that Inca kings preferably married outside of the dynastic line and received their first wives from non-Inca lords with whom they forged symmetric alliances (Table 8).[68] Although these marriages with high-ranking noblewomen aimed to ensuring political stability,

---

66 Geertz 1980: 30-33; Valeri 1990b: 48.

67 Valeri 1990a: 166.

68 The kinship terminology 'symmetric alliance' describes the nature of the marriage exchange; it does not indicate the equivalence of the spouses' hierarchical rank (isogamy). It remains difficult to identify post-marital residence patterns among Inca rulers.

Table 8. List of the Incas' principal wives, with mention of their origin.

Inca ruler	*Principal wife*
Manco Capac	– Mama Ocllo, sister (in Betanzos, Las Casas, Sarmiento, Román, Cabello, Gutiérrez, Guaman Poma) – Mama Huaco, sister (in Fernández, Cobo)
Sinchi Roca	– Mama Coca/Chura, native of *Sañu* (in Betanzos, Las Casas, Sarmiento, Román, Cabello, Murúa, Cobo) – Chimpu Coya, sister (in Murúa) – Mama Ocllo/Cora, sister (in Garcilaso)
Lluque Yupanqui	– Mama Caua Pata, native of *Oma* (in Las Casas, Sarmiento, Román, Gutiérrez, Cabello, Murúa, Cobo) – Mama Cura, sister or first cousin (in Murúa)
Mayta Capac	– Mama Chianta/Chiancha, native of *Sañu* (in Las Casas) – Mama Tancaray/Tacucaray, native of *Tancaray/Tacucaray* (in Sarmiento, Román, Cabello, Murúa, Cobo) – Chimpu Urma, sister or first cousin (in Murúa) – Mama Coca, sister (in Garcilaso)
Capac Yupanqui	– Curi Hilpay/Indi Chigia, daughter of the *Ayarmaca* lord (in Las Casas, Sarmiento, Román) – Curi Hilpay, native of Cuzco (in Cabello) – Chimpu Ocllo/Maca Curi Ilpay, sister (Murúa, Garcilaso)
Inca Roca	– Mama Micay, daughter of the *Huayllacan* lord (in Las Casas, Sarmiento, Román, Cabello, Cobo)
Yahuar Huacac	– Mama Chiquia/Ipa Huaco Coya, daughter of the *Ayarmaca* lord (in Las Casas, Sarmiento, Román, Cabello)
Viracocha Inca	– Mama Runtucaya, daughter of the *Anta* lord (in Las Casas, Sarmiento, Román, Cabello, Cobo) – Coya Mama Runtu, sister (in Garcilaso)
Pachacuti Yupanqui	– Mama Anahuarque, native of *Choco* (in Las Casas, Sarmiento, Román, Cabello, Cobo) – Coya Anahuarque, sister (in Garcilaso)
Tupa Inca Yupanqui	– Mama Ocllo, full sister (in Cieza, Betanzos, Las Casas, Sarmiento, Cabello, Cobo, Garcilaso)
Huayna Capac	– Mama Cusi Rimay, sister (Cabello, Murúa) – Mama Ragua Ocllo, sister (Cabello, Murúa, Garcilaso)

they also resulted in the birth of powerful pretenders to kingship who could mobilize active support from their maternal kin and allies when the time came to fight for leadership.

Finally, some terminological features suggest that Inca kings privileged asymmetric cross-cousin marriage, restricted to the matrilateral form. In their idiomatic usage, *kaka* designated a man's maternal uncle, his father-in-law, and his wife's brother (MB=WB=WF), while *ipa* was a woman's husband's sister and her father's sister's daughter (HZ=FZD).[69] *Kaka* also referred to the foreign nations that provided spouses for the rulers, such as the Choco and Cachona Ayllus, known in early colonial documents as "Indians Caca Cuzcos" or, under the name of their principal *waka*, "Anahuarque."[70] The following report indicates that these ayllus were exempted from the *mit'a* owing to their matrimonial connections with the Cuzco nobility:

> It came about that Pachacuti Ynga Yupangui, who was Lord of these Kingdoms, married Mama Anaguarque *aunt* of the Yncas, [who descended from the] Caca Cuzcos Anaguarquez. At that time, when the Yngas succeeded the aforenamed Pachacuti Ynga Yupangui, the said Caca Cuzcos Anaguarquez benefited from the aforementioned freedoms; they were addressed as *uncles* and called the Yngas *nephews*. This custom has survived and is still in use today.[71]

The document makes clear that the denomination Kaka Cuzcos referred to a population that provided an Inca ruler with a wife, Mama Anahuarque. Pachacuti Yupanqui's wedding was the condition for the acquisition of this statutory title whose categorical aspect is underlined clearly: the Kaka Cuzcos had the privilege of *addressing* the Incas as "nephews" and were *called* "uncles" in return. Hence, when Betanzos argues that the principal wife of a king had to be his sister or first cousin, it implies that the queen could be as much a genealogical relative as a high-ranking non-kin whose marriage thereupon entitled her to the statutory designation of *pana*. This terminology alludes to royal matrimonial strategies open to the closest possible relative as well as to the most remote candidate, which are precisely the two types of wedlock described in primary sources.

---

[69] Zuidema, 1964: 40-43; Lounsbury 1986: 122-123. These terms are glossed in Santo Tomás, Ricardo, González Holguín, Torres Rubio [1619].

[70] Rostworowski 1988a: 135.

[71] Archivo Departamental del Cuzco, Real Audiencia: Ordinarias, leg. 42 (1670 [1589]), Expediente sobre el Cacigazgo de Chumbivilcas, fol 16v. The translation and emphases are mine. I am most grateful to Prof. David Cahill who generously gave me a copy of this document. See also Rostworowski 1993a: 132-37.

Positional Inheritance

So far, this overview of the royal descent group has shown that the perpetuation of the prestige hierarchy depended as much on bilateral principles of affiliation as on marriage. Another social practice called positional inheritance completes this presentation and sheds light on the transmission of Inca historical narratives. Recently, Ramírez (2006a) conclusively showed that Inca lords followed the rule of positional inheritance whereby a person is given a name that had belonged to several of his illustrious ancestors and to which corresponded a given position in the social hierarchy. The same individual would assume different names or titles throughout his lifetime, which he transmitted in turn to his younger relatives each time he attained a higher status. His military deeds and notable achievements, as well as those of his predecessors, were therefore conflated and remembered under the name they all bore in common. In this way, both male and female names corresponded to a hierarchy of statuses to which adhered ritual responsibilities. They remained in the ownership of individual kin groups, and thus belonged exclusively to a specific moiety. In this way, they stood as kinship affiliations and mediated positions of social prestige. Garcilaso suggests that the Incas sanctioned with reprobation any person uttering his/her kin's names in public. He says that Manco Capac's name was so highly esteemed that no one dared to pronounce it publicly.[72] Several other sources provide sound support for this argument and indicate that Inca lords bore several names in their lifetimes. They received these soon after birth, on being initiated, and later on their accession to the throne. The chronicles record that three kings had been called Titu Cusi Hualpa as children before they became Yahuar Huacac, Huayna Capac, and Huascar Inca, respectively.[73] Another ruler, Hatun Tupa Inca, was later remembered as Viracocha Inca. He was bestowed with this name upon completing his initiation when, according to Sarmiento, the eponymous god appeared to him in a dream.[74] Finally, Atahualpa took over the name of the thunder deity *Q'achqa* upon receiving the royal fringe in Carangue. He became thereafter Caccha Pachacuti Inca Yupanqui.[75] The newly appointed rulers thus inherited a prestige name owned exclusively by their kin group and which evoked a tutelary god or a mighty ancestor. Betanzos indicates that the generic name *yupanqui* referred to "their

---

72 Garcilaso: Bk. 1, Ch. 23, 55.
73 Sarmiento: Ch. 20: 72; Ch. 56: 139; Ch. 63: 150.
74 Sarmiento: Ch. 24, 80.
75 Betanzos 1996: Pt. 2, Ch. 6, 221.

lineage of origin because it had been Manco Capac's nickname (*sobrenombre*)."[76] Inca noblewomen are also listed under various appellations, suggesting that they too inherited different names, such as the common Ocllo, Chimpu, or Coca.[77] Illustrating this usage is Yñaca Anahuarque, a secondary wife of Huayna Capac and "niece" of her homonym Mama Anahuarque, who had been the principal spouse of Pachacuti Yupanqui.[78] This practice may well explain why single Andean lords appear under varying names in colonial documents, and how those who had lived at different periods in time could assume the same name. This particular point already caught Rostworowski's attention in 1970 and, although she did not identify the practice with positional inheritance, she drew the first lines of a similar argument by suggesting that Tocay Capac and Pinahua Capac were generic names of the Ayarmaca chieftains, the Incas' long-avowed rivals.[79] Ironically, the sacraments of the Catholic Church allowed this usage to endure under Spanish rule so that many Inca nobles of the early colonial era still received their ancestors' names when christened. Sayre Tupa Inca, son of Manco II, was baptized Don Diego Hurtado de Mendoza Manco Capac Pachacuti Yupanqui in 1558, a name composed of the viceroy's father's name and that which Sayre Tupa had received when he was given the royal *borla* in Vilcabamba.[80]

All of these considerations concur with the rules of positional inheritance. It also throws light upon an important aspect of this social practice. As Ramírez herself observes, positional inheritance is indivisible from a mode of recording and narrating the past that reduces the deeds of many generations under the umbrella of a single figure. This is no denial of historical events, but rather a conflated presentation of the past answering to the practicalities of oral transmission. Positional inheritance implied that the Incas credited to each panaca founder the accomplishments of his most illustrious descendants. Through this process, they only perpetuated the memory of those exceptional individuals who had received their forebear's name as insignia of prestige and had thereupon acquired its corresponding genealogical position in the authority structure. As Cieza claims,

[76] Betanzos: Pt. 1, Ch. 17, 52. For a translation of the word *yupanqui*, see Cerrón-Palomino 1998: 435.

[77] See specifically Murúa and his many chapters on the *quyas* in both the Loyola and Wellington manuscripts.

[78] ADC, Real Audiencia, Ordinarias, Leg. 42 (1670 [1589]), Expediente sobre el Cacigazgo de Chumbivilcas, folio 13v.

[79] Rostworowski 1993b: 242.

[80] Fernández: Pt. 3, Bk. 3, Ch. 5, 83; Yupanqui 2006: 146-147.

the lords who did not stand out through their personal merit were condemned to oblivion. The oral material that condensed these exploits formed the life stories of the panacas' heads. They were the traditions, along with the genealogical accounts, that chroniclers exploited to compose their history of Inca kingship. In transcribing these narratives into Castilian, the Spaniards embedded on paper a genealogy of eleven kings without reckoning that ten of them were heads of the royal lineages that had provided actual rulers. Viewed in this light, the longevity of the Inca dynasty would have largely exceeded two centuries. This interpretation is consistent with the conclusion drawn by Hiltunen (1999) in his intriguing and still debated study of chronicler Fernando Montesinos' dynasty list, which records more kings than commonly accepted. This last argument still awaits the discovery of Blas Valera's full corpus, but it undoubtedly finds support in the growing archaeological database on the Cuzco area indicating the ancient and undisrupted settlement of the valley from AD 1000 onwards.[81]

The Accommodation of Historical Contingency

The various traits of Inca socio-political organization reviewed thus far are consistent with the formal model of the conical clan. Within the royal descent group, individuals were ranked in terms of kinship categories where relational distance from the common ancestor determined status. Core principles of affiliation and social stratification were: bilateral transmission of status, siblingship with relational hierarchy, and positional inheritance. These rules of social organization ensured the perpetuation of the dynastic line and stability of the prestige hierarchy. Yet, on the ground, the Spaniards observed that this system was open to a greater flexibility. Transmission of titles and land benefits, for example, not only proceeded through conventional principles of affiliation, but also depended on the magnanimity of the ruler, who was entitled to redistribute his subjects' prerogatives upon his accession to power. Upon his investiture, Huayna Capac thus allotted "farmlands of the [Yucay] Valley to the lords of Cuzco, both to the living and to the dead lords whose mummy bundles (*bultos*)

---

81 Bauer & Covey 2002; McEwan, Chatfied & Gibaja 2002; Bauer 2004; Covey 2006b: 171-173. These data contradict the earlier conclusion drawn by John H. Rowe (1944) and largely unchallenged until the late 1990s. See D'Altroy & Bishop 1990; Stanish 1992: 136-157; Morris & von Hagen 1993: 18. For criticisms of this hypothesis, see Meyers 1975; Pärssinen & Siiriäinen 1997.

were there."[82] The life stories of Hanan kings record how these lords often discharged certain ayllus of important religious and administrative positions to nominate others who had shown loyalty. Huayna Capac is even said to have appropriated for himself the office of *mayordomo* of the Sun's herds, which was a charge traditionally entrusted to a king's brother or uncle. This royal licence to redistribute privileges inevitably induced conflicts of interest and factional wars at every new nomination. Royal succession was the ideal opportunity for low-ranking lineages to align with a powerful outsider to the dynastic line who could increase their status and authority once in power. In addition, the bestowal of names and titles not only followed conventional rules of transmission, it could also result from personal merit. During the initiation ritual, for example, the young novice who had won the race to the top of Anahuarque hill received the prestigious name *Waman*, which distinguished him among his peers. Betanzos reports that, "These young men who make such an effort when they become *orejon* warriors will be renowned, for when the city goes to war, they, being the quickest, will climb on the boulders and combat the enemy."[83]

All of these elements demonstrate the great flexibility of the system, which facilitated upward social mobility through good deeds.[84] As Inca domination extended with time, the enforcement of primogeniture and close-relative marriage became less viable with the reality on the ground. The ruling elite's expansionist pretensions led different factions to regularly confront each other through bloodshed in the hope of placing their favourite in power. The various narratives of Cuzco's moiety formation evoke precisely this tension between on the one hand, a dynastic model ideally established upon the perpetuation of the prestige hierarchy through strict rules of status transmission, and on the other hand, a dynamic system open to the nomination of the royal successor who proved the most apt to govern and to maintain the prestige-good system through enduring alliances with local lords. These narratives differ firstly on the identity of the king who founded the moieties. Sarmiento, Cabello, and Murúa credit this creation to Inca Roca. Moreover, none of these authors allude to the social status asymmetry between Hanan and Hurin, or to the hierarchical organization of each division. On the contrary, Inca Roca is said to have devised

---

[82] Betanzos: Pt. 1, Ch. 43, 187.

[83] Betanzos: Pt. 1, Ch. 14, 67.

[84] The dynamics of the socio-political arena in Cuzco are certainly not specific to the Inca society. For example, Valeri (1990a, 1990b) and Petersen (1999) offer other examples of such state of affairs in Hawaii and the Caroline Islands.

this partition without shattering the pre-existing social order—so much so that this particular corpus of narratives establishes a lineal temporal continuity between the kings of the two moieties. This picture of perpetuity is all the more revealing in that Cabello and Murúa describe the Inca dynasty as an uninterrupted succession of elder sons. Murúa also belongs to a group of chroniclers, including Fernández whose texts influenced the Mercedarian friar, as well as the authors of the *Discurso*, which claim that these firstborn kings systematically married their full sisters.[85] Their rendering of Inca history also offers a harmonious vision of the panacas' relations, with each ruler begetting at his death a lineage distinct from that of his predecessor. It discusses every reign parsimoniously without effusing over conquest stratagems. Only the names of each province they marched over and each people subjugated are recorded. In these accounts, finally, the rebellious threats emanate from outside Cuzco, not from the ruling factions residing in the city.

This descriptive framework contrasts with the stories of moiety formation recorded by Betanzos, Las Casas, Román, and Gutiérrez, which detail the circumstances leading to the accession to power of an illegitimate son. These texts attribute the creation of the Hanan and Hurin divisions to a reformer king, the youngest son of an aging and submissive ruler. They exalt his individual qualities, such as endurance, charisma, or ingenuity, which enabled him to transcend the prescribed order of kingship succession and, once in power, to impose a reorganization of the Inca prestige hierarchy. This emphasis on personal merit is particularly manifest in the epic of Inca Yupanqui that Cieza, Betanzos, Las Casas, and Román put in writing. These authors relate how the youngest son of Viracocha Inca overthrew his eldest brother, Inca Urco, in defiance of his father, who favoured his firstborn as the most legitimate heir. To demonstrate his aptitude for leadership, Inca Yupanqui mobilized divine and earthly support during the Chanca attack on Cuzco. The situation, however, seemed desperate. The old king, his eldest son, and the main members of the court had fled the city to find refuge on an estate west of the city named Caquia Jaquijahuana. However, they had not succeeded in convincing Inca Yupanqui to join them in exile. Against all odds, the young man stayed in Cuzco and

85 Sarmiento's text follows a somewhat different structure as it attributes foreign wives to most Inca rulers. His chronicle intermingles the reports of several members of the indigenous elite, both Inca and provincial, and therefore telescopes several historical traditions. In fact, Sarmiento later evokes the second narrative variant of the moieties' foundation by Pachacuti.

gathered an army, which he entrusted to three of his relatives. These men were called Apu Mayta, Vicaquirau, and Quilescachi Urcon Huaranga. According to Betanzos, they were *wakcha q'uncha,* or "poor" members of the royal ayllu's junior branches, who certainly hoped to gain authority in the prestige hierarchy by supporting the young upstart's efforts to obtain royal investiture. Their strategy soon paid off. Three times the Chancas attempted to seize Cuzco, and three times they were repelled. The last attack was eventually quelled with the intervention of stone warriors (*purun awqas*) that the Sun god had sent onto the battlefield to assist Inca Yupanqui. During this final combat, the young man killed the Chanca army chief and immediately sent his spoils to his father in Caquia Jaquijahuana. With this gesture, Inca Yupanqui expected Viracocha Inca to tread upon their enemy's remains as a sign that the old king endorsed his younger son's victory and recognized his predisposition to leadership. In spite of that, Viracocha remained loyal to his eldest son and asked Inca Urco to accomplish this act instead because he was the one to inherit kingship. The story states that Inca Urco did not have the opportunity to fulfil his father's wish. Soon after Inca Yupanqui's feat, he was ambushed by his younger brother's partisans and killed.[86] His death heralded the start of a prosperous era with the nomination of Inca Yupanqui as Sapa Inca. The accession of the young king, however, not only transcended the old order, it gave way to new institutions and inaugurated the rule of a new god, the Sun deity who had helped him gain victory. Inca Yupanqui was a reformer who devised the ceremonial calendar, invented the *purukaya* ritual, and divided the city into two moieties of asymmetric values. His achievements emphasize dual antagonisms and the hierarchical pre-eminence of the Hanan sector. In Betanzos' account, Inca Pachacuti's partition of Cuzco negates the anteriority of the Hurin leadership by associating the Lower sector with the descent of the three lords who assisted the young leader in his victory against the Chancas:

> From the Temple of the Sun upwards including everything between the two rivers up to the hill which is now the fortress, he distributed to the most prominent lords among his relatives and descendants of his lineage in a direct line, children of lords and ladies of his own family and lineage. The three lords whom he sent to settle in the section below the Temple of the Sun, as you have heard, were illegitimate sons of lords, though they were

[86] See also Sarmiento: Ch. 33, 98-99.

> from his lineage. They were born of women foreign to his nation and of low extraction.[87]

As with this last narrative, Las Casas, Román, and Gutiérrez do not mention the anteriority of Hurin versus Hanan lords. Instead, they list the lineages of the Upper part first and explain moiety division as a result of hierarchical distinction.

In the course of Inca Yupanqui's accession to power, the epic story not only praises the young man's exploits and good nature, it also contrasts his deeds with the multiple flaws of Inca Urco. The legitimate heir appears "simple-minded", "arrogant", and "contemptuous of others."[88] Cieza even depicts him as "vicious and so given to evil courses" that "he went about the city drinking (...) and without shame exposed his nether parts from where he evacuated the chicha turned into urine."[89] Cieza is also original in arguing that Viracocha Inca was not particularly eager to offer the royal fringe to his firstborn. "Yet, the *orejones* and principal men of the city did not wish the laws to be broken, nor anything they had observed hitherto by order of their ancestors. Although they knew how evil the inclinations of Inca Urco were, they still desired that he, and no one else, should be the king after his father's death."[90] Also according to Cieza, Inca Urco assumed kingship for a short period and spent this time in recreation on the royal estates, away from Cuzco. The story says that he abused his regal privileges over women and undertook none of what his office required of him: he did not embark upon any war of conquest, nor did he order the construction of new edifices. Cieza finally observes that the royal narratives do not include Inca Urco in their list of Inca kings owing to his mediocre temperament. In contrast, his younger brother was "a youth of great valour, and of good conduct, resolute and fearless, and endowed with good and lofty aspirations."[91] The story discursively warns the audience that had the rule of succession been followed, the Inca state would have perished at the hands of an adverse polity. It thus recognizes and encourages personal merit as the main driving force that ensured the viability of Inca sovereignty. Far from being arbitrary, the discrepancies setting this epic account apart from the narratives of moiety formation recorded by Cabello, Murúa, and Sarmiento form two homogenous ensembles, which may be qualified

87 Betanzos: Pt. 1, Ch. 16; translation by Hamilton & Buchanan 1996: 71.
88 Betanzos: Pt. 1, Ch. 8, 31; Sarmiento: Ch. 25, 83.
89 Cieza: Pt. 2, Ch. 44, 129-130.
90 Ibid: 128.
91 Ibid.

Table 9. Dual oppositions in Inca historical narratives.

HANAN	*HURIN*
– Disruption of the established order	– Perpetuity of the prescribed norms
– Election of the most able contender	– Primogeniture
– Temporal condensation, contemporaneity of the lineages	– Temporal extension, linear succession of the lineages
– Cadets, outsiders to the royal senior line	– Firstborn sons, insiders to the royal senior line
– Extended kin/allies outside the ruling line	– Kin/allies restricted to the ruling line
– Virtuous, dynamic	– Low morality, apathetic
– Territorial expansion	– Restricted dominion

as genres following Julien (2000). The characteristics of the epic stories conform to the descriptions of the Inca historical chants that each panaca composed and perpetuated about the life of its founding ancestor. These accounts celebrated the victory of the ambitious youth who had defied the law of the older generation represented by the father and elder brother, in order to demonstrate his personal ability to rule. Even though the epic story of Inca Yupanqui is certainly the best-preserved example of this narrative genre, he is not the only Hanan lord whose reign is described in those terms. Various chroniclers, including the authors of the *Discurso*, Cabello, Gutiérrez, and Murúa, attribute similar achievements to Viracocha Inca or Tupa Yupanqui. The Hanan lords all demonstrated the qualities of young upstarts who defied the past/established order embodied by their elder brother, to conquer nations further away from the lands subjugated by their ancestors. Their disruption of the dynastic continuity is thus closely linked to their reorganization of the panacas' hierarchical status, and thereby to moiety asymmetry. As a result, their life stories invert the structural oppositions of the normative model that orders the Inca dual organization system (Table 9).

Hanan versus Hurin, or the Legitimation of Inca Kingship

Scholars have long considered that brothers' rivalries and fratricidal wars, whether real or ritualized, were commonplace in royal succession,[92] but the connections between siblingship and moiety conflicts have remained largely unstudied. Gose (1996a) laid the foundations for an understanding

---

[92] See for example Conrad & Demarest 1984: 131-132; Rostworowski 1988a: 159-167; Bravo Guerreira 1992; Pease 1991: 95-146; Hernández Astete 2002b.

of this nexus by arguing that temporal process, and thus generational difference, was a core principle of the moiety system. He claims that dual opposition "encodes a developmental cycle for the careers of Inka sovereigns". Following Zuidema, he also remarks that Inca history portrays Hurin lords as timid and senile, and as the original rulers that the younger leaders of Hanan Cuzco were required to overthrow.[93] There is more to this picture because temporal process in Inca narratives, that is, the linear perspective on dynastic succession whereby Hurin lords preceded Hanan leaders, should not simply be understood as an elaboration of Spanish chroniclers. This is suggested by the very nature of the dichotomies that placed the two moieties in opposition, which operated within a strong temporality. The kings of Lower Cuzco were associated with past action and exercised restricted power, whereas the ambitious leaders of Upper Cuzco, epitomized by the young Inca Yupanqui, enjoyed the fruits of extended conquests. These oppositions not only reflected a structural order whereby past generations of lords were relegated to a lesser status than the living ruler, they also allowed for the articulation of a linear perspective on dynastic succession, which the primogeniture rule ideally perpetuated. This undisturbed representation of the royal genealogy can be associated with the second Inca historical genre evoked in primary sources. As Julien evinced, the chronicles of Sarmiento, Cabello, and Murúa, as well as Cobo and the authors of the *Discurso*, reflect the same genealogical structure. Their dynastic accounts emphasize patrilineality and the historical continuity between Hurin and Hanan lords in a sequence of firstborn royals consistently marrying their full sister. This corpus of traditions, which Julien calls "dynastic genealogy", articulated kingship succession around the notion of *qhapaq*, that is, the divine essence flowing from the apical pair to members of its descent group through patrilineal succession.[94] This particular mode of historical recollection forged an immutable image of royal succession by emphasizing continuity in the dynastic line. It aimed to create a group identity wherein the two moieties merged to constitute a line of single ancestry in much the same way described by Valeri for Hawaii:

> Rules expressed in terms of duration allow the social group they define to magnify and thus legitimate itself by adding extent in time to its extent in space. Thus society and its rules mutually legitimate one another through

---

93 Zuidema 1964: 111-113; 138, 156-166.

94 Julien 2000: 21-48. Julien's definition of *qhapaq* is closely akin to the Polynesian *mana*, which is the divine potency acquired through descent from the gods, or directly given by them to a person they elect. It is the ultimate source of legitimacy (Valeri 1990a: 167-168).

> one single powerful image: duration as proof of greatness, potency, vitality, righteousness, divine election, predestination, historical mission, historical necessity, or whatever.[95]

Undisrupted dynastic continuity, however, was a historical ideology. Besides the life story of Inca Yupanqui's great deeds, two separate accounts illustrate the unavoidable clash between innate and acquired leadership. The first describes an event that would have taken place soon after the unexpected investiture of Viracocha Inca. According to Cieza, Viracocha's predecessor had died tragically without fathering a son, leaving Cuzco without a commander. The Inca *principales* found themselves in the midst of a heated debate to decide who would inherit the royal office, when a noblewoman stood before the lords of Hanan Cuzco and suggested the nomination of Viracocha Inca, a nephew of the late king. All agreed to the young man's investiture, seeing how "worthy" he was, and he received the *maska paycha* after the customary ceremony. His predecessor, however, had a brother called Capac, whose name indicates appropriately his legitimate link to the main dynastic line. Infuriated by the ascent to power of a distant kin, Capac resolved to take back the title he should have rightfully received. To do so, he gathered the support of the Hurin lords, those who were willing to preserve the established rule of inheritance. While Viracocha Inca was campaigning in *kuntisuyu*, Capac executed his coup. He entered the Sun temple where the king's officials had assembled, and assassinated the governor. Despite this sudden change of events, Capac's popularity was short-lived because he lacked the support of powerful allies. He ultimately killed himself after every follower had deserted him.[96] The Hanan defenders of individual merit had once again defeated the leaders of the Hurin moiety, too vulnerable to maintain the provisions on hereditary succession.

The second event followed Huayna Capac's death, as the royal fringe awaited its next owner. Huascar, who was residing in Cuzco, became enraged upon receiving the presents his younger brother Atahualpa had sent him from Quito. Understanding his sibling's intention to claim the *maska paycha*, Huascar declared that he no longer belonged to the Upper moiety:

> Because Atahualpa was of Hanan Cuzco, descended from the lineage of Inca Yupanqui, and he no longer wished to be of that lineage (...) Henceforth [Huascar] wished them to recognize him as from Hurin Cuzco because he

---

[95] Valeri 1990a: 162.
[96] Cieza: Pt. 2, Ch. 40, 118-120.

> intended to kill Atahualpa and all his kinsmen and lineage that was of Hanan Cuzco, and form a new lineage of Hurin Cuzco.[97]

Most chroniclers assert that Huascar was indeed the legitimate heir to the imperial office. The firstborn of Huayna Capac and his full sister, Raura Ocllo, he belonged to the Qhapaq Ayllu, which affiliated him with the Hanan moiety.[98] Shattered by the intrigues of the political arena, Huascar had no other choice but to fight for the *borla* he should have inherited by birthright. The two chroniclers to report his change of moiety affiliation are Betanzos and Sarmiento, both of whom appear to have drawn from the "life history" genre. Following the lines dictated by this narrative genre, both chroniclers depict Huascar with the characteristics traditionally associated with Hurin lords. He appears as the keeper of traditions and embodies the values of the past. Before the fratricidal war, we read, Huascar lived a leisurely life in Cuzco with his court and did not take part in any military campaigns. It was the rule of kingship succession by birthright that he hoped to re-establish on his accession to power by forming "a new lineage of Hurin Cuzco". Atahualpa, on the other hand, is said to have been actively involved in expanding Tawantinsuyu's territories under his father's command. His abilities as an army chief and his lower status in the prestige hierarchy made Atahualpa an archetypal upstart of Hanan Cuzco.

Yet, Atahualpa's accession to power, or that of any *wakcha q'uncha* like Viracocha Inca or his son Inca Yupanqui, would have shattered the attribution of statuses and offices in the prestige hierarchy. Despite this inevitable social disruption, primary sources reveal that the monarchy's openness to historical contingencies and individual exploits did not challenge the hierarchical structure of the descent system, nor did it question the authority structure based on divine ancestry that organized the whole society. This is because every new ruler, whatever his genealogical closeness to the senior line, resorted to at least three strategies of kinship legitimation: divine sanction, marriage, and the hierarchical reorganization of the panacas around his person. First, he had to demonstrate that the Sun god itself, the divine founder of the Qhapaq Ayllu, recognized him as his son, *Intip churin*. This condition was examined during the sacrifice called *kallpa rikuy*. The Inca-to-be began by fasting for several weeks in the Sun temple, where he presented the god with daily food. These offerings accomplished,

---

97 Betanzos: Pt. 2, Ch. 2; translation by Hamilton & Buchanan 1996: 194. See also Sarmiento: Ch. 63, 151.

98 Rostworowski 1999: 107-110.

a high-ranking officiant examined divine will by performing a *kallpa rikuy* on a llama. This practice, which literally means "seeing the vital strength", involved interpreting the entrails of sacrificed animals.[99] This was done by blowing into the lungs of a llama (or a guinea pig on other occasions) in such a way that the organ revealed its crisscross mesh of veins. If no fault or breach had damaged the inflated lung, it meant that the *kallpa* of the potential heir, that is, the vital strength infused in him by the Sun god, was powerful. He was thus destined to accomplish great deeds. If the organ was imperfect, the ritual had not been successful and more sacrifices were required. The pretender who was endorsed by the god received the *maska paycha* on the day following the sacrifice, which marked the beginning of ostentatious festivities. This divine sanction transformed the heir as "unique son of the Sun", the closest relative to the ultimate divine ancestor, so that his mythical relationship with the Sun positioned him at the top of the prestige hierarchy. He then underwent a change of name through which he acquired the rank corresponding to his new prestigious capacity.

The recognition of the heir's divine ancestry also constitutes the substance of many historical narratives. Inca Yupanqui is said to have acceded to power with the assistance of a god who appeared to the young Inca before the decisive battle against the Chancas. This supernatural being, who identified himself to the prince as the Sun, bid him not to be afraid because he was his father and assured him of an imminent victory. Animated with the vital strength infused by the god, Inca Yupanqui manifested his capacity to lead an army of supernatural warriors, the *purun awqas*. In a different narrative, the eldest son of Inca Yupanqui, called Amaru Tupa, relinquished the *maska paycha* to his younger brother, Tupa Yupanqui, after visiting his cadet in the temple of the Sun, where the young boy had lived under the protection of the god for several years. Sarmiento records that Amaru Tupa fell at his brother's feet when he saw how much magnificence and how many noble allies surrounded Tupa Yupanqui.[100] Like ranks, divine sanction to the royal office could be inherited as well as acquired through merit. In practice, it meant mustering support from the Sun's attendants, who were also high-ranking lords of the royal descent group.

---

99 Molina: 62-63; Sarmiento: Ch. 40, 112; Ch. 58, 140; Ch. 62, 148. For rituals on guinea pigs, see Bolton 1979; Morales 1995: 99-128; Stahl 2003: 471-476.

100 Sarmiento: Ch. 43, 117. The same conclusion can be drawn from the narrative of the Ayar siblings' settlement in the Cuzco valley, as Mama Huaco's violent divination act revealed the support of the potent divinity on the Incas' side (see Chapter 5).

The second strategy is described in the corpus of Las Casas, Román, and Gutiérrez on the origin of moiety division in Cuzco. In these narratives, the young ruler became the new genealogical referent from which the ranking of all individuals was determined heretofore, thereby relegating the founding ancestor of the dynasty to a remote and obsolete past. Once in power, he nominated his firstborn son as the head of the Qhapaq Ayllu, the main senior line, and distributed the highest ranks to the descendants of his direct ascendants, which formed the Hanan division. He then regrouped his most remote kin in Hurin Cuzco and bestowed on them the lowest statuses of the descent group. These social changes made him a *Pachakuti*, literally a reformer of the pre-existent time and space.[101]

Finally, the last strategy of kingship legitimation for a lowborn aspirant was to wed a woman of the highest possible rank. It ensured that his firstborn's status would be increased through the combination of his parents' ranks, and at the same time it condemned the access of other ambitious pretenders to her and to her influential kin network. The most profitable union for a secondary son was a marriage with the woman closest to the main dynastic line, ideally his full sister. It was precisely this strategy that Tupa Yupanqui adopted when he received the royal title despite being the youngest son of Inca Yupanqui, also a usurper. To sanction this investiture, Tupa married his high-ranking genealogical sister called Mama Ocllo.[102] Women were therefore fundamental actors of historical changes in the course of dynastic successions. The bilateral principle of rank/name transmission and the kindred attached to her person allowed for two different matrimonial strategies described in primary sources: the king's union with the closest possible relative or his union with the most remote candidate. These politics of affinity corresponded in turn to different perspectives on the noblewoman's role in the perpetuation of the royal descent group. The two narrative genres identified thus far relayed precisely these opposite views. The Inca genealogical accounts, recorded by Murúa and Cabello, reduce the kings' principal wives to passive conduits of rank. They linger on the eminence of these women's descent, but appear oblivious to their historical agency. The life histories, in contrast, emphasize interlineage

101 The term *pacha* (world) bears both a temporal and spatial sense. González Holguín, *Vocabulario*: 27.

102 This may also suggest that sisters held a higher rank than their brothers, as Webster (1980) observed in the modern Cuzco region. This union also explains that the most prestigious division of the Qhapaq Ayllu went to the descent of Tupa Inca Yupanqui (*qullana*), whereas the grandchildren of his older brothers, Amaru Tupa and Tupa Yupanqui, respectively held the *payan* and *qayaw* statuses.

conflicts and give prominence to influential women in the political process because they were the repositories of each panaca. The chroniclers who had access to these narratives, such as Betanzos and Sarmiento, provide numerous examples of the central role mothers and wives played in the diplomatic missions or in the election (as opposed to inherited succession) of the most able ruler.[103] These women and the heads of their panaca played a significant role in imposing their favourite on the council of Inca *Orejones* that eventually conceded the title of Sapa Inca. Cieza, as we noted, mentions a woman who successfully convinced the lords of Hanan Cuzco to elect Viracocha Inca. Sarmiento describes as well the failed intrigues set up by the female relatives of Capac Huari, an older brother of Huayna Capac, to evict the latter and place their own kinsman on the throne. Upon the death of Tupa Inca in Chinchero, one of his mistresses called Curi Ocllo hastily returned to Cuzco with the news. She spoke to her kin in these words: "Lords and kin! Know that Tupa Inca is dead and when in good health he had named his son Capac Guari as heir. But eventually, as death approached he said that Titu Cusi Hualpa, the son of Mama Ocllo, should succeed him. You must not allow this. Instead you should call all your relatives and friends and appoint and raise Capac Guari."[104] Her plan was not carried out because the supporters of Huayna Capac were prompt to uncover it. They seized Capac Huari and condemned him to exile. They later charged his mother, Chuqui Ocllo, with fomenting a rebellion and the killing of Tupa Inca with evil spells. She and Curi Ocllo were finally executed. The influence of these women, together with their ability to mobilize political support, is particularly perceptible in the accounts of the fratricidal war that set Atahualpa against Huascar upon the death of their father. They relate that Huascar's mother, who resided with Huayna Capac in Quito, left the city with an escort of noblemen immediately after her husband's death. She was able to join Huascar in Cuzco before the arrival of the funeral escort, and promptly discussed with him the execution of an election plan. This mobilization drew the attention of the opposite camp, led by several noblemen of the Hatun Ayllu, to which Atahualpa and his mother belonged. As the wind turned in favour of Atahualpa, these lords successfully convinced a number of Huascar's generals to desert their leader and join their ranks.

---

[103] This question was addressed by Rostworowski 1995: 7-8, Hernández Astete 2002a: 129-134.

[104] Sarmiento: Ch. 55, 138.

Finally, one particular event, the discovery of Cuzco's main irrigation source, offers a patent illustration of a single episode being reported through two distinctive discursive genres in which the queen holds a contradictory position. The first attributes the finding of this water spring to Inca Roca, soon after the young man had his ears pierced. Still in great pain and with his left ear bleeding, the king-to-be ascended Chaca hill, northwest of Cuzco. There, he prayed to the gods and implored their help in providing the city with water. As he appealed to them, lightning struck. It frightened him to the point that the king lowered his head to the ground, where he placed his bloodstained ear. It was from the very same place where drops of his blood fell that a spring suddenly gushed forth. Inca Roca then gave orders to have the water directed to the city through elaborate canals that supplied both the Upper and Lower parts of Cuzco.[105] In this way, the narrative focuses on the supernatural origin of the stream as a gift from the gods that only a divine being, namely, the son of the Sun, could obtain. Murúa summarises the episode as follows: "He discovered the waters of Hurin Chacan and Hanan Chacan, with which they irrigate the entire Cuzco valley until today, and it is owned by his descendants."[106] The second narrative reports that it was Inca Roca's wife, Mama Micay, who endowed Cuzco with its main water supplies. She was the daughter of the powerful lord of Huayllacan, whose lands extended northeast of Cuzco. Soon after her wedding ceremony, Mama Micay realized that the valley lacked sufficient water to irrigate the maize fields. Seeing this, Cobo writes, "she had brought there most of the water that it has now (…) In memory of the courtesy she did to this place, her family and the lineage proceeding from her stayed in charge of the repartition of water that irrigates the valley."[107] This account identifies Inca Roca's marriage as a political strategy that ensured Cuzco's access to crucial water supplies. It emphasizes human agency behind the construction of the irrigation canals and identifies Mama Micay as the founder of the descent group in charge of water redistribution around Cuzco.

Finally, this reading of primary sources sheds light on another expression of moiety antagonisms that Betanzos recorded in detail, that is, the ritual battle of *purukaya* during the second royal funerals devised by Inca Yupanqui. This splendid ceremony, which is regularly described as a

105 Cieza: Pt. 2, Ch. 35, 105-106; Sarmiento: Ch. 19, 71.

106 Murúa, *Historia general*: Bk. 1, Ch. 13, 58-59.

107 Cobo: T. 2, Bk. 12, Ch. 9, 72-73. See also Zuidema 1986b.

celebration of the Incas' glorious deeds in battle,[108] can alternatively be seen as a demonstration of male dependence on their female companions to access high status and subsequent privileges. The festivity went as follows: after the nobility had mourned the dead for two weeks at various sites outside Cuzco, several dancers regrouped on the main plaza. Four men, accompanied by two young children of both sexes, positioned themselves at each corner of the square. Each adult had attached to their belt a fine rope of gold and wool that a group of ten women held firmly in their hands. The four men then executed a movement of backwards and forwards while the women released or pulled the cord as they pleased. During this performance, the little girls escorting each ritual prisoner fed him with coca leaves and also carried a stick: "threatening him with it, [they] would go hopping to one side of the disguised man as if [they] wanted to hit him on the arm."[109] The performers eventually retreated, exhausted by the constant strain of the dance, as the Sapa Inca entered the main plaza with the rest of the nobility. There, they engaged in communal mourning before a group of female warriors reappeared, brandishing weapons of war and accompanied by men carrying slings. As they eventually left the plaza, communal mourning reached an end, giving way to the final sacrifices in honour of the dead.

Betanzos' informants identified the ten women as the "lords' will" (*voluntad*), leaving little doubt that this scenario enacted a form of male subjection. He explains: "If his will gave him a long piece of cord with which he was tied, he acted like a free man. If it was pulled up short, he did not do another thing, as he did when he was given a long piece of cord, he would say that his will had him tied."[110] The Spanish word *voluntad* only approximately translates the Quechua *munana* or *munay*, whose stem *muna-* means "to want, love, or have the will to do something", and in which the notions of appetite and desire are implied.[111] The gender relations enacted at the *purukaya* dance assumed this wide semantic dimension. They depicted as much the male dancers' conflictive will as their love and desire for the women they were tied to. How shall we reconcile this interpretation with the main lines of the *purukaya* royal funerals? The ceremony acquires further significance in light of the ritual battle that followed the dance, and in which members of the Upper moiety violently opposed those of the

108 MacCormack 1991a: 128.
109 Betanzos: Pt. 1, Ch. 31, 146-147.
110 Ibid.; translation by Hamilton & Buchanan 1996: 136.
111 González Holguín, *Vocabulario*: 249-251.

Lower moiety. Betanzos reveals that the invariable victory of Hanan Cuzco on *purukaya* represented the wars that the deceased had won during his lifetime. Yet, in this instance, he does not refer so much to the dead Inca's conquests during his royal tenure as to the victory that he, as the most capable and promising heir, had won against the established and past order, which dictated that legitimacy should be acquired by descent. *Purukaya* was one of the rare festivals when the life history of every mummified Inca was recited. These epics, like the ritual battle, celebrated kingship's openness to history as the source of imperial expansion. By the same token, the *purukaya* dances exalted the central role of high-ranking women in the perpetuation of the Inca dynasty. It is significant that no other festival like these funerals overtly portrayed the political and military authority of Inca noblewomen. What better mise en scène, though, than this celebration where they appear to ensure the viability and resourcefulness of the Inca monarchy? It cannot be incidental that there were ten women associated with the four corners of the main plaza. Like the ten panacas to which yearly ritual duties fell upon, each of the women was affiliated with one of the four *suyus*, and in all likelihood stood for the ten lineages that competed for the royal office.

The Vision of the Conquerors

Kingship succession, with its share of contingencies, lay at the heart of the various moiety antagonisms evoked in royal narratives. These accounts covered two perspectives on past events that were both intrinsically linked to the way the Inca conical clan system used articulated time to construct hierarchical relations. In this structure, time determined the relative closeness/relationship of every Inca nobleman to his common ancestor. To borrow Valeri's words, the conical clan model "both recognizes history and triumphs over it."[112] It acknowledged the role of the primogeniture inheritance rule as fundamental in the stability and perpetuation of the Inca prestige hierarchy. In this way, it regarded filiation as history understood in genealogical (lineal) terms and as a social rule confined to past practices. Because they reflected the inward perspective of the Inca noble groups, the life histories owned by the panacas featured moieties' antagonisms and internal rivalries. They exalted change as a means to legitimate power and presented the rulers who strove to maintain the prerogative of primogeniture as weak and indecisive. These lords, identified with the

---

112 Valeri 1990: 159-160.

Hurin moiety, were holders of past customs who married their full sisters to preserve the continuity of the main dynastic line. As proof of their unfitness to rule, their sovereignty did not exceed Cuzco's surroundings. Their offspring, likewise, were weakling creatures in whose hands an empire was not viable. Their younger brothers, in contrast, were endowed with divine legitimacy and represented an alternative future made of expansionist conquests. Their victory over obsolete rules was celebrated in the life histories of each individual ruler and in the ritual battle that closed the second royal funeral.

Temporal process in Inca royal narratives not only structured these oppositions around two contrasting principles of succession, it also mediated asymmetric relations by allowing individuals to transcend the hierarchical position they had been given at birth to access the highest position in the senior line. Hence, divine legitimacy, marriage strategies, and rules of rank transmission ensured that the genealogical continuity of the dynasty was preserved in spite of the reworkings of historical actors. In this way, the underlying principles of the Inca conical clan formed a social structure, a frame of reference that individuals were prompted to transgress but which they ultimately strove to restore in order to ensure social stability.

As we shall see now, the conflictive relations the Cuzco nobility forged with the populations that provided spouses for their rulers took on similar dualistic features. These nations were entitled to certain privileges and were required to attend specific festivities of the Inca calendar. They entered into a personal relationship of mutual obligation with the king that ensured the stability of the empire and the tenure of local chieftainships. The next chapter provides a systematic overview of foreign presence during Inca festivities and demonstrates how ritual context dramatized the equivocal relations that the ruling elite maintained with other native leaders. It examines the ritual strategies devised by the Inca elite to express their ascendancy over the conquered nations while establishing relations of reciprocity with them. The most obvious illustrations for this overview are the foreigners' entrance at the end of *qhapaq rayme*, at the *sitwa* purification rite, and during *qhapaq hucha*. It examines the Inca/non-Inca relations as an instance of dual opposition that took up and extended the characteristics of Cuzco moiety division.

CHAPTER THREE

INCA AND LOCAL ELITES: RITUAL POLITICS OF ASSIMILATION

Ambiguous ties connected the Inca nobility with the foreign ayllus that once dominated the Cuzco region. Historical discourses and descriptions of the metropolitan festivals depict this nexus of conflicting relations in discordant terms. They either celebrate mutual dependence and long-lived confederation, or emphasize Inca supremacy and the exclusion of outsiders. A look at the ritual geography of the initiation ceremony and the narratives associated with its origin provide a useful illustration of this ambivalent picture. *Warachiku* ("the donning of the loincloth") was publicized as an elitist celebration of Cuzco's noble youth when they were first bestowed with distinctive signs of prestige. This event required that every foreigner leave the city while the male novices had their heads shaved, had their ears pierced, and received their first loincloth as a sign of passage into adulthood. Yet, the secretive and elitist nature of *warachiku* met with several contradictions. First, the Inca probationers alternated physical ordeals with offerings to the ancestral shrines of important provincial ayllus located outside the city. Second, some oral traditions indicate that several ethnic groups of the Cuzco region observed the same initiation customs before Manco Capac reportedly settled among them. These stories thus contradict the Inca origin myths that credit the founder of the royal dynasty with the conception of the initiation ritual that he thereupon imparted to the neighbouring nations.

Warachiku was not the only Inca celebration to follow an incongruous development. During *qhapaq hucha* as well the king centralized the ritual activities around his person but sent members of the panacas and non-royal ayllus to pay special honours to several *wakas* related to the history of local nations. Likewise, the cleansing festival of *sitwa* excluded foreign presence from its ceremonial centre but nevertheless covered a ritual geography peopled with ethnic groups that shared a history with the Incas, and in some cases had been their long-standing enemies. Why, then, would the ruling elite have periodically honoured the tutelary *wakas* of these nations? Silverblatt and Niles posit that in the course of Inca expansion the rising rulers of Cuzco "reconstructed customary order" from ancient customs and rewrote the past in which the deeds of their rivals were inscribed.[1] By appropriating provincial shrines for themselves, the Incas devised an important aspect of their imperialist strategy whereby they appeared as the prime holders of elitist traditions and benefactors of their loyal subjects. In Niles' view, discordances in narration were the result of this strategy of hegemonic policy. To borrow her expression, the "reshaping of history" would have brought legitimacy to the dominion of the authoritarian leaders by erasing the memory of past leaders.[2] This rationale certainly sheds light on a fundamental aspect of Inca politics, but it does not account for the ruling elite's periodic offerings to the tutelary shrines of neighbouring chiefdoms whose deeds were explicitly remembered in the oral traditions of the *seqe* system. I argue here that the Incas did not completely annihilate the memory of past leaders, but instead held two contrasting types of discourses explaining their relationship to local *wakas*. One set of traditions depicts the *qhapaq* dynasty as a successful line of autonomous and unallied leaders who imposed their sovereignty through conquests. It traces the ancestors of outside communities back to pre-Inca times and emphasizes their early resistance to Inca dominion, under which they eventually came to be alienated subjects. Another set of narratives ties the same foreign ayllus to the Cuzco elite, either through ancestral siblingship or matrimonial alliances. This perspective on foreign affairs related to the perpetuation mechanisms of the royal descent group outlined in the previous chapter. The evidence marshalled shows that the Inca dynasty not only depended on strategies of alliances to gain political control over an extended labour force, it also opened the most strategic positions of its prestige hierarchy to "illegitimate" contenders born of foreign women. The viability—and the fragility—of this expansionist system relied on extended kin politics that were remembered in a series of oral traditions and periodically celebrated in metropolitan festivals. Like the narrative and ritual antagonisms placing Hurin and Hanan in opposition, historical discourses on the Incas' relationship to foreign ayllus emphasize assimilation on one hand and differentiation on the other. This twofold poetics articulated the sequences of the *sitwa*, *qhapaq hucha*, and *warachiku* festivals. Following this line, I address the nature of the Incas' relations with foreigners in ritual settings as instances of dual opposition that took up and extended the characteristics of Cuzco's moiety division. I suggest that the ritual implementation of matrimonial strategies (e.g., the

[1] Silverblatt 1988: 92-93.
[2] Niles 1999: 20-24.

handing over of Inca brides and enactments of marriage networks with foreign ayllus) mediated antagonisms between insiders and outsiders in these conventional contexts.

The Foreign Ayllus of the Inca Heartland

As the previous chapter briefly outlined, a vast web of shrines, interconnected via the *seqes*, encircled Cuzco and extended over lands and populations partly under the jurisdiction of non-Inca chieftains. The *Relación de los adoratorios de los indios* is the only surviving description of this system, listing 328 *wakas* distributed over 41 *seqes*. The report provides details of the shrines' locations and configurations, the sacrifices they received and, for some, their astronomical function. Most of the sanctuaries are explicitly associated with Inca leaders, kings, queens, or army commanders, while others appear to be linked to local traditions and, notably, to the deeds of provincial lords. The Ayarmacas and the Huallas, for example, worshipped *Senga* (Ch. 5:9) and *Antuiurco* (An. 1:4) as their respective places of origin, while *Vicaribi* (Ch. 9:5) is described as the sepulchre of a notorious Maras lord. The Allawillay Ayllu, undocumented in other official chronicles, also had a *waka* included in this system where its chiefs had been buried (Co. 4:5).[3] The status of these foreign ayllus was such that their sanctuaries received periodic offerings from the panacas and non-royal ayllus. The *Relación de los adoratorios* indicates that each Inca lineage was given a *seqe* and provided food and rich offerings to every shrine crossed by this particular line.

Yet, not all of these provincial nations shared similar relations with the Inca nobility. Of the aforementioned ethnic groups, the Ayarmacas were the most powerful. In the Late Intermediate Period (c. AD 1000-1400), they had subjugated nations from Pucyura, west of Cuzco, to the Chinchero area, northwest of the capital, where their interests increasingly collided with those of the Incas.[4] For this reason, the two powers long maintained fluctuating relations, ranging from bride exchanges to armed conflicts. It was Pachacuti Yupanqui who reportedly put an end to their rivalries when he captured and killed the powerful Ayarmaca warlord known as Tocay Capac.[5] The Huallas, unlike the Ayarmacas, were an indigenous

---

3 Polo, "Relación de los adoratorios": 10, 15, 16-17, 29.
4 Rostworowski 1993b; Bauer 2004: 79-80.
5 Julien 2000: 245-247; Cahill 2002: 616-618; Yaya 2008: 67-68.

population, producers of coca and chilli, who never attained regional authority. The Inca origin myths record that upon arriving in the valley, Manco Capac and his siblings usurped the parcel of land cultivated by the Huallas in Huanaypata.[6] The same traditions describe the Maras Ayllu as one of the non-royal groups to have emerged from the cave of Paqariq Tampu with the Ayars and to have taken part in their migration in search of fertile lands.[7] Little is known about this ayllu, whose name also refers to an area thirty kilometres northwest of Cuzco that is renowned for its salt-pans. This zone fell within the jurisdiction of the Ayarmacas when they were at the apex of their supremacy, but the kind of relations these two groups maintained remains unclear.[8]

For Garcilaso, some of these provincial ayllus were Incas-by-privilege who resided in the Inca heartland within a radius of up to sixty-five kilometres from Cuzco.[9] He tells us that Manco Capac peopled this zone, having settled in the valley where he established the future capital. From this time onwards, Manco Capac favoured the neighbouring nations with special honours similar to those displayed by his own descendants, such as the wearing of pendant earrings and participating (albeit separately from the royal lineages) in an initiation rite similar to that performed by young Incas. On account of these privileges, the Spaniards called these people *orejones* (large ears), although this term more frequently described the Inca elite. The type of earring ornaments they wore, which indicated their hierarchical status, distinguished these nations:

> Some were given as a token a splinter of wood as thick as the little finger, as were the tribes called Mayu and Çancu. Others were to have a little tuft of white wool which stuck out of the ear on both sides the length of the top of the thumb: these were of the tribe called Poques. The Muina, Huaruc, and Chillqui tribes were to have earplugs of the common reed the Indians call *tutura*. The Rimactampu tribe and their neighbours had them made of a plant called *maguey* in the Windward Islands and *chuchau* in the general tongue of Peru (...) The three tribes bearing the name of Urcos, Yucay, and Tampu, all dwelling down the river Yucay, were given the special privilege and favour of wearing larger holes in their ears than the rest.[10]

[6] Betanzos: Pt. 1, Ch. 4, 20; Sarmiento: Ch. 13, 60; Cabello: Ch. 10, 269; Múrua: Bk. 1, Ch. 3, 45.

[7] Molina: 106. Sarmiento: Ch. 6, 53.

[8] Santo Tomás, *Grammatica*: 128-129; Sarmiento: Ch. 11, 53.

[9] Garcilaso: Bk. 1, Ch. 20: 48-49; Bk. 1, Ch. 23, 53-55. See also Betanzos: Pt. 1, Ch. 15, 73; Molina: 97; Fernández: Pt. 2, Bk. 3, Ch. 5, 83; Gutiérrez: Ch. 64, 254; Guaman Poma: 740[754], 690. See also Zuidema 1983b: 54-69; Bauer 2004: 16-22.

[10] Garcilaso: Bk. 1, Ch. 23; translation by Livermore 1989: 57.

Guaman Poma offers a slightly different list of ethnic groups that "pierced their ears according to the law and ceremony they followed under Inca rule" (Table 10).[11] For him, "everyone who has ear ornaments is called yngas, but they are not perfect, rather they are poor Indians and low-born people, they are not *caballeros*, but commoners (*picheros*) instead. Of those who have ear ornaments, only one was the first Inca king, Mango Capac. For this reason he was called capac [powerful]; to say ynga is common, it does not mean king."[12] These foreign ayllus were tribute-payers; they received goods from other lands in exchange for their own contribution and held intermediately placed offices in the hierarchy of the imperial administration.[13] Some of them also provided the Inca rulers with principal wives, such as the Ayarmacas and the Anta, which occupied lands in the northeastern Cuzco region. During the early period of Inca state formation, these two polities contracted intermarriages owing to their strong political positions.[14] Cobo reports that the first foreign rulers to pay obedience to the Incas originated from the Huaru Valley, located "three leagues from Cuzco, which was well populated and had powerful lords."[15] Both Garcilaso and Guaman Poma mention this nation in their list of *orejones*. They too exchanged wives with the Cuzco rulers, for they are occasionally recorded as Kaka Huaruq Inca.[16] For Sarmiento, Lluque Yupanqui was the first Inca king to make alliances with the Huaru and other Incas-by-privilege, including the Quilliscache.[17] Despite the prerogatives they were entitled to, not all of these *orejones* came under Inca jurisdiction peacefully. The Muyna (Lucre Basin) and the Ayarmacas appear to have rebelled countless times against the Cuzco leaders.[18]

Contrasting with Garcilaso's version of the origin myth, some traditions suggest that these nations held equal status as that of Manco Capac's descendants during the early phase of Inca expansion and had shared similar signs of prestige since times immemorial. Cieza, for example, evokes the Cauiñas "who, before the Incas subdued them, had their ears pierced and placed their own ornament in its hole, and they were *orejones*. Manco Capac, founder of the city of Cuzco, befriended them. They wear woollen

---

11 Guaman Poma: 337[339] 310.
12 Guaman Poma: 84 [84] 64-66. My translation.
13 Guaman Poma: 740 [754], 690.
14 Sarmiento: Ch. 22, 76-78.
15 Cobo: Bk. 12, Ch. 6, 68.
16 Guaman Poma: 84 [84], 66.
17 Sarmiento: Ch. 16, 65-66.
18 Sarmiento: Ch. 19, 71; Ch. 24, 80.

Table 10. Incas-by-privilege according to Garcilaso and Guaman Poma.

	Garcilaso de la Vega	*Guaman Poma de Ayala*
Antisuyu	Poques	
Kuntisuyu	Masca Chillque Papri	Masca Chillque Papri Tampu Acos Yana Uara
Chinchaysuyu	Mayu Sancu Rimac Tampu Yucay Chinchay Pukyu Tampu	Mayu Sancu Rimac Tampu Anta Equeco Sacsa Uana Quilliscache Quichiua Lare Uaro Conde
Qullasuyu	Ayarmaca Quespicnacha Muyna Urcos Quehuar Huaru Cauiña	Quehuar Huaru Cauiña

clothes, most of them shaved and had a braid wound around their head."[19] Another text, attributed to Father Bartolomé Segovia (1553), also records that several populations occupying the Cuzco region not only pierced their ears but also wore the loincloth and occasionally tonsured their hair well before the Incas' arrival in the valley. He, like most Spanish historians, mentions the conflicts that placed Cuzco in opposition to these chiefdoms in the early period of the Inca dynasty.[20] They suggest that earlier on, these ethnic groups held a potent control over the fertile lands of the valley and

---

[19] Cieza: Pt. 1, Ch. 97, 268.

[20] Cieza: Pt. 2, Ch. 33, 100-101; Sarmiento: Ch. 14, 61-62; Ch. 17, 67-68; Ch. 20, 72-74; Ch. 23, 79.

competed against each other for a monopoly on irrigation sources and matrimonial networks:

> Some of the *orejones* were shorn, while the others grew long hair, called *chilques* today; these peoples fought each other, and the shaved ones subdued the others in such a way that these last never became leaders nor residents of the city. Today there are villages of them in the vicinity of Cuzco; yet, they were forbidden to live in the city itself, except for the commoners who served in what they were asked.[21]

By the time of the Spanish invasion, the Incas had developed direct control over the Cuzco region's resources (land, water, livestock) and labour force. Their heartland not only covered a vast area peopled with tributary allies, it also enclosed the plots exploited to benefit the deceased rulers and their lineages. These royal estates extended over large areas of the Cuzco region and provided each panaca with expedients for their ritual commitments as well as quarters for leisurely pastimes.[22] In the vicinity of these domains, the foreign ayllus that had chosen to resist Inca hegemony were quelled by military means, deprived of substantial political influence, and had their lands confiscated. Those that had accepted the jurisdiction of Inca leaders benefited from tribute reciprocation; they were granted subordinate positions in the state administration but could increase their authority by marrying their women to the king and other noblemen of the *qhapaq* descent group. This prospect was offered to foreign delegates precisely at the time when they were asked to attend certain Inca festivals.

Ritual Geography of the Cuzco Region

None of the foreign ayllus mentioned administered lands beyond the topography delimited by *sitwa*, the cleansing festival whose ritual space would have defined the boundaries of the Inca civilized world.[23] Set during the Inca month of *quya rayme* (August), *sitwa* followed an interesting structure of foreign exclusion and reinsertion. Before it started, every provincial lord subject to the empire was pledged to bring to Cuzco the effigy of the principal deity worshipped by his community. During the first stage of the celebration, these images were kept in their private temples while all foreigners, as well as disabled and deformed people, were expelled from the

21 Segovia: 72-73. My translation.
22 Rostworowski 1993a; Murra [1995] 2002; Sherbondy 1996; Rowe 1997.
23 Farrington 1992: 374-375, 378-381; 1995; Rostworowski 1993a; Rowe 1997.

city, followed by dogs, whose barking was not only considered disruptive to the ritual but also regarded as a bad omen.[24] Then, on the day of the moon's conjunction, the Inca and the main officials assembled in the Sun temple, where they waited three days for the first crescent to reappear. When the new moon finally rose in the sky, the king together with the high priest of the Sun left the temple and decreed the start of *sitwa*. This announcement was soon relayed throughout Cuzco, and all of its inhabitants now shouted: "plagues, disasters, misfortunes and dangers get out of this land!" all the while shaking their clothes to expel the evils. At that signal, a crowd of four hundred armed warriors who had assembled in the square of Haucaypata ran out of the city in the direction of the four *suyus*, carrying burning torches and exhorting the illnesses to depart from the land. All of these men were members of the Inca nobility accompanied by two or three non-royal ayllus, each one grouped according to its moiety and *suyu* division. Their specific journeys outside Cuzco each reached a secluded close situated a few kilometres away from the capital, in Satpina, Chita, Acoyapongo, and Churicalla, respectively. There, they relayed their task to groups of *mitmaq* that had been displaced from their land of origin and resettled around Cuzco to ensure political stability. Squadrons of them carried on the ritual expulsion and eventually ended their courses by symbolically pouring the pestilences into four rivers: the Quiquisana, Apurimac, Pisac, and Cusibamba, in which they bathed and cleansed their weapons. All city residents repeated the same ritual cleansing on the following day before they assembled at dawn in the streams corresponding to their respective *seqe* to purify themselves with water.[25] For the next few days, the nobility congregated in the main plaza to address their prayers and offerings to the Inca gods. *Quya rayme* finally closed as the last portions of offerings were dispensed to the exiles and foreign delegates, who were now authorized to re-enter the city with the images of their divinities. Upon doing so, they were required to declare publicly their obedience to the ruler and his tutelary gods. For two days, a great quantity of clothing was burnt and each nation was asked to perform its own traditional dances, whereupon the provincial lords received rich goods, garments, servants, and wives in recognition of their loyalty. Upon returning to their land, however, they were only entitled to take the figures of the gods that had remained in Cuzco the previous year. The images they had brought for this last occasion were kept in the Incas' possession until the next *sitwa* was

---

24 Murúa, *Historia general*: Bk. 2, Ch. 34, 425.

25 Polo, "Tratado": 23.

commemorated. In this way, the ruling elite always held several of their subjects' ancestral shrines in custody. As MacCormack observes, this practice ensured "a carefully balanced system of patronage and censure of provincial shrines."[26]

In 1617, a Jesuit missionary witnessed in the Cajatambo district a ritual that presented a spatial conception similar to *sitwa*. In a *carta annua* to his superior, he reports the existence of a cleansing festival that commenced with a race of several naked men pursued by a group of villagers who carried a figure called Aupi, or "illness". The celebrants ran beyond their community to reach the lands cultivated by the adjacent village and threw the malevolent figure over to the opposite side, thus "passing the illness over to their neighbours."[27] The four rivers that signalled the precincts of *sitwa*'s purification activity divided the ritual space in a similar fashion and most likely stood for the symbolic confines beyond which stretched the enemy nations.[28] Inside this enclave, the Incas had resettled loyal subjects from distant provinces, forged strategic alliances with powerful chiefdoms, and assigned to each of these a position in the state hierarchy corresponding to its *suyu* and *seqe* divisions. Outside this perimeter, powerful confederations threatened to overcome Inca jurisdiction. *Sitwa* exercised a performative role in weakening their influence. It not only infected the enemies' resources and labour forces with disease, but also reaffirmed the ruling elite's marriage networks and their control over the distribution of resources. It followed a centrifugal movement inclusive of the outside world where each ayllu occupied a hierarchical status determined by the central power. It was only once Cuzco and its surroundings had been purified that the provincial delegates would re-enter the city and share with the ruling elite the sacrificial offering prepared by the Inca officiants. It reaffirmed the paramount position of the Cuzco nobility on the political chessboard and simultaneously conceded a role in its religious apparatus to provincial delegates. The Incas led this festival as they administered their heartland, by controlling the allotment of land, water, and brides in accordance with their allies' position from the central authority.

Duviols (1976c) suggests that a similar centrifugal structure ordered the Inca ritual of *qhapaq hucha*, an event during which rich offerings, and occasionally child sacrifices, were offered to the most venerated *wakas* of Tawantinsuyu in temples but also in mountain peaks, springs, lakes, or

---

26 MacCormack 1991a: 104.

27 *Carta Annua* 1617, ARSI, Perú 14: f. 54, in Polia Meconi 1999: 392-393.

28 Zuidema 1986b.

other natural formations. According to Gerald Taylor, *hucha* translates as "obligation, duty, debt,"[29] so that the name of this ceremony may be best summarized as the completion of a ritual obligation to/of the royal dynasty. Depending on the sources, this imperial celebration would have been held cyclically or on exceptional occasions, such as the death of a Sapa Inca, the birth of a legitimate royal son, the accession to power of a new ruler, or the subjugation of additional nations.[30] These disparities in the descriptions of the *qhapaq hucha* led Rostworowski (2008) to suggest that the ritual assumed two distinct forms. The first involved human sacrifices, whereas the second consisted of offerings of blood carried in vessels over specific routes. When a *qhapaq hucha* was announced, the provinces sent to Cuzco their tribute of precious garments, food, and ornaments, together with virginal boys and girls aged between four and fifteen, drawn from the local elite. These children were selected for their good appearance and the perfection of their skin, which could not present any blemishes. Betanzos indicates that the boys and girls were symbolically married once they reached the city. The Inca central power thus performed ritually a control of marriage networks between the native elites.[31] For the next ten days, the provincial delegates and the ruling nobility all feasted together, drinking in honour of the Inca gods and royal mummies placed on the main plaza. The king then decided on how the offerings should be distributed among the *wakas*. It was generally the case that the divinities that had favoured the Cuzco sovereigns in battle or formulated auspicious prophecies in their favour received the richest oblations, such as human sacrifices. On the appointed day, the ritual processions departed from Cuzco in the directions of the four *suyus* that composed the realm. Walking in wing formations, each cortège of participants was composed of provincial and Inca representatives who carried offerings of llama blood and *mullu* (spondylus shell), which they dispensed along their way to the different shrines.[32] Upon reaching their final destination, the victims were intoxicated with corn beer and eventually met their death surrounded by rich offerings of vessels, maize ears, garments, gold and silver figurines, etc. Sometimes their burial sites received a cult perpetuated through several generations.

---

[29] Taylor in *Ritos y tradiciones de Huarochirí*: 283, n. 11. See also González Holguín, *Vocabulario*: 199-200.

[30] Betanzos: Pt. 1, Ch. 11, 51; Ch. 30, 141-143; Cieza: Pt. 2, Ch. 29, 87-89; Molina: 120-127; Sarmiento: Ch. 13, 59; Ch. 31, 96; Ch. 42, 116.

[31] McEwan and Van de Guchte 1992.

[32] Betanzos: Pt.1, Ch. 11, 52-53; Molina: 120-127; Cobo: T. 2, Bk. 13, Ch. 32, 222.

A vivid narrative from the Huarochirí manuscript reveals that the *qhapaq hucha* offerings were subject to reciprocal benefits and served in part to ensure the allegiance of powerful chiefdoms to the cause of the Cuzqueñan leaders. This narrative opens with Tupa Inca Yupanqui's setbacks in his battles against three rebellious populations. After twelve years of combat in vain, the ruler summoned all the *wakas* of the empire to Cuzco in order to end the rout. Assembled in the capital, Tupa Inca reminded the deities of the silver, gold, foods and other presents he regularly sent them, and implored their support in a conflict he could not resolve alone. Hearing this, a *waka* of the central sierra called Macahuisa assured Tupa Inca he would help him to victory. The deity kept his promise and in return for defeating his enemies, the king in person served offerings of *mullu* to Macahuisa. From then on, Inca celebrants serviced this *waka*, "danced for him and held him in great esteem."[33]

The Incas likewise rewarded the local lords who had given their children as *qhapaq hucha* sacrifices with a position in the state hierarchy. A famous example is the sacrifice of ten-year-old Tanta Carhua, whose father, Caque Poma, was lord principal of Ocros (Ancash). The extirpator of idolatry Hernández Príncipe (1622) reports that she was brought to the summit of Aixa, near her native town, where she was buried alive surrounded by precious vessels. After her *qhapaq hucha*, the Inca king granted Caque Poma the hereditary title of *kuraka* of Ocros, which he passed on to his descendants over succeeding generations.[34] Attached to this position were the exemption from tribute, the right to be carried on a litter, and access to prestige goods and secondary wives from the repositories of native elite women controlled by the Inca ruler.

Despite the reciprocal nature of the festival, *qhapaq hucha* respected the asymmetric relations that the ruling elite had established with foreign ethnic groups. The Inca nobility was indeed exempted from ritual donations, while at the head of this prestige hierarchy, the ruler was assigned the task of redistributing the goods prepared by his subjects and forged (symbolic) marriage networks between native lords.[35] Several scholars also suggest that the different routes followed by each group corresponded to the formation of the *seqes,* and as such Cuzco represented the concentric

33 *Ritos y Tradiciones de Huarochirí*: Ch. 23, 289-301.

34 Hernández, in Duviols 2003: 753-778. See also discussion by McEwan & van de Guchte 1992.

35 MacCormack 1991a: 103-105. See also Abercrombie 1998: 170-173.

centre from which the ceremonial necessities of the whole empire were distributed.[36] It is also apparent from the chroniclers' descriptions that *qhapaq hucha* not only aimed to coalesce populations under the yoke of a single power, it also served to (re)define local rights to natural resources and labour assignments on the lands crossed by the ritual processions. This is first suggested by some of the extraordinary circumstances for which it was held: royal succession or the conquests of additional nations always effected a redistribution of land and water rights. Communal repartition was all the more crucial to regional stability because ayllus did not occupy a cohesive geographical zone and arable lands were cultivated on a rotational basis. *Qhapaq hucha* was also destined to commemorate the prominent *wakas* of regional polities and other community-level founding ancestors. These figures received a cult for having first occupied the parcels exploited by their descendants and for having distributed their resources among them. *Qhapaq hucha* provided a stately sanction of these *wakas*' ancestral authority over the land and water that contemporary ayllus exploited. Tanta Carhua, for example, was sacrificed immediately after her father effectuated the building of an irrigation canal in the vicinity of Ocros. Second, in each province, local officials accompanied the ritual procession to the periphery of their land, where a similar envoy mandated by the neighbouring community took over for them. Rostworowski (1988b) shows that each ethnic group was responsible for carrying the offerings through the parcels they exploited and did not allow foreigners to trespass because crossing a *chakra* meant having tenure over it. Arguably, the *qhapaq hucha* local delegates held duties similar to those of the *mitmaqs* who during *sitwa* carried the illness to the limits of the land they were assigned to work, whereupon it was taken over by subjects of the next province. This emphasis on communal repartition is also suggested in the chants that the *qhapaq hucha* processions sang in honour of Wiraqucha, the deity who at the origin of time had assigned to each people the arable lands they should populate and exploit. In this way, the repartition of communal lands and their resources, the reciprocal yet asymmetric nature of the Incas' relations with native lords, and the ritual integration of foreign ethnic groups into the core of the ruling elite constituted three major features of *sitwa* and *qhapaq hucha*. These characteristics were framed by the

---

36 Zuidema 1973; 1982b: 428-435; Duviols 1976b; Sallnow 1987: 39. Zuidema (1977a) suggests that the *qhapaq hucha* was a ritual only performed in the Chinchaysuyu region and was generalized in the whole empire by the chroniclers. Archaeological evidences have since refuted his conclusions. See Ceruti 2004; Bray et al. 2005.

concentric structure of the festivities whereby the centre (ruling elite) opposed the periphery (subjects).[37]

The same twofold poetics articulated the sequences of the *warachiku* festival that largely took place outside the capital, in various locations ancestrally linked to local ayllus. It commenced with the expulsion of the foreign lords to the outskirts of Cuzco, followed by their reinsertion into the ceremonial centre immediately after the end of the novices' ordeals. The primordial migration re-enacted ritually during *warachiku* transcended the asymmetrical divisions that set the Inca nobility apart from the native chieftains. It celebrated the matrimonial alliances that the kings of Cuzco had contracted with the local ayllus whose *wakas* were visited by the *warachiku* celebrants. In this way, it commemorated the Ayars' siblingship in spite of them having reportedly founded different, and sometimes rival, ayllus. Here, this ritual will be examined in light of a narrative recorded in great detail by several chroniclers, principally Sarmiento, Gutiérrez, and Fernández, who depict the ambiguous relations the Incas maintained with the chiefdoms surrounding their capital. This story tells the abduction of the young Titu Cusi Hualpa, later known as Yahuar Huacac. In this account, the matrimonial unions that the protagonists, all foreign lords of the Cuzco region, contract with the Inca nobility take place after they successfully accomplished a series of ritualized performances traditionally assigned to the novices' maternal uncles. Through the completion of this formal procedure, the conflicting nature of the local leaders' relations with the Incas becomes one of affinity and reciprocity.

The Politics of the Initiation Ritual

Ordeals and Pilgrimage

Warachiku is certainly the most documented celebration of the Inca calendar, but it is also one for which we have many contradictory accounts. Attempting to draw a single picture of this initiation ritual is a challenge, the intricacies of which stem partly from the very nature of the festival. As with other rites of passage worldwide, *warachiku* involved secluded and secret practices entailing a certain degree of social and gender partition. Only Incas of noble status participated in the celebration, while other ethnic groups from the Cuzco region conducted their own initiation at a

---

37 Duviols 1976b: 16-24; Zuidema 1989c: 272; Sallnow 1987: 37-40.

different time of year.[38] Several sources also indicate that this festival was the occasion for the young men to receive a new name, dispensing with the one they had been given at birth. None of the accounts, however, situate this episode in the exact sequence of the events, which suggests that it took place in the privacy of the novices' houses. In this context, it would seem unlikely that any Spaniard personally witnessed the entire ritual, a likelihood that could partly explain the discrepancies observable in colonial texts.

Despite these ritual constraints, women and provincials were not entirely denied access to this drama. A group of maidens was able to perceive the young men's sufferings as they followed the procession around while carrying a jar of chicha to alleviate the novices' thirst. These girls, aged between twelve and fourteen years, probably underwent their own, albeit undocumented, initiation.[39] As for the foreigners, they were commanded to come all the way from their provinces at the very beginning of the celebration, only to wait outside Cuzco for several weeks in the knowledge that the Inca nobility was conducting the initiation of its youth.

> To those [provincials] who were leaving and entering the court, [the Incas] indicated a certain place, designated for this purpose, at the entrance of each road. In each of these locations gathered the people of the *suyu* through which this road went. There, they collected the tributes and the goods for the Religion that every province of the realm provided in this season, and those who had brought them waited there for the king's and the huacas' ministers to receive them.[40]

Although they were denied any involvement with the celebrants until the last days of *qhapaq rayme*, foreign officials were aware that the probationers' pilgrimage outside Cuzco was a demonstration of the Incas' bravery and endurance at war. In this sense, the effectiveness of the *warachiku* not only rested on the impact it had upon the initiates and the ruling elite itself, but also on those who were excluded from it. They were made aware that the young men returning from their journey and passing by the foreigners' secluded residences were wounded and covered with blood from their trials.

While the chronicles attribute the institutionalization of the Inca initiation ritual to different kings, two important traditions prevailed. The first locates the origin of *warachiku* before the Inca settlement in Cuzco,

---

38 Molina: 97.
39 MacCormack 1991a: 114-117.
40 Cobo: T. 2, Bk. 13, Ch. 25, 208. My translation.

when Ayar Uchu appeared to his brothers at the summit of Huanacauri and enjoined them to perpetuate ear piercing as a sign of nobility among their descendants.[41] It was therefore at the foot of this hill, in Matahua, that Manco Capac's young son underwent the first trials of the initiation rite, whereupon he and his relatives recommenced their journey, leaving the petrified *waka* of Ayar Uchu behind them.[42] In memory of these founding events, the novices paid their respects to this shrine and several others where their forefathers had stopped along their way towards the fertile valley. Another cluster of narratives assigns the entire conception of this ritual to Pachacuti Yupanqui who, having defeated the Chancas and taken his place on the imperial stool, instituted the numerous festivities of the metropolitan calendar, decreeing that the *warachiku* was to mark the beginning of each year.[43] Spanish authors often intermingle these two traditions and report that although *warachiku* had been inaugurated at the beginning of the Inca era, the rite of passage would have undergone various transformations under the auspices of Hanan lords.[44] Still, despite these contrasting aspects, it remains evident that the itinerary of the *warachiku* celebrants replicated the journey of the Inca ancestors within the topography of the Cuzco region.

The public celebrations of *warachiku* took up the entire month of December and came to an end with a mock battle in January. Prior to that, the novices-to-be and some of their family members were kept occupied with the preliminaries of the forthcoming ceremony. During this time, the boys left their home and settled in the countryside to fast and collect bundles of straw, which they later dispensed among their female kin. In a location close by, on a hill called Chaca, a group of maidens gathered together to spin the wool for weaving the loincloths' fringes. Cobo indicates that while the spinning process was under way, the young girls worked under the auspices of the *waka* of Huanacauri, which had been temporarily moved from its location southeast of Cuzco (Co. 6:7) to this hill in

41 Cieza: Pt. 2, 6-20; Molina: 107; Sarmiento: Ch. 13, 58.

42 Cieza: Pt. 2, 16-17; Sarmiento: Ch. 13, 57-58; Murúa, *Historia general*: Bk. 1, Ch. 2-3, 42-45.

43 Betanzos: Pt. 1, Ch. 15: 71; Polo, "Tratado": 18; Cabello: 349; Murúa, *Historia general*: Bk. 2, Ch. 38, 437.

44 Sarmiento, for instance, claims that the *warachiku* had been instituted in honour of Manco Capac's son when father and son entered the valley at the dawn of the Inca era. Later, however, he adds that Pachacuti Yupanqui built four houses in honour of the Sun around Cuzco, so that his son could undergo his initiation ritual there, and receive his new weapons in each place (Sarmiento: Ch. 63, 116-117). See also Molina: 106-107.

Chinchaysuyu (Ch. 5:7).[45] Cobo's chronicle is partly based on the *Relación de los adoratorios*, to which he had access, wherein he could have read the following lines: "Chacaguanacauri; which is a hill on the road to Yucay where the young men who were armed *orejones* went to find a certain straw they attached to their lances."[46] Also linking this site with the *warachiku* is the account of Inca Roca's discovery of Cuzco's major hydrological source. Brian S. Bauer identifies this place with the ruins of Chacan, northwest of Cuzco, the site of "an important canal system, a huge outcrop spanning the upper Saphi (or Chacan) River, several large outcrops with platforms on their summits, a number of carved stones, and many terraces."[47] Situated nearby the ruins of Chacan is the spring of Tambo Machay, whence emanated a subterranean canal leading to lands whose produce went to the Ayarmaca Ayllu. Zuidema (1997a) notes that present-day popular traditions identify the main provider of Tambo Machay's waters as Lake Qoricocha, located on native plots administrated by another ethnic group, the Huayllacans. This population had political authority over subjects living fifteen kilometres north of Cuzco, extending from the high plateaux of Patabamba down to the present-day communities of Coya and Lamay on the banks of the Urubamba River (Map 1).[48] Given the ecological diversity of this area, they had access to a wide variety of yield productions as well as pasture lands. To consolidate their position in the region, the Huayllacans contracted matrimonial alliances with Cuzco and provided Inca Roca with his principal wife, Mama Micay. She is credited with having an active role in setting the water-ditching project of the main irrigation canal supplying Cuzco, the same water source whose mythical discovery was attributed to Inca Roca. Given the chroniclers' consensus over the elitist nature of the *warachiku* rite, it is intriguing that Chaca constituted a major site visited by the Inca initiates in preparation for their initiation ordeals. However, as we shall see, Chaca was only the first stage in a series of pilgrimages that commemorated several shrines ancestrally linked to non-Inca ethnic groups.

On the eleventh day of the month *qhapaq rayme*, an elderly relative of the novice-to-be sheared the child's hair and dressed him in a tunic woven previously by his female kin. He then placed a black headband (*llaytu*) on

45 Cobo 1964: T. 2, Bk. 13, Ch. 25, 208.

46 Polo, "Relación de los adoratorios": 9. The Spaniards used the word "orejones" to refer to the members of the Inca elite, in relation to their large pendant earrings.

47 Bauer 1998: 61. From this point, the Chacan canals run southwards onto the Saqsaywaman plateau before turning eastwards towards the city.

48 Bauer 2004: 81-82; Covey 2006a: 147-155.

the boy's forehead and threw a light-coloured *manta* over his shoulders, both of which were adorned with black feathers. Several maidens from the Inca noble caste joined the participants on the main square, where each of them paired with a boy. From then on, the young girls would accompany the novices on their journey outside Cuzco and alleviate their thirst with the small jar of chicha they bore with them. Together, the celebrants proceeded to Matahua, located some twelve kilometres from Cuzco, where the Ayar siblings had settled for a short season before entering the valley.[49] In memory of this foundation act, the novices and several adults spent the night there, and at sunrise they ascended the slope of Huanacauri to reach their ancestor's shrine. Upon arriving, they presented their offerings to the *waka*, in return for which they received a sling (*waraka*) similar to the one their Inca ancestors carried when they entered the valley.[50]

Huanacauri, however, was not exclusively linked to the Inca migration story. Duviols (1997) reveals that a series of documents commissioned by viceroy Toledo offers an alternative version to this origin narrative where Ayar Uchu became petrified before the very eyes of his siblings. Between 1571 and 1572, several jurists of the Crown recorded the testimonies of three ayllus: the Sauasiray, the Antasayaq, and the Alca Uiza, who all asserted they had exerted sovereignty over ancient Cuzco before the coming of the "tyrannical" Inca rulers. The Alca Uiza brought forward testimonies alleging that their ancestor, called Ayar Uchu, settled in Pukamarka and cohabited with three other ayllus before Manco Capac's group entered the valley. The same reports stated further that the early Incas progressively subjugated the Ayar Uchu Ayllu and eventually annihilated them under Mayta Capac.[51] Betanzos also suggests that a chieftain by the name of Alca Uiza ruled over the Cuzco valley before the Inca ancestors arrived there. It was Wiraqucha who, on his way to the sea, elected this lord before disappearing. The god "also gave the name of Cuzco to the place where he named this lord. He ordered that the *orejones* should emerge after he had left."[52] Hence, as in the case of Chaca, there existed two contradicting traditions as to the origin of the ancestor petrified in Huanacauri. One depicted Ayar Uchu as a kinsman of the first Inca ruler, while the other portrayed him as the forefather of a foreign ayllu.

---

49 Bauer (1998: 110-112) identifies this site with the shrine of Matoro (Co 7:5), which may have been located on the mountain summit above the present-day community of Matao.

50 Molina: 101.

51 *Informaciones al virrey Toledo*: 135.

52 Betanzos: Pt. 1, Ch. 2, 15.

Map 1. The Cuzco region with the major sites of the *warachiku*.

The offerings completed in Huanacauri, the celebrants descended the hill to a nearby ravine called Quirirmanta, where their ancestors had initially interrupted their journey in order to dispense different functions amongst themselves, entrusting the perpetuation of their lineage to Manco Capac and the occupation of the land to Ayar Awqa, while Ayar Uchu was commanded to remain "a huaca for their religion."[53] In memory of this event, the "uncles, relatives and kurakas" who attended the initiation festival struck the novices soundly and enjoined them to follow their ancestors' example. On receiving the blows, the young men remained on their feet and stoically restrained from showing any signs of distress. Everyone then

[53] Sarmiento: Ch. 12, 56.

intoned a chant called *wari*, a term denoting undomesticated camelids, especially the vicuña.[54] According to Father Molina, the provenance of this song can be traced back to the mythical emergence of the Ayar siblings from the cave, when Wiraqucha entrusted it to Manco Capac with the directive to sing it only at *warachiku*.[55]

On their way back to Cuzco, the novices met up with a small procession led by a banner-bearer carrying the two royal insignia that the Incas had previously paraded for the festival of *inti rayme*. The first emblem was the *suntur pawqar*, followed by a white llama called *napa* (also *tupa guanaco*) chosen from the herd of the Sun god P'unchaw. This animal represented the first of its species to have emerged from the cave with the Ayar ancestors. Its ears were pierced and its back covered with a red mantle over which was laid a harness adorned with red shells or spondylus. Two *mamakuna* accompanied this procession while the *napa*'s herder blew into a large conch shell, probably a *strombus* that pastoralists traditionally played.[56] Following the sound of music, the celebrants launched into another *wari* dance before finally reaching Cuzco. The use of conch shells and spondylus on this occasion was certainly related to their alleged virtues of attracting heavy, and sometimes even apocalyptic, precipitations. In Inca times, offerings of *mullu* were presented to the rivers in order to bring rain to the young crops. Sounding the *strombus* was also believed to cause a reversal of time and social order brought about by the rise of the water level (*unu pachakuti*).[57] Accordingly, the shells' appearance at that time of year not only referred to the meteorological conditions of the month of December, when the rains are heaviest, but also created a link between the time of the celebration and the primordial Deluge that destroyed the old world to be replaced by a new order. It was following this cataclysmic event that the Ayar siblings entered the valley and saw the propitious rainbow spanning Huanacauri, announcing the advent of their reign. In a similar fashion, the novices passed through their initiatory ordeals to be reborn under a new name, thereby reaching another stage of social life.

After a day of rest the participants returned to Haucaypata, where the novices and the young maidens received a new tunic that had been woven by the state tributaries as part of their labour service. Now on the plaza, the young men held in their hands a spear (*yawri*) to which they had

54 Urbano and Duviols in Molina: 102, n. 77; Bertonio: 151; Gudemo 2005: 33.
55 Molina: 107.
56 Gudemo 2005: 36.
57 Paulsen 1974: 603; Allen 1988: 98; Allen 1993-1994: 89-90.

attached a tuft of red wool and a handful of straw (*ch'awar*). According to Cieza, these additional items represented their enemies' mortal remains. They reminded the novices of the imperatives of war, "as a sign that when they went to battle against their enemies, they should endeavour to bring back the heads and hair of those."[58] As a test of their endurance, the boys once more endured the blows inflicted by their older relatives before departing for a long march to Anahuarque hill (Cu. 1:7). This shrine, also described as "the guaca of the Indians of the village Choco and Cachona", was the birthplace of Pachacuti Yupanqui's homonymous wife, Mama Anahuarque.[59] Here, after offering food to the *waka* and being struck again, they were obliged to stand bravely on their feet, grasping their *yawri*, ready to witness the elders performing another *wari* accompanied by the trumpeting sound of a shell. After this dance, the young men were given no time to rest, as the prospect of a new ordeal opened before them.

For this new contest, the maidens first reached their allotted post some two kilometres away from Anahuarque, at the top of Rauraua hill (Cu. 2:4), where they waited for the young men to arrive. On the other side of the dale, the boys prepared to race down and up the two hills by forming a single row on the departure line. Each youth was supervised by an adult positioned behind him, and at the signal given by a noble Inca, the adults and youths all started to run in the direction of Rauraua. During this arduous race, the adults stayed behind their protégés to urge them on every time they tumbled on the wet, slippery slopes or wounded themselves on the thorny vegetation; Molina records that many contestants even died from their injuries. The winner's prize, however, was worth the torments, for he received the prestigious appellative of *waman* (hawk), and thenceforth was assigned to high-ranking military posts. According to Molina, the origin of this trial traced back to the primordial Flood from which Anahuarque escaped by running to the summit of the hill "as fast as a falcon flies."[60] It was on this site that the probationers demonstrated their agility in conditions ostensibly similar to those extant at the dawn of time,

---

58 Cieza: Pt. 2, Ch. 7, 18. For Betanzos, this practice occurred in Anahuarque on the following day. This warlike practice had been observed long before the Incas evidenced by trophy heads in Andean iconography dating from the Ancient Horizon onwards.

59 Molina: 104. Sarmiento locates these two related communities in a district (*barrio*) south of Cuzco, where the Incas venerated the petrified *waka* of Chañan Curi Coca (Cu 8:1), a woman lord (*cacica*) of Choco and Cachona who fought against the Chancas on Pachacuti's side. See Sarmiento: Ch. 27, 88; Pachacuti Yamqui: f. 19v., 220; Polo, "Relación de los adoratorios": 39.

60 Molina: 105.

exactly at the height of the rainy season. At the end of the race, all of the contestants were offered a tumbler of chicha to quench their thirst, whereupon the adults danced another *wari* that culminated in them pelting the young boys with their slings, bringing the ordeal to a close.

With the race to Rauraua hill over, the procession preceded by the white llama and the *suntur pawqar* headed back to Cuzco in the afternoon. Upon entering the main square, the novices respectfully saluted the deities and the sovereign before attending the rest of the festival, standing all the while. Another *wari* dance was performed, followed by the adults assaulting the young boys. Finally, the celebrants left Haucaypata and arrived at a place called Huamancancha (Ch. 4:5) at the foot of Picchu hill, where they spent the night.[61] The following morning they reached the shrine of Yauira, located on the summit of Picchu, which was a shrine associated with the past of the Maras, one of the foreign ayllus to have emerged with Manco Capac from Paqariq Tampu.[62] Alongside this tradition, the *Relación de los adoratorios* also links Yauira to the primordial destruction of the world:

> He had emerged from the earth with Huanacauri, and having lived a long time there heaved himself up and became a stone; one that all the ayllus went to worship during the Raymi feast.[63]

Here, the celebrants conducted the usual rituals of offering maize and coca before sacrificing several llamas and using their blood to trace a line on the boys' faces from ear to ear. They also ate a piece of raw meat sliced from the animal to remind them that, "if they were not valiant, their enemies would eat their flesh in like manner they had eaten the llama they killed."[64] This practice was preceded by a long oath in which the novices expressed their obedience to the Inca and their dedication to the worship of the Sun god. In return, Yauira's attendant offered them a loincloth (*wara*) made by the empire's subjects as their contribution to labour service. The novices also received pendants of gold and silver together with a colourful feather headband from which hung golden rings. After performing another *wari*, the novices' parents struck them while they reminded the youths of their pledge of allegiance to the sovereign and the main deities. They finally returned to Haucaypata, where the deities and the mummy bundles of the

61 Polo, "Relación de los adoratorios": 8.

62 Yauira was also located next to the shrine named *Vicaribi*, the burial place of a principal lord of the Maras people (Polo, "Relación de los adoratorios": 15).

63 Polo, "Relación de los adoratorios": 15.

64 Cieza: Pt. 2, Ch. 7, 19.

Inca lords and their royal spouses were gathered. There, they danced and drank for the next six days.

Twenty-one days had gone by since the start of *warachiku*, and the novices now headed toward the water shrine of Calispuquio (Ch. 3:8) behind the walls of Sacsahuaman in order to bathe and put on a new garment. They then walked back to the main square, where the deities' images were assembled, and there they received war weapons from their "main uncle" and other noblemen. For each present the boys received, the adults would strike them and enjoin them to be brave and faithful to their ruler. Eventually, on the following morning, the novices' families gathered in the privacy of their homes and offered their boys the four jars of chicha that the women had prepared prior to the celebration. The probationers, befuddled by the beverage, were then carried onto the fields to have their ears pierced.

The *qhapaq rayme* feasts finally came to a close when foreign governors were invited to re-enter the city precinct and were given an offering of *yawar sanku* similar to the one they had shared after *sitwa* as a sign of confederation with the Inca.[65] The closing ceremony of *qhapaq rayme* was also the opportunity to acknowledge the contribution of every subjugated nation to the imperial tribute, and in this spirit, every foreign delegate received *akllas*, wool, herds, and clothes in the ruler's name.[66] The nature of these transactions placed the Sapa Inca at the heart of the redistribution of goods, for only those who displayed obedience and loyalty towards the central power benefited. It was with this pageant display of asymmetric reciprocity that the nobility eventually closed the public demonstration of *qhapaq rayme*.

Name Transmission and the Articulation of Marriage Networks

Although *warachiku* ended with the affirmation of the Sapa Inca's paramount control over state resources, this highly restricted festival took place in several locations outside Cuzco that oral traditions associated with the past of at least four local ayllus: the Huayllacans, the Ayar Uchu Ayllu (or Alca Uiza), the Choco and Cachona Ayllus, and the Maras Ayllu. Among

---

[65] Polo, "Tratado": 18-19.

[66] Polo, "Tratado": 18; Polo, *El Mundo*: 56, 80, 84. In this last document Polo situates the collection of the tribute in February, after the *rayme* festival that lasted one month. His other account of the Inca festivities contradicts this information and places *rayme* in December, like most chroniclers, so that the re-entrance of foreign delegates would have taken place in January. This discrepancy in the writings of the same author has not found any conclusive explanation but from a comparative perspective, the most likely option for the collect of labour services would be January.

them, the Huayllacans and the Choco and Cachonas had provided principal wives for Inca rulers. Moreover, colonial sources alluding to the origin of these shrines emphasize two major traditions: one sets their emergence in primordial times and evokes their resistance to the Deluge that preceded the Ayar siblings' settlement in the valley; the other links these *wakas* to the past of foreign ethnic groups. These perspectives, however, may not be contradictory. In fact, they are reminiscent of the discursive patterns that characterize the Inca genealogical accounts and the life story genre. The former highlight continuity in time and the consanguineal link between Manco Capac and the Ayars, whereas the latter presents the initiation stations as associated with outsiders and relatives by marriage. The actual coexistence of these two storylines would also explain that the conception of this ritual was credited to either Manco Capac or Hanan kings. If this interpretation stands, how can we explain that the most elitist and secretive ritual of the Inca festival calendar required worship of acknowledged foreign *wakas*? I suggest that *warachiku* was fundamental in articulating marriage networks through the bestowal of new names because a man's name-giver was ideally his wife-provider. Following this prescription, a novice's sponsor could be as much a consanguineal kin, and thus an insider to the Qhapaq Ayllu, as he could be an outsider whose native ayllu had contracted wedlock with the royal descent group. Two types of discourse on the relations novice/name-giver, centre/periphery emerge from this picture that account for the divergent origin narratives of the *warachiku* pilgrimage shrines. At the core of this relational dynamic stood the name-giver, whose identity brings new insight into the nature of Inca affiliation. For Inca males, new names were bestowed during the festivals of *rutuchiku* (hair cutting) and *warachiku*, while there is evidence suggesting that the ruler also received a change of name at his investiture on the royal *tiyana*. Depending on which chronicle is consulted, the bestowal of the first name took place a few weeks to two years after birth, during the private ritual of *rutuchiku*.[67] On that occasion, the child's relatives gathered with offerings, and the "closest uncle" cropped his nephew's hair, whereupon the infant received his first name. In the central highlands, the same celebration was enacted in the presence of both parents' brothers. A document from Chinchaycocha (Central sierra) reveals that it was the matrilateral uncle or, in his absence, "the closest relative from the mother's side" who conducted the ceremony, cropped the first lock of hair, and

---

[67] Cieza: Pt. 1, Ch. 65, 200-201; Betanzos: Pt. 1, Ch. 45, 191; Molina: 119; Garcilaso: Bk. 4, Ch. 11, 345.

presented the child with a llama and other precious goods. The patrilineal uncle came after him to perform the same acts, followed by the rest of his relatives.[68]

Years later, the imposition of a new name and the ritual tonsure were renewed in a similar fashion during the initiation ceremony of both boys and girls. The Spanish chronicles evoke the presence of "close relatives" or "main uncles" side-by-side with the Inca novices during the *warachiku* trials. They were the ones who repeatedly assaulted the youths, performed a series of dances, and handed over the war weapons to the male novices. For Garcilaso, the young Inca heir was "sponsored" by the great priest of the Sun, the very same man who was ideally the sovereign's brother or uncle.[69] Guaman Poma explicitly identifies the initiate's relative with the mother's brother, referred to as *kaka*. He writes that on the final days of *warachiku* "they took with them certain loincloths that they call uara and the uncle called caca tied it on them."[70] In early colonial Chinchaycocha, a similar context framed the initiation of young boys whose hair was first cut by their mother's brother, from whom they received the *wara* loincloth.[71] The matrilateral uncle thus conducted the male initiate's various rites of passage and symbolically removed him from his former social status.

As for the Inca maidens, the first menstruation ritual, called *k'ikuchiku*, also involved adopting another name that the initiate received after three days of seclusion and fasting. The ceremony was held privately and required that the young girl bathe before putting on new clothes. A "most important relative", albeit a different one than the mother's brother, would then name her and instruct her on the duties of a married woman. The rest of the family and friends thereafter offered the maid the domestic goods she was now entitled to possess.[72] Molina's description of the first menstruation ritual clearly distinguishes the relative who named the boys from the girls' ritual sponsor.[73] From the context he depicts, the latter was

---

[68] Arriaga: 64-65; Polia Meconi 1999: 348.

[69] Betanzos: Pt. 1, Ch. 14, 65-70; Cieza: Pt. 2, Ch. 7, 18; Polo, "Tratado": 18; Fernández: Pt. 2, Bk. 3, Ch. 5, 83; Molina: 98-110; Múrua 2001: 437; Garcilaso: Bk. 2, Ch. 9, 199.

[70] Guaman Poma: 893 [907], 837. This author indicates that similar events took place during two other rites de passage he called "cuzmallicoc" (*kusmalliyuq*) and "anacaco" (*anakaku*). The first referred to a specific garment called *kusma* that males start wearing at a particular age, although unspecified; the second describes a female skirt, the *anaku*, which women start wearing at a mature age.

[71] Polia Meconi 1999: 354. For contemporary ethnographical studies see Christinat 1989.

[72] Molina: 119-120.

[73] One century later, Cobo did not make this distinction and used the expression "most prominent uncle" to refer to the girls' ritual sponsor. However, Cobo's information largely,

acquainted with specific female responsibilities in the household, the knowledge of which was the domain of women. This important point alludes to a situation where name-givers and name-receivers are same-sex individuals. Since affiliation was matrilateral for men (from MB to ZS), the father's sister was likely the ideal name-giver for women (from FZ to BD), as had occurred in the Huarochirí district in colonial times.[74] This practice, otherwise called cross-transmission of names, was also fundamental in articulating the bestowal of statuses with marriage practice, since there was a predilection for matrilateral cross-cousin marriage among the Inca nobility. The central figure mediating this connection was the *kaka* (MB), who was at once name-giver and wife-provider to a man. He, in turn, ideally married his name-giver's daughter in a union that consolidated the bond created in his infancy and early teenage years. On the one hand, *warachiku* was a ritual that demarcated insiders from outsiders to the dynastic line with distinctive signs of prestige and exclusion from the ceremonial centre. On the other hand, it celebrated marriage networks with local ayllus as a necessary mechanism that ensured the viability of Inca sovereignty and the perpetuation of state expansionist politics. By opening up their most elitist festivities to foreign affines, the Incas shared with them

---

if not entirely, draws from Molina's text, which indicates that he probably misinterpreted his source or simply assumed that the same relative performed this role for both boys and girls. See Cobo T. 2, Bk. 14, Ch. 6: 247.

[74] Zuidema 1980: 74-75. Name-givers also belonged to a category that included individuals positioned two ascending generations removed in a direct line from the child. This emerges with particular clarity from an archival document quoted by Pärssinen (1992) in his study of Inca political organization. Recalling an event that took place more than thirty years after the conquest, a servant at the house of Doña Catalina told how this Inca noblewoman objected to the names chosen by the Spaniards for her grandson. She was the widow of Paullu, crowned Sapa Inca in 1537, with whom she had two sons, Felipe and Carlos. In 1571, the latter became the father of a boy that Spanish officials wished to christen Melchor Viracocha Inca. Upon hearing this news, "Doña Catalina responded that they should not for Viracocha Inca is a name of the Incas Hurin Cuzco that belongs to a different *parcialidad* (moiety) from that of my husband" (Audiencia de Lima 1575-1576, fol. 2588 v., Justicia 465, AGI in Pärssinen 1992: 176). Doña Catalina's influence and key position in relation to the young boy were such that her admonition was followed. It is significant that, in order to justify the appropriate name to be bestowed, she does not mention her own son's moiety adherence, although he is the name-receiver's father, but evokes instead her husband's membership to Hanan Cuzco, suggesting that he was the kin whose name and moiety affiliation were passed on to his grandchild. As Pärssinen observed, Doña Catalina's statement that Wiraqucha was a name of the Hurin Incas is peculiar because Viracocha Inca was also a famous lord of Hanan Cuzco. This information suggests that names/titles followed a complex rule of moiety affiliation that cannot be readily established from the scant documents at hand. For Pärssinen, the answer may lie with the quadripartition of Cuzco, whereby each moiety was subdivided into a Hanan and Hurin division.

part of the secrecy surrounding their success in warfare. In return, the novices' *kaka* played an active role in emulating their nephews' martial qualities at the service of the state. The celebrants' pilgrimage to provincial *wakas* supported this relationship of (asymmetric) reciprocity. The narrative of Titu Cusi Hualpa's abduction clearly illustrates the political dimension at play in *warachiku*.[75]

The Ritual Abduction of an Inca Heir

The story opens with a conflict between two neighbouring chiefdoms of the Inca heartland. The Huayllacans, who had promised Mama Micay as wife to the Ayarmaca lord called Tocay Capac,[76] did not fulfil their pledge and instead married the woman to Inca Roca. To avenge this disloyalty, the rejected chief declared war on the Huayllacans. The ensuing conflict lasted for years, during which time Mama Micay gave birth to Titu Cusi Hualpa. As a condition for ending the hostilities, Tocay Capac requested custody of the Inca child and entrusted Mama Micay's brother, called Inca Paucar, with handing over the young boy. Consenting to this, the Huayllacan lord invited his eight-year-old nephew to his lands in Paullu on the pretext that he wished to present the child to his mother's relations and to display the *haciendas* that he intended him to inherit.[77] Titu Cusi Hualpa was received with pomp and ceremony, but as soon as the festivities were over, Inca Paucar gave word to Tocay Capac that all of the men had left home to plough their lands, leaving Paullu deserted. This was an opportune time to

---

[75] Zuidema (1986a: 57-64) summarizes the narrative's essential theme as a contest for the hero's paternity. He uses this account to defend his hypothesis on kinship groups and matrimonial alliances summarized in his well-established hierarchical diagram, beginning from the great-great-grandfather function, founder of the irrigation system, in this case Inca Roca, to the grandfather Viracocha Inca. Between the reigns of these two monarchs came that of Yahuar Huacac. According to Zuidema, these three generations would have represented the political evolution of Cuzco and illustrated the various unions contracted between the Incas and the foreign ayllus, mythically conveyed by a succession of exogamic, symmetric, and finally endogamic alliances.

[76] The identity of the figure(s) behind the name of Tocay Capac has recently given rise to an extensive literature. In the mid-colonial period, he appears to have held an elevated position in the Inca nobility. See Rostworowski 1993b: 241-259; Julien 2000: 245-249; Cahill 2002: 616-626; Yaya 2008.

[77] The haziness of the *hacienda* description precludes any identification of the status of the land/people referred to in the account. In fact, *hacienda* has a broad, generic definition that does not necessarily have anything to do with lands. Concerning the classification of the lands, see Moore 1973; Rostworowski 1993a: 105-116; Murra [1995] 2002. On the notion of inheritance in pre-Hispanic times, see also Ramírez 1996: Ch. 3, 42-86; Hernández Astete 1998: pp. 109-134; Salles & Noejovich 2006: 47-50.

"descend on the town and carry off the child, doing with him what they chose, in accordance with the agreement."[78] The Ayarmacas entered the village immediately, wounded the boy, and abducted him.

Stunned with remorse, the Huayllacans confronted the kidnappers in a town called Amaru, but they were rapidly fought off. The child was then taken away to the Ayarmaca capital, Ahuayrocancha, where he boldly reminded Tocay Capac that he was the son of Mama Micay and Inca Roca. Furious, the Ayarmaca lord sentenced the child to death, which prompted Titu Cusi Hualpa to invoke curses upon his people: should the threats declared by their chief be carried out, they would face certain annihilation by the Incas. Professing these words, tears of blood ran down his face. This retort deeply disturbed Tocay Capac, who after this event decided to rename the child Yahuar Huacac: "the one who weeps tears of blood". In order to dispel the curse, he resolved to let him starve in captivity. Thus, during one year Yahuar Huacac remained sequestered on an Ayarmaca estate. It was there that one of Tocay Capac's spouses noticed him. She was the daughter of the *sinchi* of Anta, whose lands were located a few kilometres northwest of Cuzco, at the entrance to the Jaquijahuana Valley. Faced with the boy's distress, she devised a plan to rescue the young Inca and asked him to organize a race with other children up the side of a mountain. Fleet-footed, Yahuar Huacac arrived first at the summit of the hill and was immediately taken by the Anta to their village. The Ayarmaca, seeing they had lost custody of the child, tried to fight their opponent in the region of Lake Huaypo but were defeated.

The boy remained in Anta for another year, cherished and indulged by his liberators. When the Antas finally considered that it was time for Yahuar Huacac to return home, they sent ambassadors to Cuzco to announce the good news. Inca Roca received them with joy but, still distrustful, the king delegated a "pauper" (*pobre*) to Anta to discreetly investigate the information. Once convinced that the child was alive, and upon being returned to his father, Inca Roca granted the people of Anta the right to be called "relatives". An alternative variant of this version recounts that it was the Huayllacan lord who, disguised as a pauper, entered the abductors' town and liberated the child.[79] Three singular alliances concluded the story: Yahuar Huacac married Mama Chicya, the daughter of his Ayarmaca abductor Tocay Capac, who in turn took his son-in-law's sister as his wife.

78 Sarmiento: Ch. 20, 73.

79 Gutiérrez: Ch. 49, 210; Fernández: 81.

Later, upon succeeding Yahuar Huacac to the throne, Viracocha Inca married an Anta woman.

Taking into consideration the geographical route covered by Yahuar Huacac outside Cuzco and the severe suffering he endured, an interesting structure emerges that evokes the *warachiku* stages with precision. The main plot confirms it: a young boy is kidnapped at the time of a celebration organized by his maternal uncle. He is assaulted, undernourished, and sentenced to death, only to re-emerge under another name. The narrative also indicates that the abduction coincided with an agrarian activity that required every man to leave town and attend his *chakra*. A type of work called *chaqma*, held precisely between the months of January and February, consisted of breaking the soil of the idle fields outside the city limits for over twelve days in preparation for sowing.[80] The Incas undertook this ploughing following the ritual battle that brought the novices' initiation to a close. It thus becomes apparent that the novices' descent down Anahuarque hill finds its equivalent in the race that Yahuar Huacac wins before fleeing to Anta, which is a town also located at a high altitude, on the edge of a plateau overlooking the Jaquijahuana Valley. Similarly, the tears that the young Inca heir shed in the Ayarmaca capital recall the lines of blood traced on the boys' faces after they reached Yauira, a *waka* of the Maras Ayllu whose territory was earlier occupied by the Ayarmacas. Located in the same province were several sites evoked in the narrative and undoubtedly associated with the *warachiku*. Two kilometres south of the present town of Maras is a hill named Ahuayro that the narrative refers to as the Ayarmaca capital (Map 1). Further on, less than seven kilometres south, spreads Lake Huaypo, also an Ayarmaca site, where the battle pitting the Anta against the Ayarmaca took place. It is in this location that Albornoz (1582) situates "Guaypon Guanacauri, a stone next to a lake [where] the Cuzco Indians pierced their ears."[81] A major *qhapaq ñan* (royal road) connected these different sites; remarkably, it passed through Maras, north of Lake Huaypo, down to Anta, tracing Yahuar Huacac's journey before he was rescued.

---

[80] *Chaqma*: "Plowing or other work such as breaking the sods and clearing the soil to make it fit for sowing" (González Holguín, *Vocabulario*: 91). See also Molina: 112. For colonial information, see Guaman Poma: 1134 [1144] 1031, 1137 [1147] 1034.

[81] Albornoz: 180.

The Politics of Abduction

The analogies between the abduction narrative and *warachiku* suggest that both traditions followed the same chronology of events and depicted similar types of interactions among the protagonists. The child's odyssey, the development of *warachiku*, and the ritual's associated myth of the primordial Flood all present a similar ritual scenario describing the end of an era and the advent of new humankind or a new king. One of the defining characteristics of the Inca rite of passage was to stage a nexus of relations between the celebrants (novices-relatives-priests) in relation to the route they covered outside Cuzco, to the different shrines attended by foreign ayllus. The same feature is still present in this narrative, in which the heir's journey through foreign towns justifies the concluding matrimonial ties. The successive roles of the three provincial lords alongside Yahuar Huacac recall the services performed by the novice's uncle during their peregrination outside Cuzco, so that each of them claims in turn an avuncular relationship with the child. In the case of Inca Paucar, the biological ties linking him to Yahuar Huacac are extant and ensure the boy would inherit his uncle's belongings. As for Tocay Capac, he expresses the ritual kinship reserved for the novice's "main uncle" by allowing the child to acquire the social recognition granted through the initiation. It is under his authority that Yahuar Huacac suffers the toughest ordeals and obtains a new identity that empowers him to ascend to the throne. Finally, the status of the Anta *sinchi* remains suspended until Viracocha Inca's rise to the kingship office. While he is granted the position of "relative to the Incas" at the end of the narrative, the matrimonial alliance that this position entitles takes place one generation later, when Yahuar Huacac's successor marries a native of Anta. The role of the three provincial lords in the course of these events not only establishes their cognation with the child, it also redefines the relations between all actors involved because the Incas and Ayarmacas made a matrimonial exchange. In other words, the ritual events constituting the nucleus of this account exercise a dynamic change of status on the entire group of participants. Yahuar Huacac's wife = MBD = ZHD, while Tocay Capac's wife = ZD = DHZ (Figure 4). These two unions appear symmetric not only because they constitute an exchange, as Zuidema points out, but also because both protagonists share as a result the same kin relations to each other. Besides being Yahuar Huacac's *kaka*, Tocay Capac is also his *qatay*, a term referring to the sister's husband and the daughter's (ZH = DH), while the young Inca adopts the same positions, having

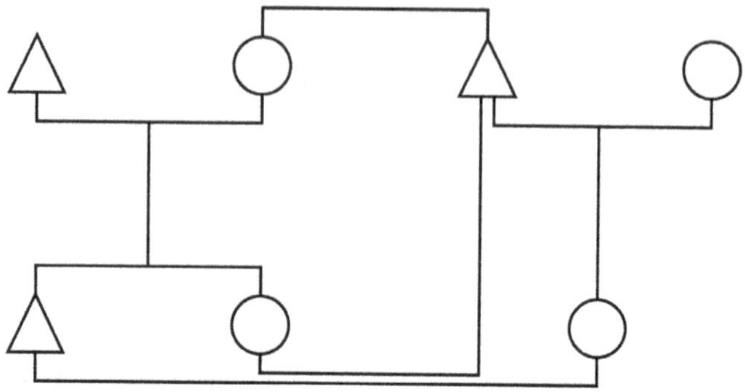

Figure 4. Model of the symmetric alliance closing Yahuar Huacac's abduction.

espoused the Ayarmaca lord's daughter (*qatay*) and being his wife's brother (*kaka*).

Finally, the threefold alliance concluding the narrative is also governed by a strong sense of temporality that serves to set different pictures of the integration of foreign ayllus within Cuzqueño power. In the first case, that of Inca Paucar, the kinship is already established and the matrimonial alliance has been concluded. The Huayllacans are therefore Kaka Cuzcos. In the second case, the kinship is actualized during the ordeals suffered by the child, ritually authorizing the Ayarmaca leader to aspire to the exchange. In the last case, that of the Anta chief, kinship is proclaimed but the alliance would only be effective in the next generation. The narrative's closing episode further stresses the hierarchical difference separating the status of the three lords. It was indeed a pauper (*wakcha*) whom Inca Roca delegated to Anta to investigate discreetly the presence of his son in this town. Gutiérrez's and Fernández's version of this narrative differs slightly but indicates that Yahuar Huacac's liberator entered the Ayarmaca town unnoticed under the disguise of a pauper in order to set the child free. Here, the image of the *wakcha* offers an interesting narrative metaphor. It refers as much to a character that does not attract attention owing to his modest outward appearance as it does to a member of the foreign nations that the Incas qualified as poor/orphans because they did not hold a trace of rich (*qhapaq*) ancestry. The *pobre* delegated by Yahuar Huacac's father can be seen as an official from one of these ayllus, who was able to investigate unnoticed the *dicta* of the Anta lord because the people of this town were themselves *wakcha*. Likewise, in Gutiérrez's and Fernández's version, the young Inca's liberator had to disguise himself as a *wakcha*, that is, as someone whose garments and ornaments were similar to those of the Anta residents. Through the semantic plurality of the word *wakcha*, the different variants of the narrative describe the same fact: if Anta was a place where *pobres* could wander discreetly, it was because the inhabitants of this town were themselves *wakcha*. Their status would only change under Viracocha Inca, when they were granted the title of Kaka Cuzcos, that is, wife-givers to the Incas.

As chronicler Pachacuti Yamqui suggests, in the eyes of the ruling elite, the Kaka Cuzcos held a higher social status than other foreign ayllus, such as the *wakcha* populations. Depicting Huascar's allies mobilized against Atahualpa, he describes the lord's war escort as follows:

> As halberdiers of his person [were] all the *orejones* of Mancop churin Cuzco <who are *caballeros*>, and *caca cuzcos* and *ayllon cuzcos* <who are specific *caballeros*>. In the frontline, he positioned the Quiguares etc., Collasuyos y Tambos, Mascas, Chillques, Papres, Quichguas, Mayos, Sancos, Quilliscaches etc, and as their halberdiers he brought the Chachapoyas and Cañares.[82]

An obvious hierarchical framework structures this enumeration, starting with Manco Capac's direct descent, followed by the Kaka Cuzcos, then the non-royal ayllus of Cuzcos, and then, finally, the disloyal ethnic groups that betrayed the Incas during the Spanish conquest. Here, the Kaka Cuzcos are differentiated and hierarchically more esteemed than the Incas' common allies. Being the Cuzqueños' statutory MB, these selected nations entered the Inca kinship nexus through matrimonial unions and constituted one of the main forces in Inca social organization and policy. At the other end of the spectrum, the *wakcha* nations, while linked to the mythical account of the original migration and observant of the custom of ear piercing, did not belong to the royal kinship nexus. Through this metaphor, the narrative offers three pictures of the foreign nations' assimilation into the Inca kinship and political nexus.[83] Hence, in Sarmiento's account, the prevailing role of the three provincial lords next to Yahuar Huacac reflects the ritual pathway of *warachiku*. Similar to the heir's journey, the initiation ceremony actualized the privileged relations between the Incas and the *wakas* of foreign populations that provided spouses for early Inca kings or whose

82 Pachacuti Yamqui: 261. My translation.

83 It is also tempting to see here a last analogy between the narrative and the general framework of the initiation ritual. The tripartite model of assimilating foreign nations into the Inca kinship nexus echoes the tripartite liminality of the initiatory rite as Turner (1969) defines it. His theory on the rite of passage distinguishes three spatial phases to the ritual: preliminal (rites of separation from the existing order), liminal (period of transition), and postliminal (rites of reintegration).

apical ancestor was mythically related to Manco Capac. By representing this link, the narrative enters the extended corpus of traditions in which foreign nations are ambiguously represented as related/unrelated, allies/enemies, and excluded from/reintegrated into the Inca ceremonial centre. In this light, rituals and mythical narratives appear as agents of social structure: by staging the characters' interrelations and thereby displaying their (classificatory) kinship links, the *warachiku* and its related mythical narrations (Yahuar Huacac's abduction and the Ayar ancestors' migration) constructed the foundations for a unified identity, serving as a powerful tool of political and social cohesion.

In the account of Yahuar Huacac's abduction, it is particularly significant that the formulation of this identity emerged from the mediation of concentric dualism with a triadic structure. As Levi-Strauss observed in his essay on dual organizations, the dynamic tension tying these two systems together is devised to regulate the fundamentally unequal nature of concentric binary oppositions. In the Inca narrative of Yahuar Huacac's initiatory journey, the boundaries that traditionally discriminated the ritual centre from its periphery enter into a dynamic process with the triadic union pattern, thereby allowing external elements to integrate progressively within the core of political power and to share its privileges. Hierarchy, nonetheless, still structures the relationship between the two poles, as the story depicts three degrees of assimilation, and thus three different positions from the centre. Importantly, this perspective places the recognition of marriage networks at the heart of the *warachiku* festival but also explains the presence of foreign officials on the edge of the celebration. As Levi-Strauss noted, concentric dualism is not a self-sufficient system, and its frame of reference is always the environment. "The opposition between cleared ground (central circle) and waste land (peripheral circle) demands a third element, brush or forest—that is virgin land—which circumscribes the binary whole while at the same time extending it."[84] In other words, binary oppositions exist because their two constituents also form a unit, which itself contrasts with an external entity. This type of settlement pattern is a widespread feature of spatial organization in the Andes. For Deborah Poole, it "opposes a central village to the fields which lie around it, while these lands are subsequently opposed to the less populated, uncultivated, and hence symbolically 'wild' *punas* (grasslands) outside of them."[85] This settlement pattern likewise translates socio-political

[84] Levi-Strauss, 1963: 152.

[85] Poole 1991: 330.

hierarchies, as in the relation between insiders and outsiders. In the example of the Inca initiation ritual, the tributary nations stationed outside Cuzco would have constituted this third element, referred to as *sallqa* (savage, uncivilized) in present-day ethnographies, to which the unity composed by the Incas and their allies was opposed. It is in this context that Turner's dumontian notion of "social totality" can be used to unravel the participants' interaction. For him, the relationship of asymmetrical predominance that characterizes dual oppositions works hand in hand with a second framework in which the two terms become a social totality by appropriating the status of its dominant member.[86] This twofold dimension describes accurately how the participants of the *warachiku* engaged with one another and their environment: on the one hand, the ritual was restricted to the descendants of royal lineages, and hence enacted asymmetrical relations; on the other hand, the celebrants paid allegiances to foreign, yet assimilated, *wakas*, and thus displayed coherency. This operation of unification, furthermore, is indivisible from the formulation of a constructed identity, forged by oral traditions and actualized by the celebrations. It is with regard to this social unity that the Incas developed a body of narratives relating their former rulers to their allies' ancestral shrines and devised the *warachiku* as a demonstration of this authority.

Marriage Alliances and Imperial Expansion

As Tawantinsuyu extended, the Incas imposed a series of stringently hierarchical institutions upon subjugated nations, co-opting the nobility exempted from tribute and labour service, to preside over their political, economic, and religious administration. Further consolidating the imperial institutions, the provincial elite, with whom the Qhapaq Ayllu established political and matrimonial alliances, assisted the Inca nobility. These native lords occupied essential positions in the imperial bureaucracy and served as intermediaries between the central power and provincial delegates. In this way, the maintenance of royal authority was partly ensured by the privileges the ruling nobility granted to their allies. In the early phase of their expansion, however, the Inca rulers maintained complex relationships with a number of provincial leaders, headmen of neighbouring chiefdoms, whose interests frequently collided with the Incas' thirst for hegemony. Historical traditions evoke these early tensions in detail and depict the Cuzco region before Inca rule as a theatre of incessant clashes

---

86 Turner 1984: 367.

between rival polities. These hostilities, however, are multifaceted: although each story ends with the brutal defeat of the resisting chiefdoms, the bloody encounters are often interspersed with political and matrimonial alliances between the combatants. A similar ambivalence characterizes other aspects of the Incas/non-Incas relationships, notably in the religious sphere. Several narratives portray a complex picture of the honours the lords of Cuzco bestowed upon foreign *wakas* and evoke the ancestral links binding the Inca nobility with rival chiefdoms. They say that the dominant groups of the region shared together a collective identity from time immemorial and observed a series of ritual practices in common, such as ear piercing and the wearing of the loincloth. Contrasting with these traditions, other accounts claim that the Inca nobility was the first to follow these ritual usages and to initiate their youth following the ordeals of the *warachiku*. They state that the Incas later granted these specific honours to their allies in recognition of their support in battle. These antagonistic representations assumed traits of dual opposition in which reciprocity was as significant an aspect as that of asymmetry and exclusion. In this light, the relations that the Incas maintained with their foreign allies disclose resemblances with the dichotomies placing Hanan in opposition to Hurin. Within the moiety structure, but also at the supralocal level of the Incas' relations with their allies, the members of the dominant group defined themselves as cognates and regarded the residents of the opposite division as affines. In both cases, the figure of the wife-giver (*kaka*) mediated the two antagonistic spheres. His position as MB articulated a structure of assimilation whereby dual asymmetry (Hanan versus Hurin and *qhapaq* versus *wakcha*) was recast onto a tripartite hierarchy of social prestige accessible to outsiders. In this way, the *wakcha* nations could aspire to privileges by becoming wife-givers to the Inca nobility, that is, Kaka Cuzcos, while moiety membership and upward social mobility were mediated by the name-giver, here again a man's MB. In this picture, triadic structures are mediating forces of dual social division.

The asymmetric nature of the Inca model and the diachronic dimension of its moiety system indisputably contradict Maybury-Lewis' view that dual organizations are relics of unshattered times, preserved from "external constraint."[87] It would appear reductive to confine gender opposition, generational hierarchy, and kin/non-kin differentiation in Inca festivals and historical narratives to a series of symbols explaining the inherent dual

---

[87] Maybury-Lewis 1979: 312. See Ortiz (1969: 131-137) for a critique of this thesis.

nature of the universe as understood by the elite. Such a portrayal ignores the dynamics of changes that nurture every culture and leaves unanswered the question of why the Inca descent group and other dual organizations, unlike others, devised binary institutions. Analysis of the *warachiku* shows that comprehension of antagonistic relationships in pre-Hispanic Cuzco is incomplete without an appreciation of Inca kinship strategies as factors of social mediation. Ritual praxis was also central to the enactment of the particular relations that brought together opposite equations. It was through ritual condensation that reciprocity and antagonisms were simultaneously depicted in Inca ceremonies. These dual ritual oppositions included respect and transgression of the established order in the form of moiety conflicts, the communal repartition of lands and resources, the assimilation of ethnic groups at *sitwa* and *qhapaq hucha*, the foreigners' exclusion from the ceremonial centre, and the celebration of their outer *wakas* at *warachiku* together with significant *wakas* of the origin myth, all of which can be added to other forms of dual oppositions documented in other parts of the Andes, such as upper/lower, centre/periphery, migrant/native, younger/older generations, and so forth. Rituals not only recalled but also shaped Inca historical consciousness by mediating hierarchical disparities and other dual antagonisms to convey another alternative, that of assimilation. The Incas, however, were a highly stratified society, so that this principle of reciprocity did not entail equilibrium between the opposite parties. Rather, it created a new social order in which asymmetry remained a core principle but gave the elite's provincial allies access to various privileges. This practice, in turn, would have enabled the descendants of foreign women and Inca lords to people the lower part of the city and to become an even more integrated part of the political, administrative, and religious system. It was through the agency of time that foreign nations could gain access to these increasing honours, rising through the ranks of social hierarchy to obtain the ultimate privilege of contracting symmetric alliances with the ruling nobility. That the Incas themselves conceived of the central role of temporal process in the articulation of social privileges is clearly illustrated in the account of Yahuar Huacac's abduction, in which the three alliances contracted by the Inca lords took place over three generations.

PART TWO

"SOMETIMES THEY HOLD THE SUN AS THE CREATOR AND OTHER TIMES THEY SAY IT IS VIRACOCHA"

CHAPTER FOUR

THE ANCESTRAL RULERS OF THE DRY SEASON

Having reviewed the social organization and politics of reproduction of the royal descent group, we shall now turn to the fundamental structures of Inca cosmology and examine the nature of the ruling elite's main divine ancestors as outlined in colonial sources. According to these texts, the heads of their religion were three: the Sun, Wiraqucha, and Illapa, who is referred to as the ruler of lightning flash, thunder, and lightning bolt. Together, they supervised a plethora of forefather-figures known under the generic terms of *waka* and *wanka*, as well as other embodiments of community-level ancestors such as the *mallkis*, *kamaqin*, and *qunupas*.[1] At the time of the Spanish invasion, the worship of these primordial beings was widely attested to in different parts of Tawantinsuyu and had long preceded its expansion. Observing their widespread influence, early missionary treatises described these cults as the "universal idolatry" of the people newly subjugated to the Spanish Crown. In this way, the roots of Inca religion were anchored in a secular past, part of its mythology bearing *inter alia* a close resemblance to pan-Andean narratives of emergence, migration, and settlement. From this common substratum, the Incas elaborated the particularities of their own belief system and devised a ceremonial calendar that reflected their specific vision of the cosmological order, with its hierarchy of deities, its play of antagonistic forces, and their influence on earthly matters.

To account for the complexity of this religious structure, I proceed by situating the lore of the Inca royal dynasty within the cultural frame that informed its oral traditions. Following this approach, the narrative threads that tied the epics of pan-Andean gods with the events of Manco Capac's settlement in the Cuzco valley emerge with greater clarity. By engaging with the founding traditions of Inca history, I also intend to unravel some of the discordance that obfuscates their different transcriptions in primary sources. This narrative dissonance relates primarily to the identity of

---

[1] *Mallkis* are the embalmed bodies of the communities' ancestors. *Qunupas* are small stones the shape of animals, or atypical fruits or tubers, which stand as household or personal attendants.

the Incas' divine progenitor, which is named either Wiraqucha or the Sun, and their place of emergence: Paqariq Tampu in the Cuzco district, or Lake Titicaca in the Collao province. These discrepancies may be harmonized not only through an understanding of colonial agendas, but also by analysing them in light of the cultural framework that informs the Inca dual organization. Anticipating this reflection, Bravo Guerreira (1992) outlined a research agenda that has remained largely unexploited. She suggested that two different, albeit interwoven, storylines composed the corpus of narrations retelling the migration of the first Incas, each starting in distinct locations. Here, I propose to further this argument and to integrate it into the wider picture of Inca cosmology. Like most dynastic accounts mentioned thus far, these traditions do not form two hermetic corpuses in primary sources that could be neatly demarcated. However, by identifying the strands of congruity running through the Inca migration stories and other Andean myths of origin, it is possible to isolate the essential features of these traditions and to associate them with the epics of two divine ancestors: Wiraqucha and P'unchaw. An analysis of their attributes and functions in the course of Inca festivals reveals the demarcation of the two seasons these gods individually ruled. Each deity arose on earth at opposite times of the year, replenished with the vital energy necessary to sustain soil, plants, and beings. In the course of his terrestrial journey, the animating force of one slowly depleted, allowing the other to claim his rule over the second half of the annual cycle.

In advancing this argument, I set aside momentarily the study of astronomical events that led several scholars to associate Wiraqucha and P'unchaw with a particular time of the year. This is not to suggest that celestial events had little impact on Inca ceremonial life, for in fact their interconnection is an important argument of this book's final chapter. However, for the time being, I wish to refocus the discussion first and foremost on recorded material, which will establish a point of entry into reassessing modern assumptions on the Inca gods, including their association with the movements of stars and planets. To do so, the following demonstration discusses the specificities of the Andean environment that primary evidence and philological analyses link to the influence of divine ancestors. Elevation, climate, natural resource constraints, and the crops' basic needs, but also the rotational organization of Andean agriculture and its impact on pastoral activities were all stringent requirements that men had to control in order to produce their staples. The regulation of environmental phenomena was therefore crucial to the subsistence of Andean communities. The periodicity and stability of these conditions were precisely in the hands of the ancestor-deities whose attributes were closely related to the characteristics of each season.

The Journey from Lake Titicaca

Fragments of our first body of narratives, penned between the early 1550s and the last decade of the sixteenth century, were recorded by Gómara, Segovia, Gasca, Zárate, Castro and Ortega Morejón, Pedro Pizarro, Atienza (c. 1575), Molina, and Gutiérrez.[2] Although these texts may disagree on the identity of the first Inca king, the same storyline runs through them. They first situate the emergence of the Inca people after a deluge that flooded the earth for sixty days and sixty nights. This event was a *unu pachakuti*, literally "a reversal of time and space by water", which destroyed the primitive humanity and marked the dawn of a new era.[3] Following this cataclysm, a man called Manco Sapaca, or otherwise Viracocha, arose from Lake Titicaca with his people.[4] Together, the newcomers headed toward the west, and after a long peregrination they settled in the Cuzco Valley to establish their rule. The chroniclers report that Manco Sapaca's most renowned descendants did not exceed six or seven and were known as Pachacuti Yupanqui, Viracocha Inca, Capac Yupanqui, Tupa Inca Yupanqui, Huayna Capac, Huascar, and Atahualpa, all of whom were Hanan lords. The unorthodox account of Gutiérrez even says that the Incas' forefather originally resided in the settlement of Hatun Collao, near the present town of Puno, where he erected his palace.[5] Manco Sapaca was a fierce warrior who had emerged with armed combatants from Lake Titicaca and had come to subdue the surrounding provinces by force. His descendants carried on his deeds, and generation after generation they engaged in repetitive conflicts against the native lords of Cuzco. One of his most valiant descendants was Tupa Inca Yupanqui, who transferred the Inca settlement next to the city of his sworn enemies, where he built the site of Hanan Cuzco. From there, the newcomers came to vanquish their neighbours and

---

2 Garcilaso also recorded these mythical events but combined them with the Paqariq Tampu narrative cycle.

3 Betanzos: Pt. 1, Ch. 17, 83; Sarmiento: Ch. 6, 40-42. On *pachakuti*, see Bouysse-Cassagne 1988: 13-14; MacCormack 1988: 966-969; Allen 1993-1994: 89-90; Salles-Reese 1997: 69-73.

4 About Manco Sapaca, see Julien 1983: 37-42.

5 About Hatun Collao (also Hatun Qulla) see Julien 1983: 89-107; Rostworowski 1999: 68-69; Stanish 2003: 75-76.

conquered the rest of the land.[6] Some earlier versions of this narrative also intermingle the identity of the ancestor-god Wiraqucha with the homonymous sovereign Inca Viracocha. An anonymous chronicler, later identified as cleric Bartolomé de Segovia, details the civilizing task of this deified-king on earth and mentions his relation with the Sun:

> These *orejones* say that the way they established lords among them was this. From a lake called Titicaca thirty leagues from Cuzco in the land of Collao, emerged the principal among them named Inga Viracocha, who was very knowledgeable and wise. He said he was the son of the Sun, and they say that he showed them the way to clothe themselves, to make stone houses, and it was he who built Cuzco, and made houses of stone as well as the fortress and the house of the Sun, which he left incomplete.[7]

These narratives of origin are certainly reminiscent of Wiraqucha's epic through the Andes, which elaborates and develops a similar narrative pattern. Like Manco Sapaca, the ancestor-god and the first humankind are said to have emerged together from Lake Titicaca in remote times. This primitive humanity would have dwelt in darkness. It practiced a rudimentary agriculture and existed in a state of perpetual conflict because its communal lands were not clearly ascribed. Dissatisfied with his people's conduct, Wiraqucha flooded the earth and annihilated the rebels with the cataclysmic waters. Some of these prototypical beings turned into *wakas* and even today still receive a cult from communities that venerate them as their progenitors. The deity then animated the sun, the moon, and the stars before forging new beings from the earth of Tiahuanaco. Remembering the conflicts that had rent asunder the first humanity, Wiraqucha divided this new race into separate ayllus, gave each one a chieftain, and indicated propitious places for them to settle. He also taught them agricultural techniques and, according to isolated traditions, endowed them with the skills of navigation and the craft of weaving. Finally, Wiraqucha provided his people with a language and laid the foundations for their customs. Las Casas, Betanzos, and Molina record that the god also forged three creatures named Imaymana, Tucapu, and Tawapaca to assist him in his task.[8] The last of these, however, had resolved to constantly defy his progenitor. Irritated by this conduct, Wiraqucha flung Tawapaca into the waters of Lake Titicaca, where he disappeared forever. The god then sent the two remaining brothers towards *qullasuyu*, *kuntisuyu*, and

---

[6] Gutiérrez: Ch. 49-50, 209-215.
[7] Segovia: 73. My translation.
[8] Las Casas: T. 105, 433; Betanzos: Pt. 1, Ch. 1, 11-12; Ch 2, 13-15; Molina: 53-55.

antisuyu to "bring forth the people" from rivers, springs, caves, or hilltops, and to allot them the lands they were to exploit for their subsistence needs.[9] The god himself set on a journey towards the Cuzco valley, travelling by the royal roads (*qhapaq ñan*) and stopping along the way to accomplish his deeds. In doing so, he animated the Canas Indians in Caccha and later instructed the ayllu of Urcos to emerge from the top of a hill where its people subsequently settled. There, the people of Urcos built a temple in honour of Wiraqucha, at the entrance of which were carved an eagle and a falcon representing the sons and brothers of Hatun Wiraqucha.[10] Having arrived on the site of Cuzco, the deity ordered the *orejones* to emerge and populate the land. He then pursued his journey along the course of the river that led to the ocean, bringing forth other nations before disappearing in the primordial waters of the western sea.[11]

The geographical area covered by Wiraqucha's demiurgic actions gives evidence that many communities of the Andean plateaux held this narrative of genesis in common. All of them traced their ancestors' origin back to natural formations. This was a conception closely linked to their cosmology, as they, like the Incas, reckoned that the world floated over Mama Qucha, the subterranean matrix of all primordial waters that surrounded the earth. Garcilaso explains:

> When the Sun set and they saw him sink beyond the sea—for the whole length of Peru has the sea to its west—they say it entered the sea, and dried up a great part of its waters with his fire and heat, but like a good swimmer he dived under the earth and came up next day in the east, whence they supposed that the earth rests on the water.[12]

Stemming from the same conception, the ayllus' ancestral founders and the prototypes of every animal on earth were believed to have taken a subterranean route that followed the course of Mama Qucha before emerging

---

[9] Greatly influenced by the Dumezilian model of trifunctionality, Urbano (1988) studied these four figures. He sees in the structure of these narratives the expression of a tripartite system of social organization in which Pachayachachiq embodied the supreme figure of the father regulator who ordained the politico-warlike domain. His first son/brother represented the knowledge of agricultural labours, while the youngest mastered the cultic and sacerdotal function. As for the fourth figure of Tawapaca, whom Urbano identifies with the southern god Tunupa, he would have symbolised the negative, reverse side of the forefather.

[10] "Hatun: the greatest, best, or the most important and superior, and most renowned", González Holguín, *Vocabulario*: 120.

[11] Betanzos: Pt. 1, Ch. 1-2, 11-15; 52-56; Cieza: Pt. 2, Ch. 5, 8-12; Molina: 84; Gutiérrez: 244-245; Sarmiento: Ch. 6-7, 40-46; *Ritos y Tradiciones de Huarochirí*: Ch. 1, 11.

[12] Garcilaso: Bk. 2, Ch. 23; translation by Livermore 1989: 120.

through the apertures, inland lakes, springs, and seas connecting the habitable earth with the underground.[13] These places of emergence were called *paqarina*, which means literally "the place from where the sun rises."[14] From there, they followed Wiraqucha to the provinces ascribed to each of them and inaugurated a new era. Remarkably, a surviving tradition from the Ocongate district (Cuzco) reports an analogous story relating how every species initially emerged from a cave or a water source together with men. This creation took place on the second day that the sun made its apparition, which echoes how Wiraqucha forged a second, more complete humankind after having destroyed the first beings. Tradition has it that before this episode occurred, alpacas had existed in darkness on earth from time immemorial. As the sun finally rose above the earth's horizon, they went to hide in the springs situated underground. It was only when the sun shed its light a second time at daybreak that all animals emerged from their refuge and spread over the land. For this reason, today people from Ocongate believe that the subterranean springs and the lagoons have command over every species of animal.[15]

This genesis conception throws light upon another variant of the Incas' origin myth, first recorded in the mid-1570s, which situates the Incas' place of birth in Lake Titicaca but differs from the previous account in its ending. Molina, Garcilaso, and Murúa document this story, according to which Wiraqucha first created every nation and then ordered the leader of the Incas, called Manco Capac, to settle towards the northwest. The ancestor-god gave him a golden rod to test the quality of the soil and told him that the appropriate site to occupy for his descendants was where the staff penetrated the ground entirely. Manco Capac and his people travelled through subterranean channels before stopping in Paqariq Tampu, where they rested overnight. On the following day, at the break of dawn they came out of the cave and walked towards the rich hillsides of the Cuzco valley, where they recognized the land assigned to them by their "father the Sun."[16]

In sum, the narratives tracing the Incas' origin to Lake Titicaca belonged to a larger body of pan-Andean traditions in which the advent on earth of the ancestor-god Wiraqucha and the nations' emergence from darkness occurred at sunrise, following a reversal of time and space (*pachakuti*). This

---

[13] Sherbondy 1992: 57-58; Itier 1992: 1039, n. 45, p. 1043, n. 53; Duviols 1993: 111.

[14] Following Doyle (1988), Ramírez (2005: pp. 274-275, n. 29) suggests that *paqarina* was both the place of emergence of a founding ancestor, and the apical figure itself.

[15] Gow & Gow 1975: 142.

[16] Molina: 52; Garcilaso: Bk. 1, Ch. 15: 37-40; Murúa, *Historia general*: Ch. 2, 39-40.

event marked the dawn of a new era ruled by the second sun after the annihilation of the archaic humanity. Wiraqucha himself is explicitly assimilated with the sun in the myth of Manco Capac's subterranean travel from Lake Titicaca to Paqariq Tampu. Later chronicles repeatedly evoke this solar identification. Moreover, the early chroniclers who penned these traditions only listed the reigns of Hanan kings, and thus ignored the rule of Hurin lords who are traditionally said to have ruled in the early time of the Inca dynasty. Bravo Guerreira notes that Gómara only makes mention of three kings after Manco Sapaca, namely, Tupa Yupanqui, Huayna Capac, and Atahualpa. For Pedro de la Gasca, "they were six or seven whose name was Tupa Inca because Tupa would have been the name of this first Inca."[17] Pedro Pizarro, finally, lists the following rulers: Inca Viracocha, Tupa Inca Yupanqui Pachacuti, Huayna Capac, Inca Amaro Inca, and two unnamed successors. Although these authors disagree on the identity of the first leader—which may indicate that their informants named their own panaca's ancestor as the founder of the dynasty—they all associate this apical figure with the Upper moiety. This material resonates with the reigning lists evoked earlier (Tables 3 to 5), which are also oblivious to the existence of Hurin leaders, even though they do not mention the Incas' origin myth from Lake Titicaca. Bravo Guerreira posits that Gómara, Gasca, and Pizarro transcribed a series of narratives that belonged to a larger corpus of traditions exclusively held by members of the Upper moiety, whose lore diverged from the oral histories of Hurin Cuzco.[18] Her demonstration ends here and leaves unexplored a logical corollary of this hypothesis: that Wiraqucha was the tutelary ancestor-god of Hanan Cuzco. This interpretation indeed differs from the conclusions drawn in modern scholarship on this issue, which associates Wiraqucha with the Lower moiety. The following discussion provides evidence contra this current view and argues instead that the lineages of Hanan Cuzco regarded Wiraqucha as their forefather who held features opposed to P'unchaw, the Old Sun of Hurin Cuzco.

The Animating Function of Wiraqucha

On account of Wiraqucha's demiurgic role in the genesis of mankind, the chroniclers dubbed him the Maker (*Hacedor*) of the Andean tradition and readily paralleled his epic with the action of the Christian Almighty god. Some intellectuals of the early colonial era speculated at length on this

17 Pedro de la Gasca, *Descripción del Perú* (1553), in Bravo Guerreira 1992: 21.

18 Bravo Guerreira 1992: 20.

point of apparent resemblance, and their works left an enduring mark on contemporary studies of the Inca religion. Key arguments to support the providential nature of Wiraqucha were his titles, Pachayachachic (*pacha yachachiq*) and Ticsi (*tiqsi*), which colonial lexicons equate with the notions of omnipresence and creative power. Early translations of the first word include "creator of all things" or "God, making something natural, creating it", in which *pacha* refers to the earth and *yachachiq* to "the one who teaches or transmits."[19] In a similar vein, Santo Tomás associates *tiqsi* with the concepts of "beginning", "foundation", "origin", or even the "end of all things."[20] Throughout the sixteenth century and increasingly after the resolutions of the Third Council of Lima, Wiraqucha's image became instrumental in the vast debate over the nature of Amerindians. To many chroniclers, eulogists of the Inca empire, the ancestor-god came to be the "First cause of the universe", a concept that ancient Peruvians would have articulated by the operation of natural reason. In Molina's words, he is the "Inscrutable God", while Cieza and Acosta depict him as the "Creator of Heaven and Earth". For a number of these early historians, Wiraqucha's work was immanent, an idea that opened the path for missionaries in South America. Mestizo writer Pachacuti Yamqui saw the works of the apostle Saint Thomas in Wiraqucha's journey over the land. Guaman Poma, although more ambivalent on the deity's importance in Inca beliefs, attributes his deeds to Saint Bartholomew. Both chroniclers believed that the extent of Wiraqucha's area of influence on the continent constituted a proof of the proto-evangelization of America.

Recent works of back-translations and historical linguistics have corrected this view. They suggest that the bilingual Quechua/Castilian material composed by colonial officials to assist in the evangelization of Peru had been instrumental in shaping the omnipotent nature of Wiraqucha.[21] At the heart of this misconstruction lay the word *kamaq* and its derivatives. This term describes the work of Quniraya Wiraqucha, a central Andean expression of the Collao deity. He is the *runa-kamaq* and *pacha-kamaq* of the Huarochirí tradition, two expressions that the author of its transcription on paper translates as "the creator of men and earth."[22] A few years before the elaboration of this manuscript, the bilingual lexicons and catechisms of the Third Council of Lima had elected this term to translate

[19] Betanzos: Pt. 1, Ch. 2, 14; Sarmiento: Ch. 6, 40; González Holguín, *Vocabulario*: 361-362.
[20] Santo Tomás, *Lexicón*: 363.
[21] See Ramírez 2006b, 2006c. For problems of translation see, Mannheim 1991: 128-137.
[22] *Ritos y Tradiciones de Huarochirí*: Ch. 1, 11.

God's action on earth, his power to create beings *ex nihilo*, and the soul with which he endowed men. Once entrenched in the language and "applied back" to the endeavours of Andean deified ancestors, *kama-* retained its Christian connotations and thereupon influenced the consolidation of Wiraqucha's image as the almighty god of Ancient Peru. However, as Gerald Taylor (2000) was the first to argue, the notion of a primordial act in the hands of a single being translates imperfectly the semantic field covered by *kama-*. The first reason is that many *wakas*, *wankas*, and *mallkis* of Andean traditions shared the power of *kamaq*. An analysis of their epics recorded in the Extirpation of idolatry reports shows that all of these ancestral figures had the power to transmit a vital force (*kamaq*) to a double (*kamasqa*), whether a living being or an inanimate object, which thereupon displayed incredible powers. The *kamaq* was "a source of vitality that animated and sustained not only man but also the entirety of animals and things so that they could *be fulfilled*, which is to say that their potential to function in a way determined by their specific nature becomes *real*, meaning effective."[23] Accordingly, Wiraqucha did not create men in the literal sense. Rather, he animated the people who had lived beneath the earth and kept transmitting this vital energy to their descendants in contemporary times. Many deified ancestors shared this faculty, and some were more powerful than others. Ecological disasters, disease, and the wars these divine ancestors fought against each other could weaken the life-sustaining force they infused into beings. To ensure the continued circulation of the vital force between themselves and their progenitors, humans had to regenerate and nourish regularly the source of their *kamaq* with offerings and sacrifices.[24]

The might of Andean ancestor-gods also depended on the expansion of their cult, which was related in turn to the martial deeds of their descent on earth. The Incas played a crucial role in spreading Wiraqucha's notoriety across the land by evincing their valour as a conquering people animated by the strength of their progenitor. In this way, spiritual power and the strength of the vital energy were closely associated with the ability to rule and subdue new populations. For this reason also, certain derivatives of *kamaq*, such as *kamachikuq*, described positions of ethnic authority.[25] The Incas, however, perpetuated the lore of an ancestor-god whose cult had

---

23 Taylor, in *Ritos y Tradiciones de Huarochirí* 1999: XXII. See also Torero 1990: 250-251.

24 Polia Meconi 1999: 123-125; Ramírez 2005: 135. For contemporary practices, see Tomoeda 1996: 199-201, 207-212.

25 Gose 1996b: 9-10; Ramírez 2005: 122, 137.

already pervaded many regions of South America prior to their emergence as a major power. Wiraqucha's deeds were widely integrated into the local traditions of the Peruvian coast, central sierra, and Bolivian Altiplano. His cult was also attested to in several regions of northern Chile and Argentina, where it continued to spread over the populations newly integrated into the Inca realm during the Late Horizon.[26] Within this large topography and despite historical adaptations, Wiraqucha's worshippers shared mutual narratives about his deeds on earth, all interwoven in the deity's oft-repeated journey toward the western seas. This main narrative thread and the traits common to the different regional manifestations of Wiraqucha reveal a strong affinity with the lore of Wari, an archaic ancestor-god of the central sierra that was heliomorph by nature. Like him, Wiraqucha held inherent solar aspects, which makes it unlikely that he had acquired features of the Sun god once assimilated into the Inca cosmology.[27] His story would have begun much earlier in the central sierra. It was in this zone that Wari's authority first rose and expanded, giving birth to separate regional adaptations that survived well into the colonial era.[28] Progressively, and through local particularities relayed by oral traditions, one of these deities called Wiraqucha came to dominate the southern Andes.

Wiraqucha's Solar Aspects

In addition to the speculations about Wiraqucha's epithets and actions, the actual meaning of his name is another point of contention in colonial and modern histories of the Inca realm.[29] Most chroniclers parsed the word as *wira-qucha*, where the first vocable stands for "fat" and *qucha* designates an expanse of water—a lake or sea, for example. With these interpretations in mind, Gómara, Gutiérrez, and Garcilaso translate the god's name as "fat of the sea", whereas Cieza and Gasca, probably disconcerted by the god's literal association with fat, offer a more lyrical reading with

---

[26] Molinié 1987: 73.

[27] Pease 1970, Cock & Doyle (1979: 57), Brundage (1963: 76-81) suggest that Wiraqucha went through a solarization process influenced by the prominence of the Sun god in the Andes.

[28] Itier (1993: 161) argues that Wari, simultaneously with his expansion in the southern highlands, also influenced the characteristics of other cultural heroes such as Quniraya in the Huarochirí region, Tunupa in the southern sierra or Huichama on the coast.

[29] An extended literature exists on this issue. Several works are particularly valuable: Pease 1970; Duviols 1977b; Demarest 1981: 1-11; MacCormack 1991a: 349-351; Szemiński 1997; Salles-Reese 1997: 58-59; Ziólkowski 1996: 38-46; Armas Asín 2002: 203-209.

"sea-foam."[30] Both chroniclers would have ignored that the native word for spume was *pusuqu*. All of these translations, however polished, suffer from a grammatical aberration: *wira-qucha* should be read as "sea of fat" because the epithet precedes the noun in Quechua. Incidentally, no particularities of Wiraqucha's mythical acts and no aspects of his cult were related to fat. The first clue for a more plausible interpretation of the word can be found in González Holguín's lexicon. It provides an original entry for Wiraqucha that reads: "the Sun's epithet, the honorific name of the god worshipped by the Indians."[31] Duviols and philologist Alfredo Torero were among the first scholars to reflect on this gloss. They suggest that *wira* was a metathesis for the name of the ancestral Sun of the central sierra, *Wari*. In their views, *wira-qucha* meant "the Sun's lake" or "the lake from which the Sun rises" in reference to the genesis of the ancestor-god.[32] Far from being speculative, this interpretation proceeds from the observation outlined in linguistic scholarship that "cases of metathesis are legion in the Andean languages."[33] This exegesis of the word, moreover, not only sheds light on González Holguín's definition, it also clarifies the resemblance of Wiraqucha's epic with the journey of the underground Sun at night described by Garcilaso.[34]

In seventeenth-century records of the Extirpation of idolatry, Wari is the subterranean deity that the Indians worshipped as the Sun itself or its offspring.[35] His name would have originated from the Pano linguistic family (Amazonian region), where the term *wari* describes the sun. According to Torero, the Pano idioms occupied an area located at the same latitude as the central Peruvian highlands before the beginning of our era. During the first millennium BC, the expansion of the Quechua I subgroup over the Andes gradually dispelled the Pano languages to the east.[36] The concurrent

30 Gutiérrez: 245; Garcilaso: Bk. 5, Ch. 21, 270; Cieza: Pt. 2, Ch. 5, 10; Bravo Guerreira 1992: 20-21.

31 González Holguín, *Vocabulario*: 353.

32 Torero 1990: 248.

33 Heggarty 2004: 7. A metathesis is the transposition of two phonemes in the same word. See also Escribens 1980; Shaver & Shaver 1989; Heggarty 2005: 80.

34 See Chapter 4, n. 12.

35 Arriaga: 30, 95, 97, 127; Duviols 1979b: 11-13.

36 Linguists distinguish two Quechua families of dialects. On the one hand, the Quechua I, also called Quechua B, is a group of dialects spoken in the central and north-central sierra of Peru. On the other hand, the dialects that belong to the Quechua A/Quechua II family extend to the department of Huancavelica to the northeast of Argentina, and are also spoken in the equatorial sierra. Although there exist morphological, lexical and phonological differences within these two groups, it is widely accepted that the dialects that composed each subfamily share a common origin. Taylor identified another subgroup,

contact between the populations of these two distinct linguistic groups would have prompted the emergence of Wari's cult in the central sierra, which then throve far into the colonial period.[37] The trial reports of the Cajatambo province (1656-63) assembled during the campaigns against idolatry illustrate this success. According to the native witnesses whose testimonies are registered in these documents, Wari was the architect of irrigation canals, the regulator of subterranean waters who lived in darkness, and the guardian of *chakras*. They referred to him as *pirqayuq* (ruler of the *chakras*' enclosures) and *yakuyuq* (ruler of the waters) because they believed he ruled over the cultivations of irrigated terrain.[38] Like the sun described by Garcilaso, he was associated with the waters of Mama Qucha that circulated below the ground. He is the ancestor travelling through her course before emerging into the light of day through earthly apertures such as lakes and seas. In the northern part of the Lima Archdiocese, his worshippers were cultivators of irrigated terrain who celebrated Wari as their *llaqtayuq* (He who protects His people/descent).[39] He was their divine progenitor who had initially divided their *chakras* and distributed the fields among his descendants. These people also called themselves Wari and "children of the Sun". They told the *visitadores* that their first ancestors on earth were giants who came from the ocean or from Lake Titicaca to settle in the central sierra. They presented Wari with offerings in the upriver streams by virtue of his relation to the subterranean waters. Even today he is given specific oblations during the ritual cleaning of irrigation canals.[40] In the Mantaro valley (Jauja), the common ancestor of the Huanca ayllus was Wari Willka, which they venerated as a spring.[41] North of this area, in Cinga (Huamalíes), Wari's underground temple would have replicated the god's dark habitat bored through with labyrinthine passages to

---

which he called "mixed dialects", in the Yauyos province. See Torero 1974: 16-36; Taylor 1990; Cerrón-Palomino 2003.

[37] Torero 1990: 246-249.

[38] Itier 1992: 1023-1024, 1027, 1031, 1036, 1042; Itier 1993: 160.

[39] He shared this office of *llaqtayuq* with *mallkis* and *wankas*. Ramírez (2006b) points out that, before the Spanish invasion, *llaqta* referred to a group of people sharing a common ancestry. Unlike today, *llaqta* did not describe a place or a settlement because *pueblos* and ayllus did not have a delimited spatial base before the Spanish-mandated *reducciones* of the 1560s and 1570s. *Llaqta* also referred to the local *waka*/ancestor as well as the community he/she protected (*Ritos y tradiciones de Huarochirí*: xxvii-xxviii). As Taylor shows (*Ritos y tradiciones de Huarochirí*: xxx), the nominal suffix /*-yuq*/ broadly translates as "who possess(es)", but also "who is protected by". Literally, *llaqtayuq* means "he who possesses/protects/animates a people".

[40] Torero 1990: 247.

[41] Cieza: Pt. 1, Ch. 84, 243-244.

allow circulation.[42] This sombre universe was also synonymous with unsettled and dangerous times. In 1656, a native commentator described the advent of Wari on earth as "a powerful and great wind", which was a condition commonly thought to announce the propagation of diseases, especially at the beginning of the wet season.[43] Today, *wayra* (wind) has retained part of this aspect, for it is "conceptualized as a localized circulatory agent of subterranean energy. He is said to rush out of his house—a high inaccessible cave—to flow like a river of air through the atmosphere before returning home."[44]

Wari's odyssey in the subterranean world finds remarkable parallels in the details of Wiraqucha's journey over the land. First of all, both evoke a close association with interior seas and water springs. Wiraqucha was believed to have bestowed his descendants with the knowledge of cultivating irrigated fields and to have delineated their arable lands. Likewise, Wari supervised the draining and ditching activities of maize farmers to whom he had given *chakras* in time immemorial. In addition, Zuidema and Demarest have argued convincingly that Wiraqucha's path over the earth can be seen as the "earthly projection" of the sun in the sky.[45] Like the star rising in the east, the ancestor-god emerged at dawn from Lake Titicaca to undertake a civilizing mission on his way to the sea. As he walked through the highlands, Wiraqucha enlightened a second generation of humankind and beings before disappearing in the ocean of Mama Qucha. His identification with the sun explains why the founders of the Inca dynasty were said to have emerged at dawn from their *paqarina*, for it was the light and vital energy of this god rising on earth that animated the group's life force. To establish that Wiraqucha embodied aspects of the sun clarifies some of the confusions that permeated the chronicles. For example, they commonly disagree on the identity of the deity—either the Sun or Wiraqucha—who appeared before Inca Yupanqui to assure him of his favours as the young contender was about to confront the Chancas.[46] The same sources also report that the Cuzco nobility traced its ancestry to the Sun, although they depict Wiraqucha's initial role in the emergence of the Inca ancestors. To account for these confusions, Pease argued that a "solar cult could be contemporaneous with one or several solar gods. The Sun can be

42 Duviols 1971: 386-398.
43 AAL, Leg V, cuad. 2, 1656 in Cock & Doyle 1979: 58 n. 14. Larme 1998: 1008-1010.
44 Allen 1988: 53.
45 Zuidema 1964: 164-165; Pease 1970: 164; Demarest 1981: 26-27.
46 Demarest 1981: 17-22.

several gods, have distinctive names and distinct 'personalities', according to its position during the day or during the year."[47] In line with this argument, Molinié suggested that Wiraqucha's four manifestations—Pachayachachiq, Imaymana, Topacu, and Tawapaca—corresponded to different aspects of the sun. She partly founds her argument on Silverman's study (1994) of the present-day community of Q'ero, north of Cuzco, where residents have developed an elaborate semiology of textile patterns. Among these motifs, the sun assumes four different designs according to the position of the star in the sky. They represented sunrise, the midday sun, sunset, and the midnight sun that Molinié associates by analogy with Wiraqucha and his three attendants. In this configuration, Tawapaca, the character who disobeyed the god, represented "the nocturnal, antithetic and hidden aspect of the solar divinity."[48] In an earlier work, Demarest (1981) had similarly postulated that the Inca sun god consisted of several overlapping aspects. He identified three of them: Wiraqucha, or the mature Sun; P'unchaw, or the young Sun; and the ancestral Sun of the Inca nobility, which he called Inti Wawqi. Transcending this trinity was a "complex, aggregate sky god" also reckoned as the "Maker", whose nature encompassed the characteristics of three major deities: Illapa, the Sun, and Wiraqucha. Like Zuidema (1976) before him, Demarest suggested that Wiraqucha supervised the Inca festival of the summer solstice at *qhapaq rayme*. According to him, the ancestor-god would have extended his tutelage over the novices' transformation into adulthood, at the time when the young sun "is transformed into its mature aspect (...) by the achievement of the summer solstice."[49] This deity would have opposed the young Sun they identified with P'unchaw, whom the Incas worshipped during *inti rayme*, the ceremony of the winter solstice.

While the days when the sun reaches its northernmost and southernmost extremes in the sky were key dates in the framework of the Inca calendar, primary sources do not offer explicit evidence supporting Wiraqucha's seniority with respect to P'unchaw, nor do they substantiate his association with the *qhapaq rayme*. Zuidema's only unequivocal source to establish Wiraqucha's link with the December festival is the late and disputed chronicle of Pachacuti Yamqui, whose discourse was embedded in the evangelization rationale that gave predominance to the "Maker" in Inca religion. There is no indication in Molina, our most detailed source, that

---

47 Pease 1970: 68, n. 29.
48 Molinié 1993-1994: 30.
49 Demarest 1981: 28. Zuidema 1964: 166-172; Zuidema 1976; Zuidema 1989: 259.

P'unchaw was the patron of *inti rayme*. On the contrary, a close reading of this text reveals that three images, those of P'unchaw Inca, Wiraqucha Pachayachachiq, and Ch'uqilla, together received offerings during the June celebration that was dedicated to the "Sun". Molina appears to have made equal use of the expression "Sun god" (*Sol*) in his description of both solstice festivals, leaving no clue as to the identity of the deity/ies behind this designation.[50] The following reassessment of primary materials on Inca festivals and the study of the gods' influence over the environment and agricultural cycle draw a different picture. They offer evidence that places a specific aspect of Wiraqucha, called Pachayachachiq, at the centre of the June solstice ceremony. This festival celebrated the deity's civilizing task at the dawn of time, when he animated his descendants and thereupon enlightened them with the knowledge of agricultural techniques. The *inti rayme* procession thus reproduced Wiraqucha's journey on earth after he had emerged from Lake Titicaca.

Wiraqucha in Inca Ceremonies

Molina indicates that during the June festivities of *inti rayme*, the Incas directed their prayers specifically to Wiraqucha Pachayachachiq so that he would guarantee the "multiplication of beings and food."[51] Itier has demonstrated that the name of this deity derived from the same stem as the term *yachakuchi-*, which described the deeds of supernatural beings from various Andean traditions. This word did not designate the concept of immanence as claimed by the Spaniards, but rather illustrated the act of bringing any object to its fulfilment, from its initial state to its plenitude. Deities exerted this power over every animated body, from humans to foodstuffs, passing through the springs of water supplying irrigation canals. Itier translates *yacha-* as "bringing to fruition", "fully carrying through", while in this case *pacha* refers to the earth as the fertile soil. As such, Wiraqucha Pachayachachiq is the paramount aspect of the ancestor-god. He is the Sun at its culminating point of fertilization, He who guarantees the full development of the earth, "the one who brings the surface of the earth to a point of maturation required (for its full agrarian exploitation) in the

50 Zuidema's argument relies also heavily on his reconstruction of the "Inca astronomic theory" and his identification of several shrines with pre-Hispanic observation points. This approach, although it reflects the Incas' undeniable interest in celestial phenomena, is based on tenuous evidence and draws from a modern and still problematic interpretation of the *seqe* system.

51 Molina: 67.

course of a founding deed that occurred in a mythical past, as well as during the annual cycles of production."[52] A review of primary sources on Inca festivals corroborates the close connection of Pachayachachiq with the earth productivity cycle. Every year, *inti rayme* marked the start of the new agricultural cycle, when men shelled the recently gathered corn and selected the grain for the following sowing.[53] On this occasion, the Inca nobility, who alone participated in this festivity, went to Mantucalla hill, north of Cuzco. There, they first burnt offerings to Wiraqucha, including figurines made of *kishwar* wood.[54] Then, they performed a dance called *wayllina* in honour of Pachayachachiq, which they repeated several times through the year to accompany the maize maturation process.[55] Two female deities attended this *taki*. The first was called Inca Ocllo, while the second, Pallpa Ocllo, bore a name that suggested her relation with sunset (*pallpani*) and tuber farming.[56] Later in the year (August), when men sowed the cultivations of irrigated fields, Wiraqucha received a white llama and other offerings. For this occasion, the Incas performed another dance in his honour because they believed it enticed the food crops to achieve full maturation. The vegetative cycle eventually ended in April with a sacrifice to Wiraqucha so "he would always bring them good years."[57]

The prayers, dances and offerings that the Incas addressed to Wiraqucha during *inti rayme* not only celebrated the first stage of the agricultural cycle, they also commemorated Wiraqucha's journey on earth simultaneously with the emergence of every species from their *paqarina*. This primordial episode was enacted in various ways, but the presence of two pairs of gold and silver life-size llama figures at *inti rayme* is a first reference to this event. Molina reveals that these images, known as *quri napa* and *qullqi napa*, represented the first of their species to have come out of the cave with the Inca ancestors. Covered with a red mantle, they headed the procession to Mantucalla hill in the custody of high-ranked Inca delegates. They remained there with members of the nobility until the month reached a close. Then, from this site part of the cortège went to the "house of the Sun" in Ch'uqi Kancha (the gold enclosure) or to the nearby sanctuary of

---

52 Itier 1993: 161.

53 Polo, "Relación de los adoratorios": 20; Cobo: T. 2, Bk. 13, Ch. 29, 218.

54 Mantucalla (An 3:6) has been identified by Bauer as the modern Salonpuncu. Bauer, 1998: 81-82.

55 Molina: 69.

56 Duviols notes (Molina: 70, n. 37) that Bertonio records *pallpallitha*: "singing [while] sowing potatoes".

57 Molina: 118-119.

Pawqar Kancha (the enclosure of rich plumage) to immolate several llamas.[58] In the meantime, a group of celebrants, all members of the Tarpuntay Ayllu, left Cuzco on a pilgrimage southeast of Cuzco "because they say that the Sun was born in this region."[59] Along the way, they laid offerings in several shrines and crossed the Pomacanchis plain and its lagoon before reaching the shrine of Vilcanota, located approximately 150 kilometres southeast of Cuzco (Map 2). Situated at an altitude of more than 5000m above sea level, the Vilcanota was one of Tawantinsuyu's main shrines, and like many high, snowy peaks of the Andean Cordillera, it fed important watercourses.[60] From there, two rivers burst forth: one flows to the Yucay valley and the other runs towards Lake Titicaca. The former was originally known as Vilcamayu (River of the Sun). It feeds the Huatanay River, which itself supplies the two streams, called Saphi and Tullamayu, that cross Cuzco. When returning from the Vilcanota sanctuary, the procession retraced the footsteps of the Sun Wiraqucha travelling through his aquatic domain. To do so, it followed the course of the Vilcamayu River, passing through several *wakas* set along its banks. Most of these sites cannot be located with precision, but two of them had been important stops on the journey of the ancestor-god to *chinchaysuyu*: Caccha (present-day San Pedro de Raqchi) and Urcos (or Tambo Urcos).[61] Caccha was located in the cool altitude of the Collao region that shelters an ecological system ideal for the breeding of camelids. Cieza, who visited the area in the 1540s, entered its temple, where he saw "a stone idol, the stature of a man with clothes and a crown or tiara on his head."[62] The chronicler also stopped in the town of Urcos, established close to a lagoon, but he made no mention of a specific shrine. What captured Cieza's attention, however, was the high wall that bordered the road entering the urban centre and which, according to locals, enclosed ingenious irrigation canals. The demiurgic origin of these *wakas* in the hands of Wiraqucha and their location along a watercourse connected with Lake Titicaca make clear references to the journey of the ancestor-god in times immemorial. The association of the Vilcamayu River with Wiraqucha is all the more remarkable in that other

---

58 Cobo: Bk. 13, Ch. 28, 216. Niles, followed by Bauer, identifies Ch'uqi Kancha (An 6:3) as Rumi Huasi Alto, uphill from San Sebastian. Sherbondy and Zuidema locate it in Cusicallanca, eastwards from the other site.

59 Molina: 69.

60 Cieza: Pt. 2, Ch. 28, 84. Reinhard (1995) locates the Vilcanota temple at the pass of La Raya, between Lake Langui-Layo and the Chimbolla peak.

61 Betanzos: Pt. 1, Ch. 2, 13-15; Sarmiento: Ch. 7, 44-45.

62 Cieza: Pt. 1, Ch. 97, 268-270.

temples dedicated to him bordered its banks northwest of Cuzco. Shrines such as Apu Wiraqucha, Urusaywa Wiraqucha, or Kurku Wiraqucha were located near Amaybamba in the present-day valley of *La Convención*. Ch'uqi Chaka was also found in the nearby Amazonian jungle. An important *qhapaq ñan*, beginning in the Collao province and following the course of the Vilcamayu, connected these sites with each other. They obviously extended the mythical route of Wiraqucha towards the western sea. Today, the Vilcamayu has been renamed Vilcanota River, but it still bears a close relation with the sun. Urton records the persistence of this belief in the community of Misminay, about fifty kilometres northwest of Cuzco:

> [The river represents] the terrestrial reflection of the path of the sun through the sky during the day and is considered the actual path of the west-to-east movement of the sun during the night (...) when it [the sun] sets in the west, it enters the 'sea' or 'other world' [*otra nación*]. After entering the sea or the other world, the sun makes a twisting motion to the right [north] and begins its journey back to the east beneath the Vilcanota River. It takes all night for the sun to move from the sea to inti seqamuna [place from where the sun rises].[63]

In the community of Kuyo Grande, a few kilometres away from Misminay, locals also believe that the sun travels from west to east through a subterranean tunnel filled with the waters of Mama Qucha, from which it drinks every day. The same belief explains why the figure of the midnight sun in Q'ero is called *inti qucha*, in reference to its journey over the primordial waters.[64] Urton corroborates the observations made by Juvenal Casaverde Rojas in a community situated on the opposite bank of the Vilcanota and demonstrates that the sun maintains an interdependent relationship with the seasonal cycles. According to local belief, the sun of the rainy season (starting in December) lives underground. It is brighter, hotter, and larger because it drinks from the watercourse while travelling back to its place of emergence. Conversely, it appears weaker during the dry season (starting in June) because it is unable to quench its thirst from the shallow flow of the Vilcanota River.[65]

It was precisely at the peak of the dry season, in June, when the Inca ruling elite celebrated the emergence of the Sun-Wiraqucha on earth and re-enacted his journey to the western sea. This event occurred six months after the ancestor-god had entered the ocean to become the subterranean

63 Urton 1981: 68-69.
64 Ibid.; Silverman 1994.
65 Urton 1981: 29-30.

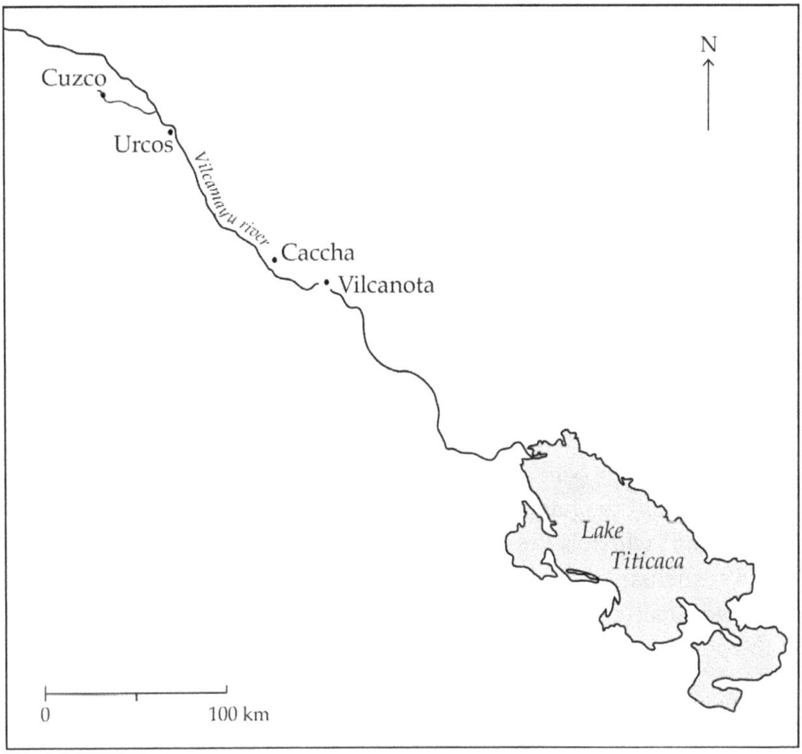

Map 2. Main shrines visited by the Tarpuntay Ayllu during *inti rayme.*

Sun who travels to its place of emergence in Lake Titicaca. In the *inti rayme* reconstitution of Wiraqucha's journey on earth, the shrine of Vilcanota represented his *paqarina,* "the place from where the Sun rises". It stood for the wellspring of the deity's aquatic matrix within Cuzco's ritual space because it fed the waters of the Vilcamayu. After Wiraqucha had arisen from this location, the river's feeble current bore him along its aquatic path. The Inca offerings laid along his route would then help him to regenerate. *Inti rayme* was therefore a celebration in honour of this ancestral Sun and his civilizing feats. It marked the beginning of a new agricultural cycle, much like Wiraqucha had initiated a new age by bestowing the mastery of agrarian techniques on humankind.

While Wiraqucha was associated with the earthly course of the Vilcamayu at the start of the dry season, he became Tiqsi Wiraqucha at the beginning of the wet season, when he was believed to enter the flow of the subterranean waters. Etymologically, *tiqsi* is a lexeme closely related to

Mama Qucha, that is, to the aquatic borders surrounding the universe from it edges down to its depths. *Tiqsi muyu* (the circumference of *tiqsi*) was a common expression still used during the colonial period to refer to the world in its totality.[66] In fact, the early colonial translations of *tiqsi* (origin, principle, grounds) certainly originated from the relation of the subterranean sea with the borders of the universe and from the genesis of Wiraqucha in these waters. Supporting this reading, Torero (1990) has shown that the etymology of *tiqsi* linked the archaic Sun with the subterranean waters. He observes that this term could not have derived from the Quechua family, but more likely proceeded from the pre-Andean family of the Arawakan languages (Amazon). Its most archaic transcriptions, *titi* in Betanzos and *tici* in Cieza, are linguistically close to the Arawakan terms for fire, light, or sun. Torero claims that the same lexeme is found in the denomination of Lake Titicaca (*tiqsi qaqa*) because this name has no connection with the *lingua franca* of the Inca empire. Titicaca should be translated: "the Fire/Sun's island", which is the name that locals still use today to refer to the island's sanctuary.

Wiraqucha received several orations under the denomination of Tiqsi at the perilous time of the *sitwa* festival, when the first annual precipitations gather in the sky.[67] August was a critical and unsettling month for the maturation of all living beings on earth. *Sitwa* was devised to repel potential threats by banishing the diseases beyond the known universe to the aquatic borders of the Inca world. To do so, four squadrons raced from Cuzco in the direction of each *suyu* and ended their courses by symbolically pouring the pestilences into four major rivers of the Inca heartland. For the next few days, members of the nobility assembled in Haucaypata with the images of their gods and the mummified bodies of their kings. They consecrated the first offerings to Wiraqucha and spent the last part of the day performing a *taki* named *arawi sitwa*, which expressed their devotion and love to the ancestor-god. The cane flute tune that accompanied the dancers was called *t'ika t'ika* (blooms) and celebrated Wiraqucha for the abundant burgeoning that followed the early sowing of the plants.[68] The Incas then addressed prayers to their deities but directed them principally to Wiraqucha Apu quchan, literally "the lord of the subterranean

---

[66] Itier 1992: 1039 n. 45.

[67] Molina: 81, 83, 84, 86, 89, 94.

[68] "Ttica: the flower which is plumage", González Holguín, *Vocabulario*: 340. In Quechua, the plural is usually inferred so that to specify an indefinite but large quantity, speakers double the noun concerned by the plural. *T'ika t'ika* refers to the numerous flowers growing after the plantation.

waters", and to two more denominations of the god whose name suggests an association with the aquatic borders of the universe: Tiqsi Wiraqucha and Qaylla Wiraqucha, in which *qaylla* refers to the "edge."[69] During this time of year, Wiraqucha was the ruler of the dangerous underground connected with the darkness of the subterranean sea. He could raise these waters above the earth and destroy the crops by bringing along a series of illnesses listed by Guaman Poma: *zara unquy* (maize disease), *pukuy unquy* (water spring disease), or *chirapa unquy* (disease of the rain when the sun shines), among others.[70] To prevent these calamities, the Incas beseeched "the one who alone is in the world below" to keep them healthy and victorious, but also to multiply the people, animals, and foodstuff.[71] They dispersed the plagues in the rivers for Wiraqucha to carry them away because "they understood that these would join the sea."[72] It is the same cosmology that underlies the *mayu qatiy* rite in January, when the Inca nobility gathered at the peak of the wet season in Pumap Chupan to thank Wiraqucha for his benevolence. On this occasion, they poured the ashes from all of the offerings of the past year in the Watanay river so that Wiraqucha would "receive it in his hands, wherever he was (…) saying that the waters will carry it to the sea."[73]

Wiraqucha and the Might of the Thunder God

Although Wiraqucha was assimilated with the subterranean Sun, several traditions link him as well with the thunder god and with the celestial rains that supplied the highlands with water springs. Like the ruler of lightning in the central highlands, Wiraqucha inhabited the upper ecological niche (*puna*) and supervised the pastoral activities of its inhabitants.[74] In Cuzco, his temple was located in the Hanan district of Pukamarka (the Red settlement). There, his anthropomorphic image was housed in the same complex as, or adjacent to, the sanctuary of the Thunder god Ch'uqilla (Map 3).[75] Wiraqucha's temple itself was the Kishwar Kancha, a name reminiscent of the wooden figurines burnt in honour of the god at the festivals of

69 Itier 1992: 1039, n. 45.
70 Guaman Poma: 253 [255], 227.
71 Molina: 82-93.
72 Molina: 75.
73 Molina: 116.
74 Duviols 1997: 295.
75 Polo, "Relación de los adoratorios": 9, 10; Molina: 59, 67, 73, 100. For a study of the compound, see Bauer 2004: 134-135.

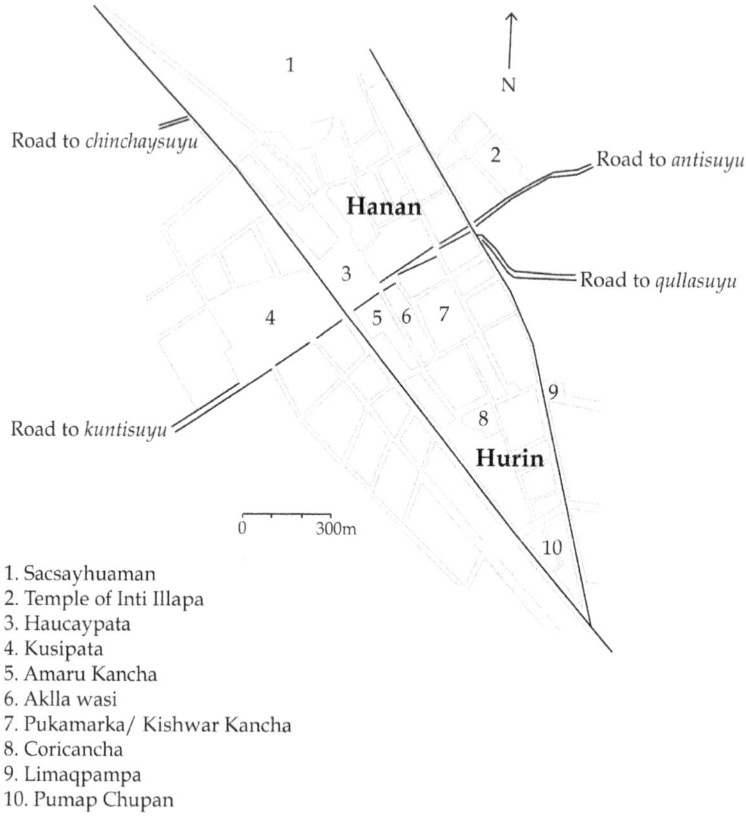

Map 3. Inca Cuzco.

inti rayme and *sitwa*.[76] The *kishwar* (*buddleia Incana*) is a native tree foresting the high plateaux of Lake Titicaca, the birthplace of Wiraqucha but also the habitat of the Thunder god. Its wood was used to make the Andean foot plough (*chaki taklla*) and to build temporary sanctuaries.[77] Although its natural environment is the lower limit of the *puna*, it adapts to higher altitudes and to extremely low temperatures. Cobo, describing the fauna and flora of the New World, writes: "in the extremely cold and bleak moors of the Collao provinces, where no other tree survives, grows the quishuar."[78]

---

[76] Polo, "Tratado": 21; Cobo: t. 2, Bk. 13, Ch. 29, 218.

[77] The Huarochirí manuscript records that, in ancient time, the villagers of the community of San Lorenzo built a *kishwar* enclosure where they worshipped their local *waka* for five days. See *Ritos y tradiciones de Huarochirí*: Ch. 7, 95.

[78] Cobo: T. 1, Bk. 6, Ch. 49: 255.

In addition to their mutual connection with the high plateaux, Wiraqucha and the Thunder god shared similar powers. Cieza, for example, attributes to the Inca ancestor-god deeds traditionally associated with the ruler of thunder and lightning. In his variant, the chronicler reports that the Cana people, who had just emerged from their *paqarina* in Caccha, raised their weapons against Wiraqucha as he passed through their province. To quell the uprising, the deity sent blazing fires from the sky, burning a range of mountains near the town.[79] Horrified, the Cana surrendered, and to commemorate this event they built a temple to the god on the site of his feat. Known in colonial times by the name of Caccha, this sanctuary is located in present-day San Pedro de Raqchi, near the most important group of inactive volcanoes of the region, the Kinsach'ata snow-topped peaks. The past eruptions of this mountain range left charred remains that are still visible today in the town vicinity and easily evoke the blazing outbursts of Wiraqucha's fury. Traditionally, however, volcanic eruptions and lightning bolts were connected to the might of the Thunder god.[80]

Archaeological excavations reveal another dimension of the assimilation process that affected these two gods. The Peruvian Instituto Nacional de Cultura and the UCL Institute of Archaeology carried out surveys in the temple of Caccha showing that the building had been constructed during the Late Horizon on the site of a more ancient occupation dating from the Formative period. The whole complex included *qullqa* storehouses, several fountains, and an artificial lake, all enclosed within high walls. Its architectonics is typical of the Late Imperial style, and the ceramics associated with its construction are Inca, Inca-Colla, and Colla.[81] In colonial texts, the site was also known as a *tampu* on the *qullasuyu* road where the Incas stored the foodstuff required to feed their army.[82] Cieza, Garcilaso, and Cobo credit either Viracocha Inca or Tupa Yupanqui with the erection of this sanctuary after they had defeated the Chancas.[83] For Polo, this triumphant outcome had been the result of the alliance Cuzco had contracted with the Canas and Kanchis,[84] two ethnic groups that were worshippers of the Thunder god and whose influence stretched 120

---

79 Cieza: Pt. 2, Ch. 5, 9-10.

80 *Ritos y tradiciones de Huarochirí*: Ch. 6, 83; Ch. 8, 109-111; Ch. 16, 215; Ch. 23, 297; Ch. 26, 351; Bouysse-Cassagne 1993.

81 Lynch 1984: 420; Niles 1999: 240-253; Sillar 2002.

82 La Lone and La Lone 1987: 54.

83 Cieza: Pt. 1, Ch. 98, 270; Garcilaso: Bk. 5, Ch. 22: 271-274; Cobo: T. 2, Bk. 12, Ch. 11, 77.

84 Polo, "Tratado": 46.

kilometres southeast of the Inca capital, over the site of Caccha.[85] Betanzos, however, claims that it was Huayna Capac on a visit to *qullasuyu* who promulgated the construction of the temple to Wiraqucha.[86] While it should not be discounted that this last ruler contributed to the extension of the complex, the Canas' allegiance to the Incas was reportedly more ancient. Betanzos himself indicates that Tupa Yupanqui displaced a great number of Cana *mitmaq* to squelch the insurrection of the Colla people, while documents uncovered by Julien attribute the creation of these Cana colonies to Pachacuti Yupanqui.[87] For Cieza, finally, it was Viracocha Inca who first made an alliance with them. He would have exempted the Canas from tribute and from attending the yearly inspection in Cuzco.[88] Cobo, who probably had access to the same tradition, writes:

> These two nations of Canas and Canchis were always held in high esteem by Inca Viracocha and his successors, who granted them special insignia of honours, because from the time they yielded obedience to this Inca, they helped and served with notable effort and fidelity in all the wars and conquest that the Incas undertook.[89]

As Bill Sillar (2002) suggested, the construction of an Inca compound on the site of the pre-existing sanctuary of Caccha would have solemnized the support of the Canas and Kanchis to the Cuzco elite. In this way, the edifice stood as an acknowledgement of the contribution of these two ethnic groups to victorious warfare. It was the material evidence of the Thunder god's alliance with the all-powerful ancestor-god of the Inca nobility.

Arguably, the late introduction of Inti Illapa as one of the three manifestations of the Inca Thunder god had a similar origin. The name of this figure evokes a composite deity that combined aspects of the sun (*inti*) with those of thunder (*illapa*). Tradition has it that Pachacuti Yupanqui chose Inti Illapa to be his *wawqi* (brother, double) after his accession to power. He ordered that a golden image be cast of it, which he carried onto the battlefield to assist him to victory. Off campaign, the effigy remained sheltered in a temple located in the Cuzco district of Tococachi (*t'uqu kachi*) (Map 3).[90] The warlike function of Inti Illapa and its introduction in

[85] Bouysse-Cassagne 1992: 134; Torero 2002: 389-401.

[86] Betanzos: Pt. 1, Ch. 45, 191-192. Niles (1999: 236-240) gives more credit to Betanzos' account which, according to her, carries a "ring of truth".

[87] Betanzos: Pt. 1, Ch. 34, 156; Julien 1983: 82-83.

[88] Cieza: Pt. 2, Ch. 42, 125-126.

[89] Cobo: T. 2, Bk. 12, Ch. 11, 77.

[90] Bauer has identified the location of this temple with the actual church of San Blas (Bauer, 2004: 137).

the Inca ritual apparatus after the annihilation of the Chancas suggest that it commemorated the alliance of the two triumphant gods, the Sun and the Thunder, like the construction of the Caccha temple.[91] The name of this site indicates as well that the cult that had thrived there before the Inca annexation was dedicated to a local thunder god, named Caccha or Caxia in the Pukina language, whose authority was previously limited to the Lake Titicaca region.[92] This name is affiliated with the Aymara term *k'aqcha*, which covers a series of glosses related to bolts of lightning, detonation, and arquebus gunfire. For Guaman Poma, *quri k'aqcha* was a synonym for the Quechua *illapa*, which in turn referred to the god of lightning flash, thunder, and lightning bolt, and came to designate the Spanish arquebus and artillery. González Holguín also writes: "Kacchanta çurcun ccacñiy: to be struck down by thunder, or to be shattered by a great noise."[93] In present-day Quechua, *q'aqcha* still describes a resounding and frightening noise, but is also the strong cracking of a whip that reproduces the roar of thunder in certain ritual dances.

In Collao, it was the thunder deity Tunupa who ruled the unpredictability and might of volcanoes. He reigned over the chaotic world of wild animals and was responsible for the draining of irrigation sources. Both him and Wiraqucha originated from the high plateaux of the southern sierra and had been labelled Creator god in colonial times.[94] This analogous status explains in part why Wiraqucha's mythical corpus possesses close resemblances with the narratives of the Aymara thunder god that some chroniclers identify with the apostle Saint Thomas. These similarities are particularly evident in the narration of Colla chronicler Pachacuti Yamqui, who credits Tunupa with the exploit of journeying from Lake Titicaca to the west sea. For Thérèse Bouysse-Cassagne, these two gods maintained an "interdependent relation of opposition and complementarity in which both entities were thought of as a single concept, which corresponds to the definition of the *allqa*."[95] Demarest has advanced an

---

91 This figure is reminiscent of another composite deity of Pukina origin: Curi Caccha (*Quri K'aqcha*) that Guaman Poma describes as the Incas' prime Thunder god.

92 Pukina is an extinct language and today's written forms of Caccha vary from *Qhaqya*, *Qhaxa*, *Cakha*, or *K'aqcha*. Seventeenth-century Jesuit missionaries observed that Caxia also designated the thunder god's attendants (Bouysse-Cassagne 1993). For studies on the Pukina language see Browman 1994; Torero 2002: 389-401; 408-456.

93 Bertonio: 44; Guaman Poma: 265 [267], 239; 405 [407], 377; 885 [899], 831; González Holguín, *Vocabulario*: 129.

94 Cieza: Pt. 2, Ch. 5, 8-10; Pachacuti Yamqui: f. 3v., 188; Ramos Gavilán: 10.

95 Bouysse-Cassagne 1997: 173. About the concept of *allqa*, Bertonio writes: "Contrario en las colores y elementos auca y de otras cosas así que no pueden estar juntas v. g. contrario

alternative perspective that conforms to his hypothesis on the interchangeable aspects of Andean gods. He proposes that "the sky god Illapa/Thunupa is an inseparable entity from the creator Wiraqucha" so that together they form an aspect of the "single, manifold godhead" that resides in the upper regions.[96] This compelling argument leaves unanswered the question as to why these two particular divinities, rather than others, were connected by mythical and historical events that reportedly brought their communities of worshippers together. Should we assume, like Demarest, that "the criteria separating the overlapping and interchangeable aspects of the gods of the upper pantheon were not consistent"?[97] Was the system of native beliefs so nebulous that there was no internal rationale behind Wiraqucha's affiliation with the Thunder god? Other aspects of the Thunder god in Andean cosmology shed more light on this issue.

From the Underground to the Firmament: Water Regulation in the Dry Season

Colonial sources generally give two denominations to the Thunder god: Libiac (*llíwyaq*) and Illapa, both of which referred to similar phenomena but originated from different regions. Documents of the northern and central highlands mention the epic of Libiac, whereas the southern provinces of the Inca empire were the domain of Illapa, whose name derives from the Quechua of the Cuzco area. Despite this geographical division, the two divinities held common features associated with the ecological level of the *puna* and the activities of subsistence practiced at that altitude: pastoralism, potato cultivation, and sling hunting. The etymology of the word *llíwyaq* also makes clear his association with the atmospheric phenomena of the dry season because it describes the cloudless and limpid (nocturnal or diurnal) sky in frosty weather.[98] In the Extirpation of idolatry material, he is often preceded by the epithet *rupay* (Quechua I), which is a specific attribute referring to the warmth of the sun during the dry season, in contrast to the sun of the wet season.[99] The same documents reveal that the

es el negro de lo blanco, el fuego del agua, el día de la noche, el pecado de la gracia". Similarly, for González Holguín, it is "Lo de dos colores blanco y negro" and "cosa blanca y negra". See Bertonio: 10; González Holguín, *Vocabulario*: 19; Santo Tomás, *Lexicón*: 233.

96 Demarest 1981: 35-39. Here, Demarest follows the lead of Brundage (1963: 47), who remarked that the Sun divinity "was often subsumed under a more generalized god of the sky and the storm".

97 Demarest 1981: 39.

98 Cerrón Palomino 1976: 94.

99 For Itier (1992), the orations beginning with *Libiac Yayanchic Rupay Yayanchic* could be translated by "Our father the sky of the dry season, Our father the sun of the dry season".

pastoral groups known as Llacuaz regarded Libiac as their forefather. They believed that in remote times, the Thunder god engendered the first of their kurakas, who then took his people on a long march towards the west. Their migration reached an end when the Llacuaz joined the Huari communities, the worshippers of the eponymous deity Wari, and built their own settlement in the Upper part of their towns. From then on, both groups "shared a common dual conception of nature, cosmos, social organization, and of their origin."[100]

Humans who had been struck by thunder or had experienced an extraordinary birth, such as twins or people born with their feet first, were elected to be Libiac's officiants.[101] The sites seared by his presence were also revered and held in great awe. Many places in the elevated altitudes of the Andean plateaux, more exposed to thunderbolt strikes and frosty conditions, housed Illapa shrines. In ancient Cuzco, the Incas would have closed indefinitely a sector of Huascar's palace that had been struck by lightning because they feared that this ill-fated omen portended tragic consequences. Following this event, the palace was named Amaru Kancha (the Serpent's enclosure) (Map 3) in reference to the animal that held close links with the thunder deity.[102] During the dry season, between May and July, snakes make their way under the earth's surface to hibernate. They only venture out of their den if they sense the imperceptible tremors announcing an imminent volcanic eruption, a phenomenon traditionally associated with the thunder god.[103] The snake, whose undulating movement was also identified with the jagged contour of a lightning bolt, was regarded as the thunder god's messenger during the dry season.[104]

In his depiction of the Inca religion, Father Cobo argues that the Thunder god bore three different names corresponding to separate images of the god: Ch'uqilla, Catu Illa, and Inti Illapa, in order of importance. All of them, he adds, had their own altars in the Sun temple.[105] Although Cobo's account is partly based on native testimonies, it also proffers a distorted

100 Duviols 1986: LVI. See also Duviols 1973.

101 Arriaga: 44, 63; Polia Meconi 1999: 165-166; Chávez Hualpa 1997: 107-118. This belief has persisted throughout the centuries in many parts of the Andes, see Tschopik 1951: 199; Mariscotti 1978b; Marzal 1988: 263-285; Fernández Juárez 2004: 20-22.

102 Garcilaso: Bk. 2, Ch. 1, 63. See also Ogburn 2004: 128-129.

103 Urton 1981: 177-180.

104 Cobo: T. 2, Bk. 13, Ch. 6, 159.

105 Cobo: T. 2, Bk. 13, Ch. 7, 160. Accounts by Cobo, Acosta and Murúa paraphrase Polo's statement. See Polo, "Tratado": Ch. 1, 6; Acosta: Bk. 5, Ch. 4, 335; Murúa, *Historia general*: Bk. 2, Ch. 28, 412.

picture of the thunder cult by suggesting that the Incas conceived of this deity as a trinitarian character.[106] Earlier sources do not support this view; rather, they outline the Incas' devotion to a single god, Ch'uqilla, whose figure was at the centre of their annual festivals. Independent of this main deity, the other two representations held a different status and a separate origin. Inti Illapa was the *wawqi* of Pachacuti Yupanqui and followed the king on the battlefield. The Spaniards discovered his golden image next to the Inca's mummy, in Tococachi, where it assisted the defunct in his after-life, just as it had supported him during the conflict against the Chancas.[107] The name of the second figure, Catu Illa, may have been a pre- or post-colonial adaptation into Quechua of the Culle noun Catequil,[108] which was an oracle of the Huamachuco region (northern Peru), ruler of thunder and lightning, whose attribute was the sling. Several documents allude to his influence during the Inca period, which is confirmed in the wide diffusion of his mythic corpus down the central sierra and along the coast, beyond its area of origin.[109] He is described as being "the most feared and honoured idol there existed in Peru, adored and revered from Quito to Cuzco", and "one of the most prominent huacas of the realm", but also "the Inca's emissary."[110] His image escorted Huayna Capac to Quito and later accompanied one of his captains back to Cuzco as he sought reinforcements to fight off the Chiriguano's incursions into Tawantinsuyu.[111] Even Atahualpa's destruction of Catequil's image, after the idol had made him an unfavourable prognostication, did not put an end to his influence.[112] His worshippers recovered part of his image, around which they perpetuated their devotion. Catequil's extended authority may have prompted the Incas to

---

[106] On discourses about the Trinity in pre-Hispanic America, see Acosta: Bk. 5, Ch. 28: 429; Garcilaso: Bk. 2, Ch. 5, 74-75; and study by MacCormack 1991a: 269-271, 312.

[107] Sarmiento: Ch. 47, 127; Polo, "Relación de los adoratorios": 5.

[108] Torero (1990: 254) suggests the translation cat(e)-: "water" from the Culle language spoken in the Huamachuco region. Sarmiento and Albornoz write his name Cataquilla and Apu Cataquillay, which is close to Cobo's spelling.

[109] Topic, Topic & Cava 2002: 303-313.

[110] Agustinos: 25-26.

[111] Sarmiento: Ch. 61, 147. The Chiriguanos were an intermixed population of Tupi-Guarani origin that migrated from the Paraguayan Amazon. They successfully defeated Inca garrisons in the Chaco region, southeast of present-day Bolivia. See Saignes 1985; Alconini 2004.

[112] Atahualpa, on campaign against his brother, consulted the oracle in order to know the denouement of the hostilities. The oracle's answer predicted Huascar's victory, driving Atahualpa to conquer the Huamachuco territory, destroy the *waka* and burn the site. See San Pedro: 173-174; Betanzos: Pt. 2, Ch. 16, 249-251; Albornoz: 186; *Ritos y Tradiciones de Huarochirí*: Ch. 20, 247-265; and also studies by MacCormack 1991b: 130; Gose 1996b: 23.

integrate him into their official cult under the Quechuized name of Catu Illa. Yet, neither he nor Inti Illapa appears in Molina's detailed description of the Inca ceremonial calendar. For Molina, the Thunder's most venerated representation was Ch'uqilla, whose image was kept in a temple located in the Pukamarka district (Ch. 5:2), next to Wiraqucha's sanctuary (Map 3).[113] *Ch'uqi* is the Aymara word for potato, and *ch'uqilla* designates the pastors and hunters of the *puna*. In Quechua, it was also a synonym for *quri* (gold).[114] In fact, the god was closely associated with precious metals because it was believed that minerals and metals came from the bowels of volcanoes and high mountain peaks. Even today, in *qullasuyu* a great number of mining operations are located in extinct volcanoes where locals had previously worshipped the thunder god.[115]

In Inca times, Ch'uqilla's anthropomorphic figure was transported to the main plaza at the beginning of each annual celebration and was placed next to Wiraqucha and Apu P'unchaw to receive the usual offerings. However, it was during the festivities of the dry season, at *inti rayme* and following the August sowing, that the Incas addressed specific prayers to Ch'uqilla.[116] In these orations, they asked the deity to keep the yields safe from hailstorms and to send his beneficial rain. During this time he was not only believed to control hail and frost that could potentially destroy the cultivations, he was also associated with the waters originating from the upper region of the cosmos, as opposed to the subterranean rivers that fed the irrigation systems. Cobo provides an illustration of this cosmology in his detailed description of Illapa, whose domain was the upper spheres. The Incas, he writes, conceived of the Thunder deity as being formed by a multitude of stars only visible in a cloudless, limpid night sky. He held a sling and a club in his hands, indicative of his association with hunting and war. They believed that when he slung stones with his weapon, the roar of thunder rumbled (*trueno*) and the sky lit up (*relámpago*) as lightning bolts (*rayos*) rent the sky. These signs announced the coming of rain that the god generated by drawing water from the river of the Milky Way (*Mayu*) and by pouring it down upon the earth. His cult was also associated with the early and unexpected rainfall of the dry season that threatens to bring hail. When such catastrophic events occurred, designated celebrants left their village for the *puna*, where they made unspecified sacrifices in the hope of

---

113 Polo, *El Mundo*: 46; Molina: 67, 73; Sarmiento: 96.
114 Itier 1992: 1019 n. 5; González Holguín, *Vocabulario*: 117.
115 Bouysse-Cassagne & Bouysse 1984: 50.
116 Molina: 67, 72.

appeasing the god. Likewise, if the rains were late to bring moisture to the high plateaux, the Indians trekked to the highest altitudes and presented offerings to Illapa.[117] For the Altiplano communities who commonly practiced rainfed agriculture, sowing depended entirely upon the regularity of the precipitation, and therefore on the benevolence of the thunder god. In this way, the Thunder deity and Wiraqucha were both associated with the atmospheric phenomena and activities of the first half of the agricultural cycle, when they contributed complementarily to the stability of the weather conditions propitious to the early growth of the crops. From his advent on the day of the June solstice, at the height of the dry season, to his disappearance in the sea, Wiraqucha had command over the waters of the underground world that supplied irrigation canals with their precious resources. It is during this period that ditches are extensively used, when the fields need to be ploughed and sown in the absence of rain. While Wiraqucha controlled the waters of the subterranean sea, Illapa reigned over the Milky Way, the river of the upper cosmos, from which he drew the substance to rain upon the earth. He heralded the wet season's first precipitations that replenish the lakes and rivers dried by Wiraqucha quenching his thirst, but also ruled over the hail, dew, and frost that were potentially fatal to the early cultivations. Together with Wiraqucha, the Thunder god regulated water circulation and ensured that it flowed from the firmament to the underground, like the animating energy that characterizes the Andean cosmos today.[118] In this cosmology, the celestial river interacted on earth with the Vilcamayu, the river of the subterranean sun that supplied hydraulic systems. This connection explains why the two deities held complementary and sometimes syncretic natures in the traditions of the Collao region. Their mutual influence over the early agrarian cycle also explains why their temples were frequently found within the same compounds in the Cuzco region, as in the Inca capital and Caccha.

The cosmological base for the assimilation of these two deities also sheds light on the identity of the divinity who appeared to Inca Yupanqui before his decisive battle against the Chancas. Betanzos records that nearly five kilometres north of Cuzco, the young man saw a crystal tablet falling from the sky into the spring of Susurpuquio (*Suksu pukyu*, An. 5:8) as he walked along the road to Cajia Jaquijahuana to visit his father in exile.[119]

[117] Polia Meconi 1999: 226-227.

[118] About the circulation of energy in Andean cosmology, see Urton 1981: 68-69; Allen 1988: 49-54.

[119] Bauer 1998: 86-87. The name of this place recalls that of Viracocha Inca's panaca.

On this slab, Inca Yupanqui saw the image of a god who addressed him in the following terms: "Come over here my son; do not fear me because I am the Sun your father, and I know you will subdue many nations. Do remember to worship me and to unite around the sacrifices you offer in my name."[120] The god then disappeared, leaving Inca Yupanqui with the crystal tablet and the promise of a triumphant future. The omen later proved to be right, and after receiving the royal fringe the new king ordered that a statue be made of "the one he had seen there, whom he called Viracocha" and whose brightness reminded him of the Sun.[121] In his record of Pachacuti's epic, Sarmiento also notes that this figure was "like the Sun", but he mentions several times that it was indeed Wiraqucha who assisted the young Inca against the Chancas, notably by sending the army of stone warriors (*purun awqas*). Like Betanzos and Molina, Sarmiento evokes the construction of a temple to the Sun.[122] These depictions are not discrepant if we consider that Wiraqucha's association with the subterranean world, his journey from Lake Titicaca to the sea reproducing the course of the sun in the sky, and the celebration of his epic at *inti rayme*—all present the characteristics of a solar deity. It was this deity's earthly manifestation that revealed itself to the young Inca in the form of a bright sun striking like a thunderbolt. The image that Pachacuti Yupanqui ordered to be made was an anthropomorphic statue "the size of a one year-old child", resting on a low seat and covered with a rich, colourful woollen garment inset with gold.[123] Crowning its head was the *llawtu* (headband of plaited wool) to which the royal fringe was attached, and its ears were adorned with large earrings made of precious stones. It wore golden *ojotas* (sandals) on its feet; solar rays emerged from its back and shoulders, while a pair of serpents curled around its arms. Lastly, puma heads appeared on its sides, or between its legs, while its shoulders were covered with a gold feline hide.[124] In the traditions of the central sierra, the puma connotes the undomesticated world of the Altiplano and is associated with the hail, snow, and early rainfalls controlled by the Thunder god. In view of these elements,

---

120 Molina: 60. My translation.

121 Betanzos: Pt. 1, Ch. 11, 50.

122 Sarmiento: Ch. 27-28, 86-92; Molina: 60-61. Cobo (Bk. 12, Ch. 12, 78) paraphrases Molina and also identifies this vision with the Sun.

123 Betanzos: Pt. 1, Ch. 11, 51-52. The seat is *tiyana* in Quechua. The rare Inca figurines of metal work can give an idea of the deity's exact position, which the documents do not explain. Seated figures were represented squatting, the two knees against the torso and the hands resting on the stomach.

124 Molina: 60-61.

the ensemble would have represented Wiraqucha in his triumphal aspect, endowed with the warlike attributes of Thunder, as he appeared before the young Inca. Together, these two deities epitomized the forces of the first half of the year that had coalesced to assist the future Hanan king in defeating his enemies. Both gods had their temples in the Upper part of the Inca capital, which is all the more significant since the early sources that trace the Incas' origin back to Lake Titicaca only record the Hanan rulers as descendants of the Sun Wiraqucha. It was precisely the panacas of these Hanan lords that supervised the maintenance of Cuzco's irrigation system and conducted the annual sacrifices directed at the ditch waters where Wiraqucha had established his domain.[125] He appears as the earthly manifestation of the vigorous Sun that generated and protected the Hanan lineages and the outsider-affines who came to power owing to their personal merit—the One who perpetuated the fertility of their lands and implemented their sovereignty. *Inti rayme* commemorated his mythical journey on earth together with the emergence of the species while celebrating the original migration of the Hanan ancestors to Cuzco. Now, did this tradition differ from the narratives locating the emergence of the Ayar siblings from a cave in the vicinity of the Cuzco valley? To provide the beginning of an answer to this question, we shall examine the Paqariq Tampu narratives in view of the chroniclers' descriptions of the god they identify as the Sun. Colonial sources evoke this supernatural being under two different names: P'unchaw and Inti.

---

[125] Zuidema 1989a: 455-487.

CHAPTER FIVE

EPICS OF THE OLD SUN

Alongside the narratives that trace the Incas' origin to Lake Titicaca, another corpus of texts stipulates that they emerged from a cave known as Paqariq Tampu or Tampu T'uqu ("the opening/resting place from where the sun rises"), presumably situated in the Pacariqtambo district, several kilometres south of Cuzco.[1] Scholars often restricted the analysis of the Inca mythical origin to the study of the Paqariq Tampu narrative cycle, thereby leaving unexamined the stories locating their emergence in the Collao province.[2] It is generally assumed that this last narrative cycle is either a Spanish elaboration assembled from elements of Wiraqucha's epic or a late Inca appropriation of the lore held by the once-renowned civilizations of the Titicaca Basin.[3] However, two distinct, albeit contemporaneous, Inca mythical cycles come to light by contrasting these narratives. Each one displayed structural dichotomies placing in opposition outsiders and newcomers on the one hand, and autochthonous and first settlers on the other. In large parts of the Andes past and present, dual organizations refer to similar dichotomies to account for the antagonisms placing the moieties in opposition to each other. Gutiérrez records a foundation story that particularly emphasizes these features. He reports that those who emerged with Manco Sapaca from Lake Titicaca settled in Hatun Collao where they lived for several generations, fighting against the great chieftains of Cuzco time and time again. They eventually migrated towards the north under the leadership of Tupa Inca Yupanqui and founded "a city very close to Cuzco, on a slope nearby a water stream, and called it Hanan Cuzco."[4] Once settled in this place, they defeated the unnamed lord kuraka of Cuzco and conquered his city. This account highlights the longstanding conflicts between the early leaders of the respective moieties and situates the native place of the dynasty founders outside the Inca

---

[1] Scholars have long speculated on the actual location of this site. See Urton 1990; Bauer 1992.

[2] See for example Rostworowski 1999: 12-15; Julien 2000: 233-244; D'Altroy 2002: 49-52.

[3] Conrad & Demarest 1984: 94-95; Urton 1990: 3; Pärssinen 1992: 61; Salles-Reese 1997: 97-99; Niles 1999: 74; Meyers 2002; Cummins 2002a: 60-61.

[4] Gutiérrez: Ch. 50, 213.

heartland. In contrast, the other body of texts, with which this chapter is concerned, locates the Incas' primordial migration within the boundaries of the Cuzco region, an area peopled by nations whose conquests were traditionally attributed to the Hurin lords.

The Ayar Siblings' Journey from Paqariq Tampu

The chroniclers who record this last tradition in its most detailed form are the authors of the *Discurso*, Betanzos, Cieza, Santo Tomás (1560), Sarmiento, Cabello, and Murúa. The main protagonists of the story are three or four siblings named Ayar who emerged together with their sister-wives and left their place of birth in search of fertile lands to settle. Sarmiento is the only chronicler to indicate that several ethnic groups, which had emerged from separate apertures, joined their enterprise. They were the Chawin Cuzco Ayllu, the Arayraka Ayllu, the Tarpuntay Ayllu, the Waqaytaki Ayllu, and the Sañuq Ayllu, which came to populate the Hanan moiety. From the other aperture proceeded the Maras Ayllu, the Quwikusa Ayllu, the Maska Ayllu, and the Uru Ayllu, which were associated later with the Hurin moiety. Together, they and the siblings stopped in different settlements along their journey: Huayna Kancha, then Tampuquiro, where they stayed for some years after Sinchi Roca was born, and finally Haysquisro. There, the siblings became aware that the strongest amongst them, Ayar Awqa (or sometimes Ayar Kachi), had become a threat. They sent him back to their *paqarina* under a false pretext and had him immured inside the cave by an accomplice. Having done that, the remaining brothers left for Quirirmanta, a town located at the foot of Huanacauri hill, where they discovered the flourishing cult of a local deity. One of them, Ayar Uchu or Ayar Kachi, resolved to take hold of this image and sat upon the *waka*. As a punishment for this sacrilege, he was immediately transformed into stone.[5] Before his fate was definitely sealed, however, he asked his siblings to pay him their respects each time they held a festival. He then decreed the course of the *warachiku* ritual and henceforth became intercessor of the Sun, the one who would reveal divine will to his descendants. Endowed with this new authority, he named Ayar Manco ruler of all the nations they would eventually subdue.

---

[5] Sarmiento is the only chronicler of this group to attribute this act to Ayar Uchu. Cabello and Murúa, like Cieza (Pt. 2, Ch. 7, 16-17) before them, identify the offender as Ayar Kachi.

After spending two more years in a nearby town called Matahua, the time came for the two remaining brothers to fulfil the wishes of the Sun god. Having reached the top of a hill, Ayar Manco indicated a site in the valley to his remaining male companion, Ayar Awqa, and addressed him in these terms: "Brother! You remember how it was arranged between us, that you would go and take possession of the land where we are to settle. Well, look at that stone, fly over there (for they say he had grown wings) and by seating yourself, take possession of the site where that boundary stone appears, because later we will come to settle and reside there."[6] Ayar Awqa complied with his sibling's command and reached the site, where he was petrified for eternity. It was in this location that the Incas later built the temple of Coricancha. Sarmiento reports that this location was known thereafter as "Ayar Auca cuzco huanca" in the *lengua particular*, which he translates as "Ayar Auca marble boundary stone of possession."[7] Now alone with his brothers' wives, Ayar Manco initiated a march to the site he was destined to occupy, at the heart of the Cuzco valley, where several ethnic groups already lived. In order to ensure the dominion of their descendants over these fertile lands, Mama Huaco, the wife of the petrified Ayar Kachi, forced her way through the dale and hunted down its inhabitants. Having caught a local Hualla Indian, she killed him in front of his people, slit open his entrails, and tore out his heart and lungs, which she put to her mouth. She then blew into these organs to inflate them, causing the frightened natives to flee.[8]

Notwithstanding the violence of this act, the newcomers' cruelty only partly explains the desertion of early settlers. In fact, Betanzos clearly reveals that Mama Huaco's deed was an act of divination, a *kallpa rikuy* sacrifice that proved to locals that a powerful deity assisted their aggressors. This divine protection gave the Ayars an incredible force called *kallpa*, which referred to the physical power of living beings as well as the might of supernatural beings.[9] *Kallpa* was the forcefulness with which they accomplished extraordinary deeds, and today it is a concept closely linked to the food substances that humans ingest.[10] The movement of vital energies, like the *kama*, fashioned the Andean cosmos. They circulated from potent spiritual entities to all living creatures on earth: men, animals, and plants.

---

6 Sarmiento: 59.

7 Ibid.

8 Betanzos: Pt. 1, Ch. 4, 20; Sarmiento: 60.

9 Taylor in *Ritos y Tradiciones de Huarochirí* 1999: xxiii; González Holguín, *Vocabulario*: 44; Santo Tomás, *Lexicón*: 245.

10 For a cotemporary study on *kallpa*, see Riva González 2005.

These essences held material qualities that were made visible in the organisms they animated. For these reasons, every creature of exceptional strength, or those who revealed unusual corporal features, such as dwarfs, twins, humpbacks, or deformed beings, were believed to be gifted with extraordinary abilities transmitted by their ancestors.[11] These vital forces circulated in a fluctuating way through living creatures, potentially withdrawing from one body to infuse another. Andean rituals were orientated towards the maintenance of these energies because they alone ensured the development of beings, crops, and the propitious alternation of the seasons. In order to verify that they had been appropriately restored after an offering, Andean ministers performed the *kallpa rikuy*, which involved interpreting the entrails of sacrificed llamas or guinea pigs by blowing into their lungs. It was a similar augury that Mama Huaco had performed when she and her siblings entered the valley, sacrificing and disembowelling a human victim in honour of the Sun, and displaying to all and sundry tangible proof that the god animated his offspring with a supernatural force. Her violent act was as much a demonstration that the migrants had come to Cuzco under the aegis of a prominent divinity as it was a display of their warlike abilities. Following this event, Manco Capac and his sisters gradually subdued the local peoples, and several months later they finally settled on land between the rivers Tullumayu and Saphi.[12]

Appropriating People and Resources

At different stages throughout their journey, the Ayars appropriated (*tomar posesión*) ancient cults and lands for their own benefit by "seating" themselves on them, whereupon they were changed into stone. The same practice is evoked in different terms in a document of the Jauja province (1582), which states that "in remote times, the valorous ancestors who left in search of new arable lands and had won them at war, marked these with a stone, different from the others, and asked their successors to remember them [in that location]."[13] Ramírez observes that these stone *mojones* were not, strictly speaking, markers of territory or boundary units, but were commemorative monuments that preserved the memory of ancestral warlords.[14] Sarmiento, in his record of the Incas' origin, calls these sculpted

[11] Arriaga: 63; Acosta: Bk. 5, Ch. 25, 260; Hernández Príncipe, in Duviols 1986: 485-507.

[12] Cieza: Pt. 2, Ch. 6-8, 13-22; Cabello: 260-264; Sarmiento: Ch. 10-13, 50-61.

[13] *Relaciones Geográficas de Indias*, in Duviols 1979b: 14. In the sixteenth-century, the word *tierras* referred to either land or peoples.

[14] Ramírez 2005, 32-46.

idols *wanka*, which was a form of ancestor cult still widely extant in late colonial times. A review of the Extirpation of idolatry materials shows how the Inca narrative falls within a pan-Andean tradition about first settlers. In these documents, the *wanka* is a powerful and valorous man who came from another realm to protect a village, its inhabitants, and their land resources. Once petrified, the *wanka* held two main functions according to its location in the community's territory. When situated in the middle of a *chakra*, it ensured the fertility of its soil and received offerings at the beginning of every agrarian work.[15] Father Arriaga (1621) notes that the Indians "also call it *chakrayuq* which means lord of the *chakra* because they think that this *chakra* belonged to the huaca [here idol] and that it is in charge of its growth, and as such they revere it and offer sacrifices specially in the sowing season."[16] When located within the urban centre, the *wanka* was worshipped as the protector of the people and lineage he founded. He was known as the *markayuq*, a function he shared with other representations of the ayllus' forefather, such as the *qunupa* and the mummies.[17]

The monolith *wanka* thus represented both the warlike and fertilizing functions of this ancestor whose boulder received particular attention during the main agrarian labours and, in certain communities, in mating rites and female defloration.[18] The same aspects—fertility and martial valour—were at the core of the Inca initiation festivities that the Ayar brother petrified in Huanacauri had decreed and supervised every subsequent year. The function of this deity, as well as its origins in the era of early Inca settlement, suggest that Huanacauri primarily held the status of a *wanka*. In that regard, its far distant provenance and its link with the Sun are reminiscent of another narrative recounting the coming of the first *wanka*-ancestors in the town of Maray (Chancay). This late seventeenth-century account, published in part by Guillermo Cock and Mary Eileen Doyle, tells us that the main *markayuq* of Maray had originated from a distant land. In his lifetime, this *wanka*, known as Huacra Yaru, had travelled with his brothers and one

---

15 Arriaga: 128.

16 Arriaga: 37.

17 Arriaga: 128; Duviols 1986: 279, 359-362, 383-384; Polia Meconi 1999: 176. For Duviols, the *markayuq* is the mythical founder of the clan. Mills (1997: 59-60) reports that in the eighteenth-century, in the town of La Ascención de Guanza, "an annually elected official called a *marcayuq* was said to have led a ceremony that involved dancing and drums". Ramírez (2006b: 362-263) suggests that *markayuq* and *llaqtayuq* were the back translations of "dueño de indios" and referred to "the caregiving individual responsible for his followers who owe him labor".

18 Duviols 1979b; Griffiths 1996: 107, 206-208, 318; Mills 1997: 289.

sister before settling in Maray, where he officiated as the Sun priest. Upon dying, he turned into a monolith and became the intercessor of the Sun deity for his people, much like the shrine of Huanacauri in the Inca tradition. One of his brothers, called Punchao Jirca (the Daylight Sun Rock), paired with him and became the *wanka* to receive the main offerings at the start of the sowing season.[19] Their brother, Rupay Jirca (the Rock of the dry season Sun), and their sister, the spring of Chuchu (the Twin dedicated to the Thunder), were both associated with the god of lightning. In the colonial period, the four ayllus of Maray venerated these ancestors as the offspring of the Sun and the Thunder. They believed that they had generated antagonist moieties among human communities and ruled over the agricultural cycle in contemporary times.[20] Therefore, it was not uncommon for the same community to worship several *wanka*, which were transcended by a more prominent one regarded by all as their common ancestor.[21]

The shrine of Huanacauri appears to have held a similar status among the Incas, even though the early conquistadors proved oblivious to its importance when they first plundered the site. During this pillage, they seized the lavish gold and silver offerings but left intact the rough monolith they did not suspect was the object of this cult. Scholars also have paid little attention to the function it held in the Inca pantheon.[22] This may be in part because the Spanish themselves perceived Huanacauri as a heathen idol of minor importance, one they did not credit with a high degree of conceptualization, and therefore a figure which did not conform to the same classification as the major divinities worshipped by the Inca elite, such as the "Maker", the Sun, or the Thunder. As opposed to these gods, the influence of Huanacauri had been restricted to the Cuzco region. Located almost eleven kilometres southeast of the city, the lore and belief surrounding his cult faded away rapidly once the Spanish eventually laid hands on his image. Despite these circumstances, slivers of information offer a different picture of Huanacauri's importance, starting with his presence during the main celebrations of the Inca calendar. We learn from Molina that his effigy was warmed and fed with other major deities and the royal mummies at the purification ritual of *sitwa*.[23] It also watched over

---

19 In the Quechua dialects of Ancash and Huanaco, *jirca* means mountain, stone. Present-day rituals dedicated to the local *jirca* involve the headcount, shearing and marking of the cattle.

20 Cock and Doyle 1979: 59-62.

21 Duviols 1979b: 10-12.

22 Exceptions are Szemiñski 1991; Ziółkowski 1996: 65-71; Julien 2000: 276-286.

23 Molina: 76-77.

the fruitful outcome of the novices' trials on *warachiku*, when the young boys presented him with offerings and orations in exchange for war weapons. The presence of his image at war, "especially when the King went in person,"[24] confirms his distinctive warlike aspect. Cieza, Molina, and Sarmiento all claim that Huanacauri was a *waka* of prominence in the Inca sacred landscape. To illustrate this point, they evoke the precious offerings that all newly appointed rulers offered to the shrine and argue that he was one of the rare divinities to have received human sacrifices at *qhapaq hucha*.[25] Cieza even lists his temple as the empire's second shrine of importance after Coricancha. According to Julien, his prestige was such that there existed various eponymous hills scattered around Cuzco that the Incas visited during *qhapaq rayme*.[26]

According to the Inca tradition, Huanacauri's prestigious position originated from the primordial events that brought the Ayar siblings to the valley, when they appropriated, distributed, and protected the newly conquered lands in the name of the Sun. In this story, both he and Ayar Awqa evince a particular relation with the Sun and with the geographical division of Hurin Cuzco: the former because he was petrified on the site of the future temple dedicated to the Sun in the Lower part of the city, and the latter through his position as the particular intercessor for this god. Many chronicles insist on this last point. They describe Huanacauri as the messenger of the deity, the one who, in Cieza's words, always "prayed to God on behalf of his siblings"; he who "would remain as an idol [and] ask their father the Sun to protect them, increase their number, give them children, and send them good weather."[27] Namely, the authors of the *Discurso* identify this Sun with P'unchaw and present Huanacauri as his messenger and *segunda persona*.[28] In their variant of the myth, two priests of P'unchaw who are tutors of the young Manco Capac enter the Cuzco valley to announce to its residents the imminent coming of the man fathered by the Sun. They enjoin the Indians to receive him as their new lord and to worship P'unchaw as their god. If they show resistance, "the Sun will send a great plague on them, which will kill them all, leaving no one alive; because he is resolved to destroy the world and to punish the mankind like the

---

24 Polo, "Relación de los adoratorios": 31.

25 Cieza: Pt. 2, Ch. 28: 83-84; Molina: 126; Sarmiento: Ch. 31, 96-97; Ch. 56, 139.

26 Julien 2000: 276-286.

27 Cieza: Pt. 2, Ch. 7, 16; Betanzos: Pt. 1, Ch. 4; translation by Hamilton & Buchanan 1996: 15.

28 *Discurso*: 24, 26-27.

Deluge, [and] then to re-populate the earth with new people."[29] The priests conclude by saying that P'unchaw had sent his servant and friend Huanacauri to the people of the Cuzco valley so they could pay homage to the Sun through him.

Despite the peculiarity of the *Discurso*,[30] it is possible to relate its account of the Inca migration to the small corpus of texts transcribed by Sarmiento, Murúa, and Cabello, which mention the apparition of a rainbow (*k'uychi*) over Huanacauri hill as the siblings entered the valley.[31] For the small group of migrants, this apparition was a propitious sign. It foretold that "the world would not be destroyed by a flood."[32] Andean people have long recognized that the rainbow presages the simultaneous appearance of the sun with the precipitations and have thus associated it with the rainy season. Garcilaso explains that the Incas naturally "came to the conclusion that [the rainbow] proceeded from the Sun."[33] It was also believed to augur a period of transition between two eras or seasons, as with the advent of the Incas in the Cuzco valley, because it only forms under damp but sunny conditions.[34] In colonial times and even today, it is considered to be a dual figure in the shape of a two-headed snake, either white or multicoloured, propitious or malevolent.[35] When harmful, it may worm its way into human bodies and unsuspectedly propagate rainbow diseases (*chirapa unquy*), which derive from the dangerous association of the rain with the sun.[36] In the Inca cosmology, the serpent (*amaru*) was also a dichotomous being. On the one hand, it was closely linked to the Thunder god. It was his chthonic envoy at the beginning of the dry season, when it penetrates into the ground to hibernate, only to resurface if Illapa creates an earthquake.[37] Urton reports that today, communities of the Cuzco region recount that the dark cloud constellation of the snake lives underground during this period, in the subterranean world inhabited by the sun of the dry season. On the other hand, the *amaru* was the celestial emissary

---

29 Ibid.: 27.

30 See Chapter 1, n. 7.

31 *Discurso*: 24, 26-29; Sarmiento: Ch. 12, 56; Cabello: 260-264, Murúa, *Historia general*: Bk. 1, Ch. 2, 41; Pachacuti Yamqui: f.6v., 194.

32 Sarmiento: Ch. 12, 56.

33 Garcilaso: Bk. 3, Ch. 21, 174-175.

34 Molinié 1986: 154; MacCormack 1988: 998-1003; Robin 1997: 378, 384.

35 Muñoz Bernand 1999: 150-165; García Escudero 2007.

36 Polo, "Tratado": Ch. 5, 13-14; Murúa, *Historia general*: Bk. 2, Ch. 34, 426; Guaman Poma: 255 [257], 229; 280 [282], 253; Cobo: Pt. 2, Bk. 13, Ch. 38: 233; Duviols 1986: 193, 228, 269; Allen 1988: 16; Larme 1998: 1008-1009.

37 Allen 1988: 53; Urton 1981: 88-90, 177-180.

of P'unchaw when the new Sun brought the rains with his coming. During the rule of the Daylight god, it was a multicoloured *k'uychi* replenished with water. Its celestial counterpart, the serpent dark cloud, ruled the sky of the wet season at night before disappearing from the firmament in February. A similar ideology shapes the lowland Peruvian lore of the cosmological serpents, *Yaku Mama* (Mother Water) and *Sach'a Mama* (Mother Forest), which travel through the different plans of the universe.[38] The former embodies water regulation during the first half of the year. It is said to emerge from the undergrounds in the form of a river and to rise to the heavens as thunderbolts. On the other hand, *Sach'a Mama* is linked to the fertility of the rains. It is conceived as a bicephalous snake that transforms itself into a rainbow when it rises to the heavens. Likewise, the rainbow rising above Huanacauri hill at the arrival of the Ayar siblings not only foretold the beginning of a new era when the sun showed through the rain clouds, it specifically announced the start of the second half of the annual agricultural cycle, and with it the advent of P'unchaw's reign on earth. The cosmology emerging from this picture opposes P'unchaw, the daylight Sun ruler of the fertilizing rains, against Wiraqucha, the subterranean Sun of the dry season.

The Sun God of the Chronicles

P'unchaw was not the only appellation of the Sun deity. Primary sources also commonly use the term Inti, although neither of these names appears in the documents recorded during the first decades following the conquest. With the exception of the *Discurso*, in which P'unchaw is clearly identified as the Sun, early sources refer to this god by its Spanish designation: *Sol*. The same texts, however, mention several terminologies that describe objects and beings in relation to the god, which are constructed around the vocable *inti*. For Betanzos and Segovia, for example, the Sapa Inca was *Intip churin,* son of the Sun, while they called the high priest *Intip yanan*, the Sun's servant. Cieza also indicates that the Coricancha temple was known as "Indequaxi" (*Intip wasin*), which was adjacent to the *aklla wasi* that sheltered the organization of *Intip warmin*, the Sun's wives.[39] For

---

[38] Valcárcel 1967: 140-141. In the Ucayali basin, these two cosmological serpents are primary sources of disease alike the rainbow. See also Gow 1991: 79, 248; Thomas & Humphrey 1994: 94-95.

[39] Betanzos: Pt. 1, Ch. 17, 83; Segovia: 74, 76; Sarmiento: Ch. 29, 93; Cieza: Pt. 2, Ch. 27, 79; Santillán: 113.

Arriaga, Inti was the solar god's "proper name,"[40] and an entity closely related to P'unchaw in the transcripts of the Inca orations to the Sun and in the late documents of idolatry. In this material, the two names are often recorded in semantic pairs (e.g., *Inti yaya, P'unchaw yaya*).[41] Early Quechua lexicons, however, clearly distinguish between the two words. For Santo Tomás and González Holguín, *inti* is "the sun, the planet". Its related glosses refer to the location of the star in the sky, to its warmth and luminosity.[42] Garcilaso's *Comentarios* backs up this definition and only makes use of the term in reference to the star in a chapter discussing Inca astronomy.[43] On the other hand, dictionaries translate *p'unchaw* as "the day and the sun". It corresponds more specifically to the diurnal light, to the time of day during which the sun diffuses its rays as opposed to nighttime.[44]

Records of the prosecutions against idolatry add another dimension to this issue. In these documents, P'unchaw is depicted as an ancestor who travelled in search of fertile lands and eventually joined an existing community whose members worshipped him thereafter as their forefather. The *wanka* and *mallki* obliterated by missionaries often bore the name of P'unchaw but were never called anything after Inti, who did not become the hero of any epic narrative as such. Throughout the seventeenth century, the cults of *wanka* and *mallki* persisted among the populations of the archdiocese of Lima and further south, in the Huamanga and Charcas dioceses. Church archives describe with profusion the regional cults of the anthropomorphic image of P'unchaw but make no mention of an idol representing Inti. Similarly, none of the major chronicles of the sixteenth century suggests that there existed a statue of the Sun called or dedicated to Inti in Cuzco. The word appears instead as the designation of the founding Sun, the origin and matrix of the star's manifestation on earth embodied in P'unchaw. In the early sources that depict Inca celebrations, the Sun's imaging is only evoked under the name of P'unchaw. Molina mentions the existence of three figures representing the Sun: P'unchaw Inka (the Inca Daylight), Apu P'unchaw (the lord of Daylight), which featured "the principal they had in their temple", and Wayna P'unchaw (the young Daylight).[45] Molina makes no reference to a representation of Inti. Around 1600, the

40 Arriaga: 26.
41 Molina: 91; Itier 1992: 1034, 1037.
42 Santo Tomás, *Lexicón*: 301; González Holguín, *Vocabulario*: 369.
43 Garcilaso: Bk. 2, Ch. 21, 108-109.
44 Santo Tomás, *Lexicón*: 343; González Holguín, *Vocabulario*: 295.
45 Molina: 67, 77, 110.

Jesuit Antonio de Vega still attested that the main Inca cult in Cuzco was that of P'unchaw, the "lord of the day and creator of light, of the sun and the stars and of all other things" whose image was kept in Coricancha.[46] In those years, the only discordant voice came from Acosta, who claimed that the Incas worshipped several idols of the Sun god, P'unchaw being only one of its representations. His statement, however, should be approached with caution, given Acosta's implication in the elaboration of the material of the Third Council of Lima, in which the miscellaneous beliefs of Andean people are regrouped under the common denomination of idolatry.

From these testimonies, it is almost impossible to ascertain which one of P'unchaw's images attended a specific festival. Even Molina does not linger over this fundamental aspect of the rituals, and only in three instances does he provide the much sought-after information. He first mentions the offerings that P'unchaw Inka, Wiraqucha, and the Thunder received at *inti rayme*, the "celebrations of the Sun" that, he avers, opened the Inca ceremonial calendar. Later in the year, the deeds of the same figure were celebrated during the cleansing ritual of *sitwa*, held between August and September. For this event, the Inca nobility directed several orations to P'unchaw Inka in which he is evoked with aspects of the warlike and triumphant Sun. He is the one who "assured victory to the Cuzco and the Tambo", and he who protects the city residents from ever being displaced.[47] *Sitwa* opened and closed with several offerings to Wiraqucha, the Thunder, and the Sun called Apu P'unchaw.[48] This aspect of the Daylight god attended the festival in the company of his priesthood and his two godly wives, Inca Ocllo and Pallpa Ocllo, who had previously escorted the procession of *inti rayme* to Mantucalla. A third wife and sister of the Sun god also sat beside him in the person of an Inca noblewoman called *quya paqhsa*. She was later sacrificed in the name of her divine husband. Finally, during the initiation of young Incas at *qhapaq rayme*, an unidentified image of the Sun was carried to Haucaypata, where it received the usual offerings. The novices then received rich garments in the name of this deity before they swore allegiance to the Sun, to Wiraqucha, and to their king. Finally, on the

---

[46] Antonio de Vega, in Mateos: 8-9.

[47] Molina: 67, 91-92. This oration evokes the Incas under the names of Cuzco and Tampu. Ramírez (2005: 13-56) argues that the term "cuzco" referred to the Inca king in power. Tampu may well designates the Ayar siblings' *paqarina*. If these interpretations are correct, these two words can be associated with the two moieties given that each sovereign belonged to the Hanan moiety once he attained power, while I associate Paqariq Tampu with a tradition of the Hurin moiety.

[48] Molina: 77.

twenty-third day of the same month, the image of Wayna P'unchaw was carried to the temple of Puquín at the top of a hill "some three arquebus shots from Cuzco", where the Incas went about their sacrifices for the multiplication of beings and foodstuffs.[49]

An overview of P'unchaw's attributes *per se* is limited because Molina is the only one to evoke the different representations of the Sun god in the context of Inca rituals. Material descriptions of these images are also very scant. Only a few witnesses testified having seen a particular effigy of the Sun before it was sent to Spain in the early seventeenth century to be melted down.[50] Among these chroniclers, Antonio de Vega is the most prolix. He reveals that the main image of P'unchaw was an anthropomorphic figure made of gold and adorned with large pendant earrings. The god sat on a stool (*tiyana*) of the same precious metal, in which was concealed a casket in the shape of a sugarloaf. A gold representation of the Sun's heart was enclosed within, together with remains of the dead kings' hearts.[51] When the image left its temple to attend the ceremonies, a set of two spears along with a club and a golden axe flanked it.[52] It may have been the same image that conquistador Ruiz de Arce saw in the Coricancha only a few years after Pizarro's troops had entered Cuzco for the first time. He describes the image of a god placed in the middle of the courtyard, beside a golden *tiyana*. Each day at noon, the Inca celebrants uncovered the stool where the Sun was believed to sit and laid their offerings at its feet, burning meat and pouring *chicha* onto an *ushnu*.[53] It was not long before the Incas endeavoured to hide the statue, which remained in the possession of the Vilcabamba rebels until 1572. At this date, Francisco de Toledo took hold of it once he had suppressed the rebellion led by Tupac Amaru. The statue was deprived of a portion of its ornaments that the viceroy dispensed among his men in reward for their hardiness in quashing the Inca insurgence. The booty was "a form of golden medallions that would shine when struck by the sun, in such a way that one could never see the idol itself, but only the reflected brilliance of these medallions."[54] Such adornments, the number of which is unclear, share several aspects in common with another description of the Sun's image that primary sources mention for the

[49] Molina: 110.
[50] Julien 2002: 709-715. See also Duviols 1976a.
[51] Levillier 1924, IV: 344-345; Acosta: Bk. 5, Ch. 12, 236-237; Antonio de Vega, in Mateos 1944: 8-9.
[52] Pizarro: 67.
[53] MacCormack 1991a: 65-66.
[54] Toledo, in Hemming 1993: 306.

first time at a date coinciding with the Spanish seizure of P'unchaw's statue. In those years, three chroniclers, Gutiérrez, Garcilaso, and Cobo, evoke the existence of a singular representation of the Sun that had hitherto never been listed. Their descriptions present such close similarities that they leave little doubt about the identity of their common source: the conquistador Mansio Sierra de Leguizamo, who played an active part in quashing the rebellion of Tupac Amaru. The controversial drawing of the Coricancha altar by the late chronicler Pachacuti Yamqui also evokes this image.

According to our three chroniclers, the image of the Sun was a round or oval plate, the size of a shield and a finger thick, without any relief, and perfectly polished. In its centre was etched a human face invested with rays. It allegedly hung on the wall of the deity's altar in such a way that it reflected the dazzling light of the rising sun when it shone through the door of the room: "the rays were mirrored and reflected with such brightness, that it looked like the Sun [itself]."[55] The first mention of this figure appears in Mansio's will (18 September 1589) in which he pretends to have owned the gold image.[56] Gutiérrez, Garcilaso, and Cobo accept his declaration, but also report that Mansio gambled his booty and lost it on the very same day he had acquired it. It is inconceivable, however, that the conquistador owned or even saw this figure before the 1570s, for two reasons. First, as John Hemming notes, Pizarro strictly forbade his soldiers to retain precious objects before they were melted down.[57] In the unlikely event that Mansio had overlooked this directive, he would not have been so foolish as to mention his bypass in his will. Second, the first Spaniards to enter Cuzco all admitted their great disappointment at never having discovered the idol of the Sun. Segovia notes in 1553 that "the Indians hid it in such a way that up until now it has not been discovered; they said that the Inca rebel has it with him."[58] If Mansio ever saw or owned one of the Sun's figures, it would have been after the year 1572, the date at which P'unchaw was seized and its golden disks distributed to the viceroy's men. It is no coincidence that these events coincided with the first descriptions of the Sun's image in the form of a disk, a description endlessly repeated in later documents and scholarly works. Mansio's account, which many historians deem fictitious, can be associated with the Vilcabamba events. He was one of just

---

55 Gutiérrez: Ch. 50, 216; Garcilaso: Bk. 3, Ch. 20, 172; Cobo: T. 2, Bk. 13, Ch. 5, 157.
56 Stirling (1999) includes an English translation of Mansio's will.
57 Hemming 1993: 131.
58 Segovia: 75; Pizarro: 184.

three conquistadors of the first generation to have participated in the triumphant expedition that put an end to the Inca rebellion in 1572, and was among those who had the privilege of seeing P'unchaw's statue before it was sent off to Spain. It is most likely that he was among the soldiers who were given the solar disks in reward for their collaboration. If this is true, Mansio's declarations were founded on reliable grounds. Although he did not own a specific idol of the Incas, he came to possess an object that had adorned the Sun's image.[59]

The only late chronicler to mention the existence of three *bultos* of Inti in Coricancha is Father Cobo. For him, these images were kept not far from the Inca main deity, P'unchaw, which was "their principal simulacrum and major devotion."[60] The three images would have represented Apu Inti (the Lord Sun), Churi Inti (the Sun's son), and Inti Wawqi (the Sun's brother), each of which had its distinctive cohort of ministers. They were covered with thick woollen cloths, had large earrings, and wore the *llawtu*. Despite their precision, these data call for critical examination. They are late and unprecedented descriptions that Cobo could only have collected either from a native informant who had held original material that no other chronicler had uncovered in the century following the conquest, or from the lost manuscript of an earlier historian. Cobo himself indicates that he extracted part of his information from the work of Polo, whose only surviving text evokes the solar triad in almost similar terms. It tells that three figures of the Sun—his Father, Son, and Brother—were brought out in Cuzco's square in order to receive offerings at the *qhapaq rayme* festival.[61] Polo did not speak Quechua and his text does not provide native names for these representations that could have supported their assimilation with the late denominations of Apu Inti, Churi Inti, and Inti Wawqi. Polo's report seems more to be approximate translations of Molina's description

[59] For Duviols (1976b), another interpretation on the provenance of the oval imaging of the Sun is possible. He mentions the work of Dominican Reginaldo de Lizárraga who benefited from first-hand information from his companions of the convent of Santo Domingo in Coricancha to write his *Breve Descripción del Perú* (c. 1603-1609). In this temple, he claims, was a large and deep octagonal basin in which the Incas poured the chicha they offered to the Sun. "The opening of this basin was covered by a golden plate on which the sun was engraved. When the Spaniards entered this city, fortune awarded it to one of the conquistadors that I knew, a man named Mansio Sierra" (Lizárraga: Ch. 80, 61). Here again, the solar plate does not refer to the main figure of the Sun, although it is unlikely that the veteran owned it at such an early date because of Pizarro's proscriptions. Yet, Mansio may have seen this object and P'unchwaw's disks at different times during his service in Peru, which then inspired the oval description of the Sun idol.

[60] Cobo: T. 2, Bk. 13, Ch. 5, 157.

[61] Polo, "Tratado": 18.

of Apu P'unchaw, Wayna P'unchaw, and Inka P'unchaw. These two chroniclers were contemporaries. They gathered their information on Inca religion around the same years, but Molina had the advantage over Polo of being fluent in Quechua, which gave him a more rigorous insight into the practices he described.

Cobo reports a tradition on the origin of the three *bultos* of the Sun that casts more light on their identity. He writes: "Some say that these three figures were made because once they saw three suns in the sky; others say that one was the Sun itself, the other the Day, and the third was the virtue to create."[62] The first reference on that list is reminiscent of Arriaga's translation of Inti as the "proper noun" of the Sun. The second corresponds faithfully to the function of P'unchaw as diurnal light. The third description, finally, can be associated with Wiraqucha: he is the Sun that animates people and cultivations on earth, the *pachayachachiq* that possesses the "virtue to create" (*kamaq*). Cobo's description of Inti's figures takes on a different dimension in light of this explanation. The Sun's triad as he describes it is a reconstruction from earlier primary sources, valid but fragmentary, which actually refer to the three expressions of the Inca Sun: Inti, P'unchaw, and Wiraqucha. These, in turn, echo the functions of the sun's three images as evoked by Molina, in which Inti paired with the Lord of Daylight (Apu P'unchaw), P'unchaw with the Young Daylight, and Wiraqucha with the Inka Daylight.[63] This last association is all the more relevant because the Collao deity was believed to have animated the founder of the royal descent group. It explains why Inka P'unchaw's only apparition occurred at *inti rayme*, when Wiraqucha's deeds on earth were commemorated together with the emergence of the Inca ancestors. Inti thus applied to the deity's generic name, its proper noun as appropriately stated by Arriaga, to the Sun to which the Incas directed offerings and tokens, and a concept that enclosed the two antithetic expressions of the solar star on earth. P'unchaw was the first of these antagonistic manifestations. He was the god who animated the Ayar siblings out of the cave of Paqariq Tampu, the entity who accomplished his deeds on earth under the eyes of the living communities to become their *wanka*. He represented the function of the diurnal sun over human activities, the one that dispenses its light over the cultivations as they ripen during the wet season. His task was directly opposed to the workings of Wiraqucha, the god of the subterranean

---

62 Cobo: T. 2, Bk. 13, Ch. 5: 157.

63 Demarest (1981: 22-25) already perceived these associations, but his conclusion diverges from my argument.

darkness and the second manifestation of the solar divinity. Wiraqucha is associated with the rising of the Inca ancestors from Lake Titicaca, and he is the one who presided over the circulation of the subterranean waters that fed irrigated agriculture.[64]

The Moon

While chroniclers do not link Wiraqucha with any female deity and assert sometimes that he embodied both genders,[65] they often pair P'unchaw with several wives, including Inca Ocllo and Pallpa Ocllo. Besides these two figures, for which no information has been recorded, the Moon certainly predominates at his side. She and the Daylight Sun were intrinsically linked through their common role as regulators of the rainy season, and via a mythological corpus that describe them as husband and wife. Pre-Hispanic beliefs about lunar eclipses illustrate the close relation uniting these two gods. When this phenomenon occurred, Andean people reckoned that the moon would perish, and with her the light of day, thus bringing about an eternal realm of darkness on earth. Obviously, a mutual destiny bonded daylight with its nocturnal counterpart, the moonlight. As Cobo records:

> They said that a lion or a snake attacked it in order to tear it into pieces; and for that reason, when it started to eclipse, they raised loud yells and cries and they flogged the dogs so that they would bark and howl. The men stood ready for war, sounding their horns, beating drums, and screeching

---

[64] Inti also embodied aspects of this last god in contexts where Wiraqucha's cult was absent. Cock and Doyle (1979) report that, in colonial times, the town of Maray (Chancay) situated in the high altitudes of the *puna*, regarded Inti as their creator god. He received sacrifices four times a year, in August before the sowing of "dry farmed" potatoes, and in October when the tubers were sowed in irrigated lands. The other two offerings were presented when the rains were either too abundant or too scarce. The same community also worshipped the *wanka* called P'unchaw Jirca who received their sacrifices specifically "when the men came down to sow maize". For these reasons, Cock and Doyle suggest that the Inti cult was associated with the altitude and subterranean agriculture of potatoes while P'unchaw was linked to the cultivation of maize, the maturation of which depended on an abundant diurnal light. In this context, Inti represented the subterranean Sun who, like Wiraqucha, presided over the waters supplying irrigation systems. His association with humidity and the underground warmth very likely explains his influence over dry farmed cultivations, requiring that the soil maintain a constant level of humidity. In the central highlands, Inti embodied aspects of the nocturnal Sun and P'unchaw represented the diurnal sun of the rainy season. In Inca Cuzco, however, Wiraqucha took upon characteristics of the Thunder, the prominent god of the Collao region, whose nature contrasted with P'unchaw. In that sense, this last deity was associated to the lower ecological area and maize agriculture in opposition to Wiraqucha's association with the upper plateaux.

[65] Pachacuti Yamqui: 208. See also Pease 1970: 169; Silverblatt 1987: 41; Dean 2001: 148-149.

> loudly they shot arrows and threw rods at the moon, and made wild gestures with their spears, as if they meant to wound the lion or the snake (...) The reason why they did it was because they believed that if the lion accomplished his design, [the world] would remain in obscurity and darkness.[66]

Despite her outstanding status, early sources seldom mention the role of the Moon deity in Inca festivals. This absence is all the more surprising in that seventeenth-century documents of the Extirpation of idolatry indicate that her cult was among the most widespread of all "heathen" practices throughout the Andes.[67] Scholars largely ascribe this lack of information to the male-oriented perspective of chroniclers who showed little interest in this cult.[68] They, in any case, would have had limited or no access to it because only women presided over it. Molina provides one of the earliest mentions of the Moon's native name in his description of the Inca calendar. For him, she was Paqhsa Mama, or Mother Moon. This word was closely connected to its equivalents in Aymara and Cauqui, *phaxsi* and *pajshi*, respectively. It had no link with the word *killa* of the general language that later documents employ extensively to refer to the worship of the Moon, mainly in the central sierra. In Molina, *killa* describes only the months and the lunar phases that set the start of certain festivities in Cuzco. González Holguín also lists several expressions constructed with *killa* in which the gloss refers primarily to the months and to the celestial body of the moon. These translations differ from his entries for *paqsha*, which applied more specifically to the beaming light of the moon, often in the context of a sky lowering with precipitations. *Paqhsa Mama* becomes "the resplendent Mother", and *paqsha* is "the clarity of the moon that breaks through the clouds". The verbal forms also suggest a close association of the word with the wet season: *paqsha-* or *paqshari-* is "glowing without rays of light in reference to the moon in the clouds", and *paqhsarqu-* is "the moon brightening between the clouds or when it ceases to rain."[69]

---

66 Cobo: T. 2, Bk. 13, Ch. 6: 158-159. My translation.

67 Arriaga: 27; Polia Meconi 1999: 350; Rostworowski 1988a: 79-81.

68 Silverblatt 1987: 47; MacCormack 1991a: 115. The same rationale may explain the scarcity of information about the cult of Pacha Mama (Mother Earth) in the Cuzco region. Although today she is a prominent figure of Andean religious life, little is know about the status Pacha Mama held in the Inca cosmology. Molina suggests that she received offerings at *sitwa*, but she was absent from other yearly celebrations. Polo's texts, which influenced generations of chroniclers, rhetorically mention the tribute she received in Inca time, but they do not detail her cult (1990: 46, 81, 84, 100). Cobo dedicated a small paragraph to her (Cobo: T. 2, Bk. 13, Ch. 7, 161). Ultimately, this lack of data could well reflect her little prestige in the Inca belief system.

69 González Holguín, *Vocabulario*: 271.

Betanzos made an earlier use of this word in reference to the Inca queen, whom he calls *paqsha inti ususi*, "the moon, and daughter of the Sun."[70] She was the highest-ranked female character of the royal descent group. In the same way that her husband presided over the general sacrifices to the Sun god, she would have officiated in a temple erected in honour of the Moon located at the junction of the rivers Tullumayu and Saphi, in the Pumap Chupan district.[71] Throughout the Andes, women have directed specific offerings to the Moon, particularly during pregnancy, because she was believed to be the *kamaqin* of the female gender. A narrative from the central sierra recorded in the seventeenth century says that the Sun created humankind in the Lake Titicaca region, but ordered women to worship the Moon as their mother who provides them with food and clothing.[72] In the sacred site of Lake Titicaca, the Moon had her own sanctuary on Coati Island, not far from her husband's temple located on the Island of the Sun. Ramos Gavilán, an Augustinian missionary in Copacabana at the turn of the seventeenth century, put on paper an account of the unusual activities that occupied the ministers of both deities.[73] We learn through him that the priests navigated daily between the two temples, bustling around with prayers and offerings. The *quya*, "wife of the Sun, being at the same time the Moon, sent him her gifts, and the Sun returned them with caresses of equal affection."[74] The priests drank to the deities and invoked them to protect the health of their king and to ensure the successful deployment of his imperial affairs. In Cuzco, the mummified bodies of the deceased queens reportedly surrounded the image of the Moon in Coricancha. A female priesthood attended her needs and carried her statue to Haucaypata during the few public celebrations she attended. On *qhapaq rayme*, the novices offered their sacrifices to the royal mummies and to all

---

[70] Betanzos: Pt 1, Ch. 16, 78. See also Anónimo [c. 1551]: 412.

[71] Guaman Poma: 263 [265], 236-237.

[72] Duviols 1986: 151. I choose to translate *criadora de las mugeres* as "*kamaqin* of the female gender" rather than "creator of the female gender", because the narrative clearly indicates that women already existed on earth when the Sun ordered them to worship the Moon.

[73] Before Inca occupation the temples had operated regionally as powerful religious centres. Archaeological excavations have revealed that the structure of these buildings was much anterior to the expansion of the Cuzqueñan lords, and had received large additional extensions during the Late Horizon. The instigator of such a policy has often been identified with Inca Tupa Yupanqui who, following a visit to the Island, is said to have offered a life-size golden statue of the moon to the Coati temple. The upper half of the figure's body was made of gold while the lower part was made of silver. See Bauer and Stanish 2001: 14-17; 98-157; Sallnow 1987: 22-23.

[74] Ramos Gavilán: Bk. 1, Ch. 28.

major gods, including Paqhsa Mama, before the start of their *warachiku*. She would have received similar honours on the festivals of *muru urqu* and *mayu qatiy* following the initiation ceremony,[75] and on the month of *quya rayme* (festival of the *quya*) that late chroniclers and scholars alike have described as a time of rejoicing for women.[76] On this last occasion, a specific day was dedicated to Paqhsa Mama and Pacha Mama during which the celebrants satiated with offerings the two female deities and lauded them in their orations. It was during this festival that a female member of the ruler's close family was called *quya paqsha* and was sacrificed to Apu P'unchaw as his wife.[77]

The pair of the Moon and Sun god was also closely related to the morning and the evening aspects of Venus, known in Quechua as *ch'aska quyllur* and *chinchay quyllur* (or *ch'isin ch'aska*).[78] As an inferior planet, Venus is always seen in the daytime close to the sun and at a low altitude above the horizon.[79] An Inca narrative tells that "the Sun, as lord of all stars, ordered it to come near him, sometimes before him and sometimes behind, because it was more beautiful than the rest."[80] The planet's atmospheric conditions make it the brightest object visible from earth at sunrise and sunset, so that Venus was and is still believed to escort the moon at night and to follow the daylight lord on its ascendance in the sky.[81] Perhaps because of this symbiotic relation with the supreme male and female deities, the chronicler Pachacuti Yamqui associates each aspect of the planet with a specific gender: the morning star with the masculine, and the evening star with the feminine. Aside from these data, our information on this double

---

75 It is likely that the Moon belonged to the group of deities worshipped at *muyu qatiy*. About this ceremony, Molina says: "and in the same way they brought out all the huacas and embalmed corpses, there, they did the accustomed reverence". This ceremony closed a cycle that started with the rituals of the *warachiku*. It was during *muyu qatiy* that the ashes of all the sacrifices made during the year, were thrown into the river, before the suspension of public festivities during several months (Molina: 100-114).

76 Guaman Poma: 253 [255], 226-227; Silverblatt 1987: 66; Rostworowski 1988a: 79-80; MacCormack 1991a: 305.

77 Molina: 77, 94.

78 Ziólkowski (1996: 55-58) posits that Venus held a close relation with Illapa because of the beliefs that identify the latter's body with constellations. *Ch'aska* and *quyllur* both mean star, but the former also describes snarled or frizzy hair. *Ch'isi* is the period of twilight, between sunset and night. Arnold and Espejo Ayca (2006: 188) suggest "Silver southern star" for *chinchay quyllur*. It should be noted that Cobo (Bk. 13, Ch. 6: 159) describes the latter as a star protector of all felines, although without identifying it explicitly to Venus. See also Gonzalez Holguín, *Vocabulario*: 98, 570; Pachacuti Yamqui: f. 13v., 208.

79 Urton 1981: 156, 166-167; Bauer and Dearborn 1995: 120-125.

80 Garcilaso: Bk. 2, Ch. 23, 113.

81 Arnold & Espejo Ayca 2006: 187.

entity is extremely fragmented. Two late chroniclers, Garcilaso and Guaman Poma, mention the altar or temple that the Incas would have erected in Cuzco to house a cult consecrated to Venus and some constellations.[82] Documents from the Extirpation of idolatry also make summary allusions to the planet. Known as *waraq* in the central sierra, it held an outstanding position among the religious beliefs of this region for being the deity that presided over the novices' initiation, like Huanacauri in Inca Cuzco.[83]

Hierarchy and the Coalescence of Solar Cults

Although early Europeans in Peru did not utterly understand the concepts of divine function and ancestorship in Inca religion, their writings enclose snatches of information and a variety of myths that, coupled with ethnographic and philological works, offer a coherent picture of the relations ancestor-gods maintained with each other and with their descendants on earth. These data also assist in framing the dynamics that underlay the dualism of Inca cosmology. In the ancient Andean world, the deities that shared a common physical envelope or inhabited the same domain, whether celestial or underground, did not necessarily interact harmoniously with each other, nor did they naturally hold complementary positions. Spanish observers often overlooked the diversity of expressions that a natural phenomenon alone could embody. The Sun of the chronicles, for example, encapsulated a plethora of antithetical aspects, such as Wiraqucha, P'unchaw, or Rupay, under a single denomination. The function and attributes of gods had their foundation in the seasonal phenomena and the vital element they were believed to herald and control. It was the dual division of the agricultural cycle, which was itself closely linked to the alternation of the wet and dry seasons, that shaped their divine nature and instigated their antagonistic relationships. In this cosmic world, the subterranean Sun Wiraqucha had established his reign in the primordial sea of Mama Qucha that surrounded the earth. Together with Illapa, whose domain in the firmament replicated the underground world, Wiraqucha ensured the regulation of water from the cosmic river, the *Mayu*, to the irrigation canals that supplied human needs. The influence of these two divinities over earthly matters extended from the dry season to the early

---

[82] Garcilaso: Bk. 2, Ch. 21, 108-109; Bk. 3, Ch. 21, 172; Guaman Poma: 263 [265], 236.

[83] Arriaga: Ch. 6, 65. Polia Meconi 1999: 354-355; Duviols 1986: 227. See also Santo Tomás, *Lexicón*: "Aranyac, o guara, o chasca: Venus of the morning", 236.

rainfalls. Halfway through the year, the solar god P'unchaw and his wife Paqsha Mama took over the task of regulating water by ruling over the late precipitations of the wet season and providing the heat and light necessary to dry the crops at the end of their maturation cycle. This cosmology did not ground the relations of opposition and complementarity between gods on an abstract verticality along a lower and upper spatial division. Instead, it anchored the gods' attributes in the environmental constraints and the requirements of the agricultural cycle.

Each component of this dual cosmological division was associated in turn with one of the two contrasting traditions retelling the origin of the Inca ancestors. In fact, to argue that each moiety held a different narrative of the primordial migration casts light on an otherwise obscure comment by Betanzos, who claimed that "[the Incas] sometimes hold the Sun up as the creator, and other times they say it is Viracocha."[84] Sarmiento also distances himself from the general bulk of chroniclers that credit the Incas' initial emergence to the Sun. The Incas, he writes, "said they were the sons of Viracocha Pachayachachi, their creator, and that they had come forth out of certain windows to rule the rest of the people."[85] This dual conception of ancestorship extended to the general makeup of the Inca religion, which is best defined as a heliolatry dominated by two entities: a subterranean expression of the Sun and a daylight manifestation of it. Wiraqucha presided over the agricultural activities and communal work of the dry season, starting in June with the shelling of the recently yielded maize. P'unchaw was the daylight sun of the wet season. He regulated the precipitations during the last phase of the cultivation growth, starting in December. In that specific month, the ruling nobility celebrated the initiation ritual of its youth that Huanacauri, the messenger of P'unchaw, had decreed in time immemorial. This dual reconstruction of the Inca cosmological order clarifies major discrepancies in primary sources by establishing that the Spanish chroniclers were not thoroughly responsible for the disparate versions of the Inca origin myth. They recorded a series of migration narratives that proceeded from two distinct clusters of traditions, each of which was individually owned by the moieties. The Hurin leaders claimed to have originated from the cave of Paqariq Tampu, whereas the Hanan lords traced their ancestry back to the Collao region, therein establishing a link with Wiraqucha, the prominent deity of that area. These migration stories also display structural features that are traditionally

84 Betanzos: Pt. 1, Ch. 11, 49.
85 Sarmiento: Ch. 10, 51.

associated with each sector. On one hand, the Hurin moiety locates the origin of its lineages within the vicinity of Cuzco, where their kings had restricted their conquests. On the other, the Hanan moiety assimilates its ancestors to outsiders by setting their birth in Lake Titicaca.

This solar dichotomy is a distinctive aspect of Inca religion, which contrasted in that regard with the belief systems of the central sierra dual organizations. There, Libiac, the god that the Spaniards loosely identified with the Thunder, held a prominent position as divine ancestor of the Llacuaz ayllus. He was the traditional opponent of the archaic sun god Wari. The mythical advent of Libiac's descendants on earth also displayed features different from those of the Inca foundation narratives because the Llacuaz believed their ancestor came to the world by lightning strike. In spite of these cultural distinctions, the mythological corpus of the central sierra and the pre-Hispanic traditions of the Cuzco region shared in common a dual cosmology that revolved around the alternation of seasons. The philological analysis of the word *lliwyaq* reveals that the god was first and foremost ruler of the atmospheric conditions of the dry season. His opposition with Wari, who was ruler of the waters (*yakuyuq*), replicated the antithetic relations of the sun gods Wiraqucha and P'unchaw.

This chapter closes with the discussion of a major issue that has fuelled many conflicting debates among scholars and at the heart of which lies another discrepancy. On the one hand, Spanish texts portray Wiraqucha as the supreme god of the Inca religion, while on the other they describe the worship of the Sun as the official state religion. If Wiraqucha was indeed the most revered god, why did the Incas not spread his cult through Tawantinsuyu instead of compelling their subjects to build temples and *aklla wasi* to the Sun? Or was Wiraqucha's prominent status a mere elaboration by the chroniclers to answer theological and legal issues on sovereignty in the Andes? The Spanish have repeatedly portrayed the Inca state religion as a belief system revolving around the single figure of the Sun. They claimed that the ruling elite enforced the worship of this prime devotion upon every nation they had incorporated into their jurisdiction. Their writings describe extensively the provincial institutions that the central authority established in honour of the Sun and the lands it pre-empted to support his religious service. Yet, the same sources often concur in describing Wiraqucha as the Creator god of the Inca nobility and as the foremost being who presided over their polytheism. These inconsistencies were unsustainable to many chroniclers, who promptly developed a series of exegeses embedded in theological concerns to account for them. One of the first chroniclers to enter this discussion was Molina, who accommodated the Platonic argument of the First Mover to the Inca context. For him, the reformer king Pachacuti Yupanqui had understood through natural reason that the sun was subordinated to the will of the Maker.[86] Garcilaso, who perhaps had been acquainted with this argument through his reading of Oré, Cabello, Murúa, or Acosta, suggests that the Incas had worshipped only one visible god, the Sun, who ruled over a series of adjunct beings that embodied different aspects of his sacredness.[87] In this polyptych of the heavens, Garcilaso claims that the threefold representation of Illapa was not an object of worship in itself, nor was it even a simulacrum of the Trinity, but was instead a figure attending the solar god in his work. To account for the coexistence of heliolatry with the worship of a supreme being, Garcilaso argues that the nature of the Creator god was inscrutable, whence the Incas did not engage with him through any representation or temple. The Mestizo chronicler identifies this omnipotent god with Pachacamac, which was an important oracle of the central coast. In doing so, he demarcates himself from most other colonial writers that assimilate the Maker with Wiraqucha, the "Eternal Light" who had bestowed part of his divinity upon his offspring, the Sun.[88]

The general approach to these colonial elaborations considers them to be products of a selection process that colonial thinkers had intentionally orchestrated to feed a prolectical argument. In this view, the chroniclers would have privileged the writing and diffusion of information relative to Wiraqucha's cult because the position of this god at the heart of the Inca ceremonial life helped to consolidate the foundations of their missionary agendas. Indeed, philological and historiographical studies have conclusively evinced that the god's names, his epithets, and his miraculous deeds were given an evangelistic resonance beginning in the sixteenth century. Accordingly, Wiraqucha's status as the paramount deity of the Inca religion would have been subsequent to the colonial process of reinterpretation that transformed him into an acculturated being. Yet, it may be reductive and even misleading to equate this material to an outright colonial construction. The general lines of the Inca cosmological order suggest instead that the divinity's foremost position in the native belief system predated the Spanish conquest and would have been conducive to his assimilation

86 Molina: 58-60.

87 MacCormack 1991a: 258-261; Duviols 1993: 61-64.

88 Garcilaso: Bk. 2, Ch. 2, 64-66; Blas Valera: 133.

with the God of the Christian faith. If the pastoral reading of Wiraqucha's attributes and the reappropriation of his epic in the hands of fervent Christians—Spanish, Mestizos, and Indians alike—cannot be disputed, certain features of his nature and other elements of his oral tradition shaped his hybrid portrayals in colonial times. The Inca dual religious system provided the auspicious ground for this assimilation because it accommodated two cults of unequal prestige that replicated the status asymmetry of the moieties, where the Sun of the upper part, Wiraqucha, prevailed over the Daylight god of the Hurin moiety. In this way, the metaphysical argument of the subordination of the Sun appears grounded in a dual pre-Hispanic framework associated with the rise to power of Hanan lords at the expense of the Hurin lineages. This transformation of political authority led to the divine supremacy of the vigorous Sun Wiraqucha and to the subsequent decline of the tutelary deity of the Lower moiety. As every version of the narrative tells, it was a leader of the Upper moiety, Pachacuti Yupanqui or Tupa Inca, who articulated this change of divine hierarchy and who ordered the construction of a temple to Wiraqucha in Kishwar Kancha.[89]

Now, if Wiraqucha was the most highly revered god of the state religion, why did the Incas impose the worship of the Sun on their subjects, rather than that of Wiraqucha? The answer may lie in the nature of the entity known as Inti. As we noted, Cobo provides the first explicit mention of a statue to Inti in his *Historia*, where he draws a portrait of three representations of the god, two of which fit Wiraqucha's and P'unchaw's profiles, while the third would have stood as the "Sun itself". Even though this information is a wide reinterpretation of authentic material on the Sun religion, it should not be dismissed entirely. The common use of the word Inti to describe the shrines and the festivals dedicated to the Sun shows that it was known as an expression of the solar god among the residents of Cuzco. Inti would have represented the encompassing manifestation of the two Suns revered by the ruling elite—the "overlapping" aspects of Wiraqucha and P'unchaw, to borrow Demarest's wording. It would have been the cult around which the two Inca moieties united—the imperial religion the rulers imposed on their subjects. Whether there existed a separate image of this specific entity remains an unresolved question, but this argument accounts for the contradictory material presenting Wiraqucha as the ruling elite's supreme divinity but describing the worship of the Sun as the official

---

[89] Molina: 58-60; Cabello: 349; Murúa, *Historia general*: Bk. 2, Ch. 36, 413.

state religion. A pre-Hispanic substratum based upon dual ideology facilitated Wiraqucha's assimilation with the omnipotent god of the conquistadors, which the evangelizing rationales reclaimed in the colonial depiction of this god's mythical actions.

CHAPTER SIX

THE INCA CALENDAR AND ITS TRANSITION PERIODS

Inasmuch as temporal orientations are fundamental to the makeup of social behaviour, the final part of this book examines the ecological, cosmological, and politico-economic principles that regulated the sequences of the Inca ritual calendar. It pays particular attention to patterns of dual oppositions, which similarly affected the composition of other calendars used in the area until the present day. In this regard, the chronicles attest to the significance of two meteorological seasons in the organization of the calendar: *puquy mit'a* (rainy season) and *chiraw mit'a* (dry season). Not only did they regulate agrarian activities, they also ruled the movement of populations and goods, which was crucial to the stability of an empire that depended on tributary policies and military conquests. In practice, the rainy season is a short period that lasts approximately four and a half months, starting around mid-October and ending in late February. Early precipitations, however, appear in September and indicate the start of planting activities, so that *puquy mit'a* is sometimes said to begin in that period.[1] Today, many Andean communities continue to divide their yearly activities into two cycles that reflect to various degrees the seasonal changes of their ecosystems. In accordance with these, human actors organize their activities of subsistence and frame the significance of the religious and socio-political events that shape their society's economy, identity, and beliefs. Within this time construction, the festivities of the wet season are often associated with food scarcity and the appearance of diseases, death, and dependence, whereas those of the dry season commonly reflect abundance, asymmetric labour cooperation, and separation from the household.[2] Yet, the impact of ecological transformations on human enterprises should not overshadow the fact that, in any one calendar, the annual celebrations do not follow a single temporal construct. Different cycles, which may or may not be interconnected, can regulate ritual and communal activities. They may even be dissociated entirely from the time when a

---

[1] Isbell 1978: 55.

[2] Isbell 1978; Bastien 1978; Wallis 1980; Van Den Berg 1989; Gose 1994; Harris 2000; Stobart 2006.

particular occasion was commemorated. For example, a planting festival or the anniversary of a deceased person does not invariably occur on the actual day of agricultural labour or precisely a year after the person's death. It might be ruled by astronomical events or/and ideological motives. Naturally, these aspects are by no means cultural particulars. Christian festivals, in all their diversity, constitute another example of the same reality. Many are set on a date unrelated to the factual event they recall (e.g., a patron saint's day); others follow different sequences, such as the Gregorian calendar (GC) or the solar cycle (e.g., Christmas), which also may be combined with the computation of the lunar phases (e.g., Easter). The ruling elite of Cuzco observed a calendar that bore similar characteristics. It was composed of a series of cycles, including the growth of crops, the reproduction of herds, the seasonal changes, the completion of communal works, or the development of human relationships, all of which were organized according to different modes of time-reckoning that operated independently or sometimes overlapped in order to generate ritual associations. Hence, the llama-birthing period concurred with the celestial appearance of the dark cloud constellation (*yana phuyu*) representing the animal and to the *warachiku* that enacted the symbolic rebirth of young Inca males. These three events, concretely unrelated, were remarkably interlocked during the initiation festivities as the novices adopted the characteristics of young llama victims.

In this picture, astronomical observations played a fundamental role in compensating for the unpredictability of the Andean climate and for the fluctuation of animal and vegetal cycles. People progressively developed an expert knowledge of the lunation periods and of the solar activities and, importantly in the high plateaux, they acquired an understanding of the firmament with the movement of its celestial bodies.[3] Astronomical observations flourished in pre-Hispanic times with the development of precision instruments. To estimate the passage of the sun at different times of the year, the Incas used solar pillars, which dwindled into ruins soon after the Spanish conquest. The calendar became increasingly less dependent on seasonal changes and a new construction emerged, still bearing connections with the natural environment but now encompassing the realities of Andean social life under state control. This Inca edifice of knowledge,

[3] Zuidema (1983a) suggests that the Inca calendar inherited traits of a solar calendar that would have been in use in the Middle Horizon (AD 600-1000). Technical points concerning Inca astronomical knowledge cannot be fully developed in this chapter for lack of space. For further information, see Urton 1978; Zuidema 1977b, 1981, 1982a; Zuidema & Urton 1976; Ziółkowski & Sadowski 1989; Aveni 1989; Bauer & Dearborn 1995.

forged on the legacy of previous civilizations, is what this final part sets out to understand, with a focus on the dual division of the metropolitan calendar. It examines the ethnographic, historical, and astronomical evidence that indicates the bipartition of the annual cycle into two equal sequences responding to the social and religious structure of the Cuzco moieties outlined thus far. It starts by questioning the role of astronomical observations and investigates their relation to specific sequences of the Inca calendar. Ethnographic documents and colonial records provide the backdrop against which the structure of the argument unfolds. They provide information about the limitations imposed by the natural environment on human activities, while offering a range of creative solutions that Andean societies adopted to interact with their milieu. Within this literature, the records of idolatry are particularly helpful in expanding our understanding of pre-Hispanic lore. Although it is undeniable that the Christian calendar enforced by Europeans affected native conceptions of time,[4] this material demonstrates the remarkable adaptability and ideological uniformity of certain Andean beliefs.[5] Approached with the required historical rigour, it provides a point of entry into understanding the Inca festal calendar. The conclusions drawn heretofore on the dual nature of the Sun religion complete this analytical approach and, taken together, reveal some distinctive aspects of Inca time construction. They show that each dual division of the Inca year was a temporal construct devised via astronomical means. The two ancestral Sun gods of the royal descent group presided individually over one specific division, which opened with a different commemoration of the Inca mythical settlement in Cuzco. These two divisions alternated with transition periods that marked the defeat of one social and cosmic half in favour of the opposite one. In parallel with this analysis, the chapter dwells upon the broader significance of each celebration and emphasizes how they contribute to a general understanding of Inca cosmology.

Computations and the Inca Calendar

In advancing a coherent picture of the sequences ordering the Inca calendar, the first obstacle to be surmounted is the scant and somewhat

---

4 MacCormack 1998.

5 Zuidema & Urton 1976; Doyle 1988; Zuidema 1992, 1996; Salomon 1995; Gose 1996b; Ramírez 2005; Gareis 2005. Almost all these authors stress the limitations of drawing analogies between the belief systems of the central sierra, as recorded during the Extirpation of idolatry campaigns, and that of the Incas. I concur with these views.

Table 11. The Inca calendar in primary sources. Words of uncertain or unidentified spelling are not italicized.

	Betanzos	*Polo Ondegardo*	*Anónimo* (*c. 1570*)	*Molina*	*Fernández*	*Gutiérrez*
January	*Hatun puquy killa*	*Kamay*	*Hatun puquy*	*Hatun puquy*	*Puray upyay killa*	*Puray upyay killa*
February	*Alla puquy killa*	*Hatun puquy*	*Pacha puquy*	*Pacha puquy*	*Chaqmay killa*	*Chaqmay killa*
March	*Pacha puquy killa*	*Pacha puquy*	Aylliua *killa*	*Pawqar wara*	*Pawqar wara killa*	Rura pawqay *killa*
April	Aylliua *killa*	Aylliua *killa*	Aucay *kuskiy*	Aylliua, *Aymuray*	Aylliua *killa*	Aylliua *killa*
May	Aucay *kuskiy killa* Yahuayracha *aymuray*	*Hatun kuskiy Aymuray*	*Aymuray killa*	Aucay *kuskiy Inti rayme*	*Aymuray killa*	*Aymuray killa*
June	*Hatun kuskiy killa*	Aucay *kuskiy Inti rayme*	*Hatun kuskiy*	Cauay (or) Ch'awa ruray	Aucay *kuskiy*	Aucay *kuskiy killa*
July	Caua *killa*	Ch'awa ruray *killa*	Ch'awa ruray	*Muru paqhsa tarpuy killa* Yahuayra	Ch'awa ruray *killa*	Ch'awa ruray *killa*
August	*Qarpay killa*	*Yapa killa*	*Tarpuy killa*	*Quya rayme, Sitwa*	*Sitwa killa*	*Sitwa killa*
September	*Sitwa killa, puray upyay*	*Quya rayme, Sitwa*	*Sitwa killa*	*Uma rayme, Warachiku*	*Puskay killa*	*Puskay killa*
October	*Uma rayme killa*	*Uma rayme Puskay killa*	*Chawpi kuskiy* Kantaray *killa*	*Ayarmaca rayme Warachiku*	Kantaray *killa*	Kantaray *killa*
November	Kantaray *killa*	*Ayarmaca rayme* Kantaray *killa, itu*	*Rayme killa*	*Qhapaq rayme*	*Rayme killa*	*Rayme killa*
December	*Rayme Puquy killa rayme killa*	*Rayme Qhapaq rayme*	*Kamay killa*	*Kamay killa Chuqanaku, Mayu qatiy*	*Kamay killa*	*Kamay killa*

Table 12. The Inca calendar in primary sources (suite). Words of uncertain or unidentified spelling are not italicized.

	Cabello Balboa	*Acosta*	*Murúa*	*Guaman Poma*	*González Holguín*	*Cobo*
January	*Puray upyay killa Kamay*	*Kamay*	*Kamay*	*Kamay killa, Qhapaq rayme Qhapaq hucha*	*Quya puquy*	*Kamay*
February	*Hatun puquy*		*Hatun puquy*	*Hatun puquy, Pawqar wara Chaqmakuy killa*		*Hatun puquy*
March	*Pacha puquy*			*Pacha puquy*	*Pawqar wara*	
April	Aylliua *killa*		Aylliua *kuskiy*	*Inka rayme*	Aylliua *killa*	Aylliua *killa*
May	*Hatun kuskiy Aymuray*	*Hatun kuskiy Aymuray*	*Hatun kuskiy Aymuray*	*Aymuray killa*	*Hatun kuskiy Aymuray killa*	*Hatun kuskiy Aymuray*
June	*Aucay kuskiy Inti rayme*	Aucay *kuskiy Inti rayme*	Aucay *kuskiy Inti rayme*	*Kuskiy killa, Inti rayme Qhapaq hucha*	*Inti rayme*	Aucay *kuskiy Inti rayme*
July	Ch'awa ruray *killa*	Ch'awa ruray *killa*	Ch'awa ruray *killa*	*Chakra qunakuy*	Anta *sitwa*	Ch'awa ruray *killa*
August	*Yapa killa*	*Yapa killa*	*Yapa killa*	*Chakra yapuy killa*	*Qhapaq sitwa*	*Yapa killa,* Yahuayra
September	*Quya rayme, Sitwa*	*Quya rayme, Sitwa*	*Quya rayme, Sitwa*	*Quya rayme*	*Uma rayme*	*Quya rayme, Sitwa*
October	*Uma rayme Puskay killa*	*Uma rayme Puskay killa*	*Uma rayme Puskay killa*	*Uma rayme killa*	*Ayarmaca*	*Uma rayme Puskay killa*
November	*Ayarmaca rayme* Kantaray *killa Itu rayme*	*Ayarmaca rayme* Kantaray *killa Itu rayme*	*Ayarmaca rayme* Kantaray *killa Itu rayme*	*Ayarmaca killa*	*Qhapaq rayme*	*Ayarmaca Itu*
December	*Rayme Qhapaq rayme*	*Rayme Qhapaq rayme*	*Rayme Qhapaq rayme*	*Qhapaq Inti rayme Qhapaq hucha*	*Rayme*	*Rayme Qhapaq rayme*

discrepant data that primary sources enclose on the computation of time. For aside from elusive references to the Julian, and later Gregorian months in which the festivities occurred, the chroniclers do not record them in their precise time setting or in their exact length. Most sources proceed by mapping Inca time within the frame of the European calendar, by equating the pre-Hispanic months with the divisions of the Christian year (see Tables 11 and 12).

Molina's account is certainly the most detailed and insightful source of all, but even though it regularly documents the day-by-day unfolding of ritual events, it does not indicate with precision the opening date of each festival. Still, some chroniclers were not indifferent to the issue of time recording and have attempted to discuss astronomical matters. The resulting literature often proposes divergent interpretations of the mechanisms ordering the official calendar, perhaps in part because not every chronicler was familiar with the technical questions involved. Polo, for example, surprisingly confounds the moon's cycle with the revolution of the sun when he indicates that the Incas determined the beginning of their lunar months by observing the passage of the sun through twelve pillars built outside Cuzco.[6] Fortunately, his predecessors provide more reliable clues to unravel the mechanisms of the state calendar. They concur in dividing the year into twelve months of thirty days,[7] although a handful of late chroniclers suggest that only eleven of them ordered this construction.[8] Some also report that the calendar was primarily divided according to the moons because the word *killa* designated the months.[9] Cieza is the first known chronicler to describe the main aspects of Inca astronomical knowledge about the stars, the computation of the Sun's rotation, or the waxing and waning cycle of the moon. He explains admiringly that the ruling elite had reputable augurs who observed the sky and predicted the future in a

6 Polo, "Relación de los adoratorios": 16-17.

7 Betanzos: Pt. 1, Ch. 15, 71-74; Polo, "Tratado": 16; Molina: 66-119; Fernández: Pt. 1, Bk. 1, Ch. 10, 86; Gutiérrez: Ch. 65, 255; Herrera y Tordesillas: Bk. 4, Ch. 5, 279; Guaman Poma: 235 [237]-260 [262], 210-234, Cobo. T.2, Bk. 12, Ch. 37, 142-143.

8 Albornoz: 178; Murúa, *Historia general*: Bk. 2, Ch. 38, 436-439. An anonymous chronicler (c. 1570: 156-157) reports as well that "the moon of March always reaches the one of April".

9 Ibarra Grasso (1982: 9) observes that there may be no actual·link between the word *killa* (moon) and the period it describes. In the same way, the term for month in our language originates from the Greek for moon, but the time it refers to no longer aligns with a lunation. *Killa* may have designated an earlier reality, before the Incas developed new calendrical devices to calculate the tropical year.

fashion reminiscent of Roman practices.[10] For him, they reckoned a calendar of lunar months and used in addition solar pillars to determine the appropriate time for agrarian labours. An anonymous source (c. 1570), particularly well informed about the Andean agricultural cycle, corroborates this last piece of information. The same source follows Betanzos, Sarmiento, and several others in attributing the conception of the ritual calendar to Pachacuti Yupanqui.[11] Molina dwells particularly on this point. He records that the reformer king "eliminated and added cults and ceremonies, divided the year into twelve months, giving each one its name and describing the festivities to be held each month."[12] Molina also acknowledges the prior role of the dynasty founder, Manco Capac, in establishing major celebrations such as *warachiku*.

All the aforecited accounts were written before the introduction of the Gregorian calendar in the Spanish colonies. The events they describe, therefore, lagged ten days behind the tropical year. In 1582, European countries under the influence of the Roman Catholic Church moved their calendar forward the corresponding number of days to solve this issue. In doing so, they brought the start of spring back to the crucial date of March 21, which was essential for fixing the feasts related to Easter. The decree, formulated under the supervision of Pope Gregory XIII, contained two additional reforms: one regarded the regulation of the leap year, and the other accommodated the Paschal date to the new calendar. In the provinces of the Peruvian viceroyalty, these adjustments were put into effect in 1584. All statements about the sequences of the Inca festivities recorded prior to that date need to be added ten days in order to be adjusted correctly to the Gregorian year.

Garcilaso's *Comentarios*, which were written after the calendrical reforms, provides the most detailed, albeit late information on calendar calculation. It includes several chapters on Inca "astrology" that deal with the ancient knowledge of celestial bodies, particularly of the morning star (Venus) and the Pleiades. They also draw attention to the Incas' observation of astronomical phenomena, such as the equinoxes and solstices, to fix the date of the three most important festivals of the state calendar: the initiation ritual at *qhapaq rayme*, the Sun festival at *inti rayme*, and the cleansing ceremony at *quya rayme*. Garcilaso not only knew of Polo's

10 Expanding further this parallel with Ancient times, Cieza (Pt. 2, Ch. 26, 77-78) associates the formulation of Inca laws to astronomical knowledge and the calculation of time.

11 Anónimo [c. 1570]: 150; Betanzos: Ch. 14-15: 65-74; Sarmiento: Ch. 31, 96-97.

12 Molina: 58.

account of the solar pillars, he also reinterpreted it. He explains that there were eight of these "gnomons" outside Cuzco, which served to fix the winter and summer solstices.[13] Perhaps he did consult the "Relación de los adoratorios", presumably authored by Polo, in which the author lists several solar markers that reportedly fixed the beginning of the winter and summer seasons.[14] Garcilaso would have deduced from this information the Incas' usage of the solstices and equinoxes to divide their calendar, which is similar to the fashion in which European seasons are demarcated. His interpretation, however, has little empirical support because the span of the two Andean *mit'a* differs entirely from that of the four seasons of the Western world and, naturally, does not concur with the same astronomical events. The Inca observations of the sun movement in relation to the pillars were more realistically connected to seasonal agricultural works, as the anonymous source (c. 1570) suggests. They would have regulated an annual cycle composed of solar months that Zuidema identifies with the *rayme* because this term refers to divisions of the Inca calendar that overlapped with lunar months. *Uma rayme* and *ayarmaca rayme*, for example, coincided with the Inca lunar months of *puskay killa* and *kantaray killa*, around October and November (GC), respectively. The names of these two *rayme* originated from those of local ayllus, the Uma and Ayarmaca, who performed a rite of passage similar to the Inca *warachiku* at a time when the ruling elite made preparations for their own youths' initiation ritual. For Zuidema, the coexistence of these two time divisions suggests that a solar calendar composed of *rayme* months coexisted with an annual cycle regulated by the moon's conjunctions. He thus identifies different computation measures at the service of two coexistent, albeit independent, calendars. The first one would have been divided into twelve solar months of 30 to 31 days, while the second would have comprised twelve sidereal months of variable length to which the Incas arguably added a period of 37 days that corresponded (only in part) to the disappearance of the Pleiades from the night sky. This cycle was also interconnected to the *seqe* system and its hypothetical observatory-shrines so that each *waka* corresponded to a specific day of the sidereal year.[15] In this way, the celebrations of *qhapaq rayme* were linked to the *chinchaysuyu* quarter and to

---

[13] Garcilaso: Bk. 1, Ch. 21-Ch. 22, 108-111.

[14] Polo, "Tratado": 16; 1917: 11.

[15] Zuidema 1981, 1982a, 1989b. A sidereal month is the time taken for the moon to complete a revolution around the earth in relation to the background stars, which are fixed. It lasts 27, 32 days.

the Hanan part of the valley, while the *inti rayme* festival corresponded to the *qullasuyu* section of the *seqe* system and to the Hurin half of the valley. These associations extended to Zuidema's interpretation of Inca religion in which *qhapaq rayme* celebrated the head deity of the Hanan moiety, while *inti rayme* honoured the tutelary god of the Hurin moiety.[16] The anthropologist has not been alone in attempting to draw a clearer picture of the confusing information that primary sources document about the Inca calendar. For several decades now, scholars have focused considerable attention on envisaging the astronomical calculations and observation techniques elaborated in ancient Peru. Among them, Ziólkowski (1987, 1988) has argued that the chroniclers collected data about at least two different calendars in use in the Cuzco area at the time of the Spanish conquest. The first was an old luni-solar or luni-sidereal calendar of twelve synodic months, which was adjusted to the tropical year by the observation of the solstices.[17] The second was a late elaboration of the ruling elite and a purely solar calendar that fixed the dates of the official religious ceremonies but also regulated the administrative and economic tasks, such as levying tribute or agricultural labours.[18]

Despite the diversity of interpretations raised in these technical works, they have all attempted to account for the discordant descriptions of the Inca calendar. They notably offer to elucidate a major disagreement in primary sources about the date setting the start of the year, which chroniclers situate either at *qhapaq rayme* with the male initiation ritual,[19] or half a year later with the festival of *inti rayme*.[20] Today, most knowledgeable studies on this issue suggest that this duplicity arose from the coexistence of overlapping systems of time computation. Only two chroniclers adopt a divergent position concerning the start of the year. The first is the anonymous author of the "Discurso" (c. 1570), who begins the Inca calendar in the conjunction of the lunar months of March and April, an event that

---

16 Zuidema's propositions have raised several criticisms. Sadowski (1989), Ziólkowski (1989), Bauer & Dearborn (1995: 64-65) observe that sidereal observations depend on more significant factors than those raised by Zuidema and that no civilization in world history ever actually devised a calendar based on such computation.

17 A synodic month is the time taken for the moon to orbit the earth. The average length of a lunation is twenty-nine and half days, resulting in a shift of approximately eleven days with the tropical year.

18 See also Ziólkowski and Sadowski 1982-1984; 1989.

19 Betanzos: Pt. 1, Ch. 15: 71; Polo, "Tratado": 18; Cabello: 349; Murúa, *Historia general*: Bk. 2, Ch. 38, 437.

20 Molina: 66; Fernández: Pt. 1, Bk. 1, Ch. 10, 86; Gutiérrez: Ch. 65, 255.

marked harvesting time.[21] His description of yearly activities in Cuzco is actually mostly concerned with the different stages of land cultivation and with tribute levy. Little if any reference is made to official rituals, which suggests that the calendar he records is dissociated from the abstract construct documented in other chronicles. Guaman Poma also suggests an unusual date for the beginning of the Inca year, in January, which was probably a deliberate attempt to echo the sequences of the Roman Catholic calendar. As MacCormack (1998) points out, Guaman Poma aimed to produce a continuity between the social past of Andean communities in pre-Hispanic times and their Christian existence after the conquest. The rest of the primary sources discredit overwhelmingly the isolated statements of these two chroniclers and offer consistent data that two separate dates set the beginning of the Inca ritual calendar. Without entering into the technical issues of calendar design, I suggest that both dates can be seen as the start of a new cycle that marked the beginning of the respective rules on earth of Wiraqucha and P'unchaw. In this way, the Inca year was halved into two equal periods of time, one starting in December and the other in June, which corresponded to an elaboration around *puquy mit'a* and *chiraw mit'a*.

Astronomy and the Dual Structure of the Calendar

Zuidema already showed that Guaman Poma's reconstruction of the Inca calendar opposes the features of state activities held in the dry months with the characteristics of the rituals observed in the wet season. He also established that the Inca year had been divided into two seasons fixed by astronomical events (1996). His latest and magisterial work (2011) develops this position to its fullest articulation. Although I depart from the cosmological construction that Zuidema attaches to the Inca calendar, the general lines of his argument on calendrical computations provide an interesting point of entry into understanding the mechanisms that articulated the dual division of the ritual cycle held in Cuzco. His identification of stellar observations as instrumental in the design of the Inca calendar is particularly fundamental. This contribution, formulated in collaboration with Anthony Aveni, receives strong support from Urton's contemporary ethnography of Andean astronomy, in which he establishes the central role of celestial formations in regulating ritual and communal activities.[22]

[21] Anónimo [c. 1570]: 150.
[22] Urton 1978, 1980, 1981. See also Aveni 1977.

Today, the correlation between stellar and terrestrial events lies partly in the belief that every species on earth is animated by a celestial double or prototypical being incarnated in the constellations. These duplicates correspond in turn to two different types of star formations. The first one conforms to our definition: it conceptualizes a design by assembling stars together. The other is a cluster of stellar dust that appears in negative against the brightness of the Milky Way. Called *yana phuyu* (black cloud), it forms a shape resembling an animal and, occasionally, an object.[23] A great number of colonial sources attest to the pre-Hispanic origin of these two stellar conceptualizations and to the belief of the celestial double. Polo, for example, records that "all animals and birds on earth were believed to have their equivalent in the sky which were in charge of their procreation and multiplication."[24] In this way, the *urqu chillay* cluster represented a multicoloured llama watching over the livestock, the *chuqe chillay* had a feline appearance and was in charge of the "tigers, ocelots and lions", and the snake-shaped dark cloud formation called *mach'aqway* protected the reptiles on earth.[25] The Pleiades, which were called *qullqa* (storehouse) and, more rarely, *qutu* (batch), were believed to be the matrix of all prototypical stars that "animated and formed [men and things]."[26] Cobo writes that all ayllus universally worshipped this star cluster and scrutinized its course more attentively than any other celestial body because it influenced the abundance of the terrestrial granaries. When the Pleiades appeared early in the year, large and shining brightly in the sky, the year was going to be fertile, but when they were almost invisible to the naked eye, Andean people feared that poor harvests were to come. In this way, the course of every cluster of stars was closely interconnected with human activities and the life cycle of the fauna insomuch as its vital force had an impact on the prosperity of its earthly doubles. Garcilaso explains that fishing was plentiful only when the primordial and progenitor fish that resided in the upper world bestowed this abundance upon men by displaying its visible strength (*kallpa*).[27] The fortune of one depended upon the fate of the other. This interdependence between the terrestrial order and the firmament also extended the Inca conception of the Milky Way because the *Mayu* was seen as the celestial projection of an earthly

---

23 Urton 1981: 95-111.

24 Polo, "Tratado": 5.

25 Cobo: T. 2, Bk. 13, Ch. 6, 159-160.

26 *Ritos y Tradiciones de Huarochirí*: 379. See also Cobo: T. 2, Bk. 13, Ch. 6, 159; Urton 1981: 113-127; Itier 1996.

27 Garcilaso: Bk. 1, Ch. 10, 29.

watercourse that supplied irrigation canals by flowing down under the earth to join the subterranean sea.[28] Interdependence and complementarity regulated these two worlds in the same way that the *waka* ancestors were believed to be dying of thirst if the lands suffered from drought.

In view of this cosmology, two elements appear central to the making of the Inca calendar. One was the observance of ritual tasks in accordance with astronomical events, and the other was a strong emphasis on duality so that what occurred in the sky was duplicated in the world below. To understand which cosmological characteristics the Incas attached to the alternation of earthly cycles, we must identify the solar and stellar phenomena that fixed the start of each dual division of the calendar. The observations they made thanks to the pillars erected on high-altitude sites around Cuzco provide a first clue in addressing this question. Early Spanish sources tell us that Pachacuti Yupanqui ordered the construction of four columns, or pyramids, called *pacha unanchaq* to fix the days for sowing and harvesting. On these occasions, an observer standing at a specific location in the city, presumably on the *ushnu* of the main square,[29] would notice the sun rising and setting in between the pillars.[30] González Holguín notes that *unancha-* refers to "fixing a moment to set about a task,"[31] which suggests that the stone pillars were not merely instrumental in establishing the length of the tropical year, but regulated specifically some important stages of foodstuff production. The word *pacha* in this expression may have referred as much to a time period as to the earth as fertile soil. Polo also makes reference to the agricultural cycle in his description of the pillars, although he mentions different names for each pair of them. According to the Spanish jurist, they were called *puquy sukanka* and *chiraw sukanka*, which refer to the two seasonal divisions of the year and, by extension, to periods of the cultivation cycle. In the central sierra and the southern highlands, *puquy* describes the months of the rainy season between October and late February, when the cultivations mature. Primary sources and

---

[28] Urton (1981: 172) notes that in the community of Misminay (Cuzco) today, the Vilcanota River is conceived as a "mirror reflecting the Mayu".

[29] The *ushnu* was a large stone erected in the centre of several Inca cities for ceremonial purposes. In Cuzco, it was covered in gold and would have reached two metres high. At its base was a basin in which the attendants poured their offerings. Betanzos: Pt. 1, Ch. 11, 53; Molina: 74, 79; Albornoz: 176. Modern studies on the subject include Zuidema 1989d; Hyslop 1990: 69-101; Meddens 1997; Staller 2008; Meddens et al. 2008.

[30] Betanzos: Pt. 1, Ch. 15, 73-74; Cieza: Pt. 2, Ch. 26, 77-78; Polo, "Tratado": 16; Polo "Relación de los adoratorios": 13-14; Anónimo [c. 1570]: 151.

[31] González Holguín, *Vocabulario*: 269.

present-day studies record the words *hatun puquy* (high maturation), *alla puquy* (the maturation of tubers), and *pacha puquy* (the maturation of the earth) in relation to the sequences of this period.[32] Conversely, *chiraw* describes a period dominated by the sun of the dry season, extending from March to November.[33] Zuidema (1996, 1997b) convincingly relates this term to carhua (*qarwa*), used in the colonial Archbishopric of Lima to refer to "something withered, shrivelled or turning yellow."[34] In relation to crops, *qarwa* describes the stage of the maturing maize when it dries and becomes suitable to reap. Accordingly, the festivals of *puquy mit'a* and *qarwa mit'a* in the central sierra celebrated two different phases of the cultivation cycle. The first ceremony was held "when the rain starts, before the Indians begin preparing the chakras, and the other [occurs] at the time of the Corpus, when the maize begins to mature."[35] These two major events, Zuidema argues, had their equivalents in Cuzco, where they were celebrated at a time determined by the solar markers. Building on his argument, I argue that both ceremonies not only divided the Inca calendar, but also respectively introduced a period of transition in the course of the year, which corresponded to a change of power in the cosmos.

Disruption of the Cosmological and Social Order: The Transition Periods

First Transition Period: The Harvest Season

According to all accounts, the maturation of the crops ended in April. It was in that month when the Inca elite presided over the ceremony that inaugurated the harvesting period in the Cuzco valley.[36] On the first day of the festival, the king and the members of the panacas gathered in an open square outside the city called Limaqpampa (*rimaq pampa*), where the

---

32 Santo Tomás, *Lexicón*: 340; Betanzos: Pt. 1, Ch. 15, 71; Polo, "Tratado": 19-20; Anónimo [c. 1570]: 160; Molina: 117; Cabello: 350; Murúa, *Historia general*: Bk. 2, Ch. 38, 438; Guaman Poma: 238 [240]-241 [243], 213-215; González Holguín, *Vocabulario*: 141, 291-292; Cobo: T. 2, Bk. 13, 214; Beyersdorff 1984: 76.

33 González Holguín, *Vocabulario*: 113; Dedenbach-Salazar Saénz 1985: 64.

34 Santo Tomás, *Lexicón*: 250-251.

35 Duviols 1986: 179, 198, 206-207; Mills 1994: Ch. 5, 84-104.

36 Following MacCormack (1991a: 74-77), a few scholars have contended that the ceremony witnessed by Bartolomé de Segovia in April 1535 was "the last inti rayme" to have been performed by the Inca nobility (Salomon 1995: 336; Isbell 1997: 46). This statement is not only discordant with the primary sources that all situate *inti rayme* in June, but also overlooks the two very different settings of these festivities.

deities' effigies and the mummy bundles had been placed (Map 3). Turning towards the rising sun, they intoned a chant that lasted from dawn until dusk while the gods' attendants continually burnt offerings on a large pyre. These activities were repeated for eight days, whereupon the Inca, in person, broke the earth with a foot plough (*chaki taklla*) to initiate the harvest.[37] The *chiraw sukanka* would have served to fix the opening day of this festival that inaugurated, concurrently, the beginning of the dry season. This event occurred on April 26 (GC), when the sun passes through the local nadir on the latitude of Cuzco. At that time, the sun set precisely between the *chiraw* pillars before disappearing. The observation thus held a metaphorical significance because it indicated to earthly observers that the season ruled by the Daylight god had ended and was about to be substituted by the reign of the subterranean deity, Wiraqucha, during the dry season. The ritual calendar of the Huarochirí district also commemorated the beginning of the dry season in April with a celebration called *awkisana*. This major event was dedicated to the ruler of the dry season, the thunder deity Pariaqaqa. It was held at a time indicated by the high priest (*yanaq*), when the sun rose behind a hill and its light hit a "wall constructed according to very specific laws."[38] Alongside the descriptions of the *sukanka* pillars, this report is the only explicit information indicating that pre-Hispanic Andean people had elaborated artificial structures for solar observation.[39] They both fixed a celestial event occurring in the same month and very likely on the same day, that is, April 26. Another piece of evidence taken from contemporary data supports this connection. Descriptions of the Inca festival inaugurating the harvesting period indicate that the solar observation fixing its start occurred eight days before the labourers actually began reaping the maize. If the sun passage through a local nadir really determined the start of the celebration in Limaqpampa, the agricultural work would have begun on May 3. In present-day Peru, the important feast of *Cruz Velacuy* (the Wake of the Cross) ends on this exact date, following three days of continuous celebration around the boundary crosses of the valley communities. In the Cuzco region, these crosses are called *Taytacha* (Christ) and are erected on the communities' *chakras* at

[37] Segovia: 81-83.

[38] *Ritos y Tradiciones de Huarochirí*: 125-145; Polia Meconi 1999: 275-277.

[39] This does not imply that Andean people did not use other implements to calculate the occurrence of astronomical phenomena. However, the solar observations reported by the chronicles undoubtedly fixed major events of the calendar. Concerning the hypothetical identifications of shrines as astronomical observatories, see Zuidema 1981; Dearborn & Schreiber 1989; Bauer & Dearborn 1995.

the end of November in order to protect the crops from frost.[40] Just before the harvest, on the first day of *Cruz Velacuy*, they are taken from the fields and carried in procession to a central chapel, where they remain until the movable feast of Pentecost.[41] On the last night of the festivities, everyone keeps vigil in the attendants' house while many individuals avail themselves of the momentary sexual licence permitted during this transition period. *Cruz Velacuy* also occurs when the *puna* communities descend to the valleys to assist with the maize harvest, occasioning a significant population movement, which Murra associated with the concept of "ecological complementarity."[42] As Platt (1986) observed amongst the Macha of Bolivia, the workforce required for the harvest generates an intensive interchange of goods between the residents of the two ecological levels. The inhabitants of the high plateaux convey caravans of llamas transporting salt, dry potatoes (*ch'uñu*), wool, fabrics, and clay from their localities in order to trade them in the *qhishwa* zone with maize, chilli, honey, or wood. In the district of Puica (Arequipa), the llama breeders coming down from the *puna* remain two months in the service of farmers who remunerate them with the produce of their land.[43] The meteorological conditions of the dry season facilitate altogether this movement of the population and the interchange of goods between different ecological niches.[44] It is precisely the complementarity of the ecosystems that Andean people celebrate on the first three days of May, during which *Cruz Velacuy* is held. They do so by offering products of both the *puna* and the valley to their boundary crosses. Ultimately, this festival infringes intentionally on the social rules that order the Andean world by blurring the frontiers that ordinarily divide the communities living in separate ecosystems. The tensions between the people of the *puna* and of the valley are forgotten for the time of the celebration, allowing a promiscuity that is otherwise condemned outside the ritual context.[45] This disruption of prescribed rules is also closely associated with unsettling celestial events. *Cruz Velacuy* is conducted at a critical time of the year, when the Pleiades are absent from the night sky. Their disappearance occurs at the end of the rainy season (April 18), only a few days before the passage of the sun through the anti-zenith

---

40 Urton 1986.

41 The earliest date of Pentecost is May 10 while the latest is June 13.

42 Murra 1972. See also Flores Ochoa & Núñez del Prado 1983; Masuda et al. 1985.

43 Inamura 1981: 70-71.

44 Concerning the trade system in use in present-day pastoralist communities, see Flores Ochoa 1979: Ch. 3, 87-109.

45 Molinié 1997; Isbell 1978: 145-151.

on the latitude of Cuzco. The constellation then ceases to be visible for approximately forty-six nights, before its heliacal rise on June 3. The present-day celebration of the crosses and the harvesting itself thus occur as men wait for the reappearance of the celestial granary that will predict the next agricultural year.[46] In this regard, the month of May is a period of transition during which the order of the social world and the cosmos undergo a threatening disruption.

The chronicles report that similar events took place in Inca times during the period concerned with harvesting. At that time, the sovereign collected and redistributed the yields, food, textiles, and other rich products taken from his subjects as concessions to the empire. This economic system, which had prevailed for centuries before the Spanish conquest, had been adapted by the ruling elite for its own benefit. In that month, the official delegates from the four parts of Tawantinsuyu gathered in Haucaypata to hand over their tributes to the Inca. After discussing the government of their chiefdoms with them, the king distributed part of the benefits among his relatives and "dispensed the rest to the same kurakas who had come to this council; to the ones who originated from Collao he gave the items brought from the Andes. To the ones from the Condesuyo he gave what had been brought from other parts, things that they lacked in their lands; what had been given by some was offered to others."[47] Finally, the Sapa Inca sent inspectors to the provinces in order to verify the records of productivity provided by local officials and to list the quantity of crops and herded animals alienated to the empire.[48] In this way, the sovereign positioned himself at the centre of a system of reciprocity and redistribution that Murra has called "vertical economy" and that is still being celebrated today at *Cruz Velacuy*.[49]

The rest of the Inca festivities performed in that month were propitiatory rituals destined to divert the unsettling time between harvest and the start of the new agricultural cycle. After several weeks of work on the *chakras*, the nobility performed the closing festival of the harvest called *aymuray*, during which the crops left on the field were finally garnered. On the first day of this festival, the young Incas who had been initiated most recently donned tunics resplendent with gold and silver ornaments. They

[46] Urton 1981: 113-127; Urton 1986.

[47] Anónimo [c. 1570]: 157-158.

[48] Guaman Poma: 254 [247]-247 [249], 219, 221.

[49] Following Murra's (1982, 1995) definition of vertical economy, several studies have stressed the ideological significance of the figure played by the sovereign in the centre of the structures of Andean economy. See D'Altroy and Earle 1985; Morris 1986; LeVine 1987; Stanish 1997; Hastorf & Johannessen 1993.

then complemented this ceremonial outfit with bracelets and a headdress of yellow feathers before heading to the *chakra* of Sausero (Co. 2:3), on the outskirts of Cuzco, to collect the cropped maize now gathered in sacks. From there, they carried the crops to the storehouses while singing a chant named *arawi*, which narrated the deeds of the ancestor to whom the ritual was dedicated.[50] These recitations may have been directed at Mama Huaco, one of Manco Capac's sisters, because her mummy bundle was the only one to receive the chicha made of the yield of Sausero.[51] The first day over, the people of Cuzco joined the young men in their task before attending the *chakras* dedicated to the other Inca deities. When all of the crops had been stored, they assembled a small figurine made with certain parts of the maize that had been carefully selected and adorned it with fine garments. These *zara mama* figurines were then placed in the storehouses to protect the crops during the year.[52]

More agrarian work soon followed *aymuray*, when several Inca noblemen wearing their martial attire returned to Sausero and ploughed the land to prepare it for the next sowing. This particular ploughing, called *kuskiy*, initiated the new agricultural cycle and occurred exclusively in the dry months.[53] After this arduous labour on hard soil, the young Incas returned to Cuzco and stopped at Limaqpampa, where the Inca and the ruling elite had previously opened the harvest season in late April.[54] On this open plaza they joined the adults to encircle four vicuñas trapped in its centre. The young boys then attempted to catch one of the wild animals barehanded, enacting in this way a traditional hunt similar to the one the pastoralists of the high plateaux performed.[55] Those who successfully captured a quarry shared its flesh amongst all of the attendants and were regarded in high esteem because, according to Cobo, "this was the sacrifice they made for victories."[56]

---

50 González Holguín, *Vocabulario*: 152.

51 Molina: 118.

52 Betanzos: Pt. 1, Ch. 15, 71; Molina: 118; Polo, "Tratado": 20-21; 1917: 27; Fernández: Pt. 1, Bk. 1, Ch. 10, 86; Gutiérrez: Ch. 65, 256; Cabello: 350; Cobo: T. 2, Bk. 13, Ch. 27: 214-215; Arriaga: 37-38.

53 González Holguín, *Vocabulario*: 72. See also Tables 11 and 12 and the use of *kuskiy* in reference to April, May and June.

54 Cobo, who wrote more than a century after the Spanish conquest and did not witness these events, suggests that this rite was held in Haucaypata. On the other hand, Betanzos reports that the celebrations held between May and June all took place in Limaqpampa, which presents an undeniable coherence with the descriptions of the opening of the harvesting, dedicated to the sun, and also conducted in Limaqpampa.

55 Cobo: T.1, Bk. 9, Ch. 58, 368.

56 Cobo: T. 2, Bk. 13, Ch. 27, 215.

The warlike connotations of this last ritual characterized another ceremonial hunt that the ayllus Huari and Llacuaz performed at the same time of year in the Cajatambo province (Huarochirí). It was on June 3, as the Pleiades reappeared in the sky, that the communities of the central sierra celebrated a new season called *unquy mit'a* (the season of the Pleiades). This date marked the end of a period of transition and uncertainty for the cultivators, as the sky was seen to display the propitious signs that foretold the coming agricultural year. On this occasion, the Llacuaz ayllus commemorated their ancestors' victory over the aborigines of the valley, known as Huari. They fasted for five days to honour the Thunder god and then proceeded to the high altitudes of the *puna* to capture several camelids destined for general feasting. On their return to the village that same day, the hunters were received amicably. The Huari ayllus proffered to them chicha, maize, and potatoes, but they also performed a dance in acknowledgement of the Llacuaz's offering of meat.[57] Duviols (1973) suggests that this ritual of confederation commemorated the bygone warfare that had forged the social and spatial order of the two ayllus, while the mutual offerings reflected the reciprocity binding the Andean ecosystems.

The festivities performed in Inca Cuzco at the time of the transition between the two agrarian cycles displayed similar features. They involved a hunt of wild animals that recalled past warfare and celebrated ecological complementarity through the redistribution of tribute. In addition, in the central highlands as in the Inca heartland, the observation of the Pleiades was instrumental in fixing the time for agricultural tasks, so that their reappearance on June 3 would have equally determined the appropriate time to conduct the first ploughing of the season in Sausero. This parallel is all the more significant in that the *kuskiy* had preceded the ceremonial hunt performed in Limaqpampa, where the opening of the harvesting season on April 26 had been conducted. The two Inca celebrations were interconnected in much the same way that the feast of Pariaqaqa (*awkisana*), held in the district of Huarochirí on a date fixed by the solar observation in April, was related to the festival of Chawpi Ñamca, which was an event held forty days later.[58] According to colonial records, it was dedicated to Pariaqaqa's sister and frequently coincided with Corpus Christi, the movable feast held on the Thursday after Trinity Sunday.[59] Yet, if *awkisana* was held on the day of the Sun's passage through the nadir, as Zuidema suggests, the

---

[57] Duviols 1986: 161, 177.

[58] Polia Meconi 1999: 277; *Ritos y Tradiciones de Huarochirí*: Ch. 10, 147-159; Ch. 9, 133.

[59] The earliest possible date for Corpus Christi is May 21; the latest is June 24.

festival of Chawpi Ñamca fell exactly on June 3, when the Pleiades reappeared in the sky.[60]

Analysed in light of the central sierra festivals, the closing ritual of the harvest in Inca Cuzco takes on a more explicit meaning. The redistribution of goods and wealth that followed the visit of foreign officials to the capital, the ritual hunting of wild vicuñas and the apportioning of their sacrificial flesh amongst all of the celebrants embodied an enduring pattern of Andean culture on an imperial level. In other words, they staged a temporary alleviation of the antagonisms that traditionally placed the upper and lower social and cosmic divisions in opposition. In the central sierra, these rituals commemorated the victory of the Thunder god's worshippers, herders of the high plateaux, over the valley farmers, and their subsequent settlement with the vanquished. In a similar fashion, the ritual ploughing of Sausero, carried out after the crops' garnering, involved the members of the nobility dressed in their war apparel to perform the usual sacrifice conducted in times of victory. These indications suggest that in the month of *aymuray*, the city of Cuzco also honoured the subjugation of one of its moieties to the other. This change in power began on April 26, forty days before the heliacal rise of the Pleiades, when the *chiraw* pillars marked the end of the *mit'a* presided over by P'unchaw. As the weak Sun of the rainy season made way for the vigorous god of the dry season, Wiraqucha, his vital force and capacity to animate similarly declined. The cosmic confrontation of these two divinities impacted in turn the social order of their terrestrial offspring whose might and forcefulness depended on their ancestor's *kamaq*. Accordingly, when the change appeared in the firmament, the Llacuaz and the Huari, just like the Hanan and Hurin in Cuzco, ritually enacted the victory of the upper moiety over the worshippers of the Sun god P'unchaw. Although the period during which these events occurred, between late April and early June, was a critical time in the agricultural calendar for the maize cultivators, the alternation of the two seasons was also deemed a requisite condition to guarantee the healthy maturation and appropriate conservation of the crops. Thus, while the universe underwent a complete disruption of the cosmological and social order, Andean communities celebrated simultaneously the complementary aspect of this transformation. In Inca Cuzco and in the central sierra, this transition period ended with the apportionment of the sacrifices and the redistribution of goods symbolizing the communion of the two

---

60 Zuidema 1996: 199-200.

moieties and their settlement together. Once the celebrations of reciprocity ended in June, the ruler of *chiraw mit'a*—Wiraqucha in Cuzco and the Thunder god in the central highlands—could assume power over a conciliated world. From that point, his reign would last for approximately five months before P'unchaw rose again to claim his due. It was when the first rains of *puquy mit'a* started to fall that the universe underwent another disruption, which mirrored the astronomical and ritual events of this first transition period.

Second Transition Period: The Height of the Wet Season

Another set of astronomical observations disclosed by Zuidema and Aveni provide the first indication of a recurring time pattern. Discerning the crucial importance of the sun's passage through the local nadir, the two scholars suggest that the star's passage through the zenith on the latitude of Cuzco marked the beginning of the rainy season. On this occasion, the *puquy sukanka* fixed the exact date of this phenomenon as when the Sun rose between the two pillars. For an observer located in the Inca capital, this event occurred on October 30, only a few days before the Pleiades reach their culmination period from November 5. For fourteen nights the cluster stands at its highest point in the sky and appears at its maximum visibility for a ground witness.[61] This occurrence and the foregoing solar phenomenon were obvious manifestations opposing the celestial events opening the dry season, which were marked by the disappearance of the stellar constellation and the sun's entrance into the subterranean world. Confirming this dualistic construction, the pre-Hispanic ceremonies and the communal activities performed at this time of year reciprocated the rituals conducted from late April to early June, but presented opposing social paradigms.

November and December correspond to the birthing period of llamas, when the dark cloud constellation of the animal rises in the sky.[62] At that time of year, Inca officials visited the four provinces of Tawantinsuyu for the second time. They inspected the *aklla wasi* with their depositories of ceremonial items, and made an inventory of the herds and crops handed over to the empire. Simultaneously with the execution of these duties came the inauguration of a ceremony dedicated to the livestock allotted to the king and to the *wakas* of the whole empire. The anonymous author of the

[61] Zuidema 1981; Zuidema 1982a: 92-93.
[62] Urton 1981: 187.

"Discurso" (c. 1570) also indicates that "in areas where there are wild herds, they organized *chacos* and large hunts of guanacos and vicuñas."[63] Zuidema and Urton (1976) note that similar rallies were conducted in the district of San Damián (Huarochirí) at a time of year that coincided with the apparition of the Llama constellation in the sky of the rainy season. This custom conveyed a ritual significance comparable to the hunt performed by the Llacuaz at Corpus Christi. More specifically, it recalled the uniting of the upper and lower populations, called Checa and Yunga, after the mythical intervention of one of Pariaqaqa's sons on earth.[64] Tradition recounts that in remote times, the thunder deity descended from the high altitudes upon the valley in the form of menacing rain and summoned the worshippers of the Sun god to mix with the upper communities as brothers would. Every year, on the day of San Andrés (November 30) when "they asked for the rain to come", the residents of the high plateaux followed the mythical route of their ancestor and performed a ritual hunt in his honour. Those who captured a guanaco or a taruca (Andean deer) would crop its tail and attach the trophy to their headdresses before executing a dance. The rest of the animal was offered to their ayllu's minister ("huacsa") to be sacrificed.[65] In the present-day Cuzco region, San Andrés is still an important celebration marking the end of the sowing period. Urton indicates that in Pacariqtambo, most seeds have to be planted before this particular day when the two moieties of the community erect the *Taytacha* crosses on top of the hills. In this way, November 30 opens a season protected by the roods, which ends in early May with the celebration of *Cruz Velacuy*.[66]

In pre-Hispanic times, November was a period of ritual preparation for the Inca initiation ritual that started at the turn of the month. During these weeks of anticipation, each novice-to-be accompanied by a male relative

---

63 Anónimo [c. 1570]: 159; see also Molina: 110.

64 The yunga refers to the mountain foots of the Andean highlands, situated at an altitude between 500m and 2500m altitude above sea level. In the eastern side of the Cordillera, it constitutes a transitional region between the Altiplano and the rainforest where the climate is extremely humid and warm, adequate for coca cultivation. In Pre-Hispanic times, the Yunga designated the ethnic groups living in the central coast of Peru, between the littoral and the highlands. They worshipped Pachacamac who, in several traditions, opposed Pariaqaqa, the main deity of the Checa, a population of the highland altitudes that lived in the district of San Damián.

65 *Ritos y Tradiciones de Huarochirí*: Ch. 11, 161-171. According to Salomon and Urioste (in Avila 1991: 18) the "huacsa" is a non-hereditary priesthood office that rotated among the ayllu members.

66 Urton 1986: 54. Regarding the celebrations of San Andrés in the Bolivian highlands, see Van Den Berg 1989: 65-68.

celebrated the *itu* ceremony in Haucaypata. This event, performed both periodically and on extraordinary occasions, was restricted to the descendants of noble lineages. It witnessed the representatives of the two moieties of Cuzco sacrificing eight llamas so as to incite an abundant downpour of the first rains. Polo observes that the dances performed on that occasion were also enacted at the time of Corpus Christi, the Christian celebration coinciding with the closing of the harvesting period.[67] Other evidence of the symmetrical nature of the May and November celebrations can be found in the collection of tribute carried out at both times. For, following the *itu* festival, provincial officials brought to Cuzco the refined garments that their people had prepared as a contribution to labour services.[68] However, contrary to the reciprocity that concluded their previous visit after harvest, these goods were intended exclusively for the ruling elite, who dispensed them among the Inca novices during their initiation. Then, the last activities closing *warachiku* echoed once again the festivities that closed the agricultural cycle: the Hurin and Hanan probationers fought each other in a mock battle, thus replicating the hostilities that placed the two cosmological moieties in opposition at the turn of the rainy season. Following this bloody encounter, the men and women of Cuzco left the city to perform a ploughing task called *chaqma,* consisting of weeding and breaking the soil of the idle fields that had been softened by the rains.[69] Contrasting with the earlier *kuskiy*, this labour was conducted on the wet ground of the rainy season before the foreign delegates could re-enter Cuzco and receive "maize bread with sacrificial blood of which they partook as a sign of confederation with the Inca."[70]

On the whole, the ritual and administrative activities of this period of transition mirrored the events held between late April and early June, starting with the ritual hunt and the collection of labour services, followed by the moieties' confrontation and the completion of agricultural labours. Here again, the symmetry of this structure was not specific to the Inca calendar. In the central sierra as well, the colonial festivals of *puquy mit'a* shared common patterns with those of *qarwa mit'a.* Kenneth Mills observes that it was only during these two celebrations that the residents of the Lima Archdiocese secretly confessed their wrongdoings to their local *hechicero,* in the hope that their crops would be preserved from

---

67 Polo, "Tratado": 26; Cobo: T. 2, Bk. 13, Ch. 30-31: 220-222.

68 Guaman Poma: 257 [259], 231; Anónimo [c. 1570]: 159.

69 Guaman Poma: 860 [874], 806; González Holguín, *Vocabulario*: 91; Beyersdorff: 22.

70 Polo, "Tratado": 18-19. See also Betanzos: Pt. 1, Ch. 14, 65-70; Molina: 98-112; Anónimo [c. 1570]: 159-160.

calamities.[71] The first confession occurred "around Christmas or a little later", "when [the Indians] prepared their chakras, before cleaning and ploughing them,"[72] while the two other confessions opened and closed the transition period between the harvesting and the heliacal rise of the Pleiades.[73] It follows that this description not only parallels the chronology of the Inca rituals, but also supports Zuidema's assumption that the practice of confession in the central sierra echoed the rituals of communion performed in Cuzco at similar times.[74]

Another point of comparison between the festivities of the central highlands and those of Inca Cuzco concerns the periodization of the *puquy* and *chiraw* seasons. Colonial documents reveal that, in both locations, the two celebrations opening these *mit'a* were held before the completion of agricultural works. Obviously, soil preparation before and after planting could take place several times annually and depends on the altitude and irrigation methods, but the particular labours conducted after the *warachiku* were called *chaqma*. They specifically consisted of ploughing ridges. Today, this labour is traditionally conducted by "three men together, working aligned: the one in the middle, *chawpi*, gives the rhythm for the two others (...) Two women or children, *rapa*, squat on each side of the drill in order to turn over the clods" and break them into two or three hunks.[75] The same scenario was documented in the mid-1570s by an anonymous chronicler who reveals how the ruling elite performed this labour on the maize fields allotted for ceremonial purposes:

> [The Inca] placed himself for all to see on the chakra he wished to prepare then sat on his gold seat. Later, they put before him a gold *taklla*, which is a swing plough, and just behind him sat the four counsellors of his empire with the swing ploughs for their use and behind them sat the caciques who were in charge of ten thousand men, and following them came the provinces' ambassadors, each one with a swing plough (...) When the Inca deemed it was time to start working, he alone stood up followed by his wife and her retinue (*dueñas*), and took the swing plough in his hand and jabbed eight or ten times in the soil removing clods of earth, and his wife with her retinue broke them into pieces before being seated; the four counsellors of the empire would then stand up and with their wives started to plough where the Inca had left off, and did twice what he had done before turning back to take their seats; and later the caciques who were in charge of the

71 Mills 1994: 84-104; see also Acosta: Bk. 5, Ch. 25, 256-261.
72 Duviols 1986: 181.
73 Arriaga: Ch. 5, 57-60; Duviols 1986: 179.
74 Zuidema 1996: 190-193.
75 Gade & Rios 1972, in Morlon 1992: 61.

> ten thousand men stood up with all the men and women present and they worked until it was time to lunch (...) and after they had eaten, they all got up to work except the Inca, the four counsellors of the empire and the caciques who were in charge of the ten thousand men who played a game called "pisca", which is similar to playing dice, although it is larger and made of wood.[76]

The *pichqa*, which means "five" in Quechua, is a game well documented throughout Tawantinsuyu that has remarkably survived centuries of Spanish occupation. Also known as *wayru*, it was and is still today used as an augury device in ceremonial contexts, particularly for post-mortem rites.[77] In pre-Hispanic times, not only was the game believed to communicate the deities' wishes, but it was also the common pastime of the living during the five nights of wake held after the death of an individual.[78] Attesting to its persistent association with death, Isbell witnessed in 1967 a *pichqa* divination and purification ritual in the town of Chuschi (Ayacucho) that was conducted on the fifth day of a deceased's wake.[79] In Inca Cuzco, Guaman Poma wrote that noblemen played the *pichqa* and other games, such as one called *ayllus* (also known as *riwi*), at harvest time.[80] The first was played inside a sacred enclosure with dice in the shape of a truncated pyramid, while the second took its name from the weapon called *ayllu*, composed of three cords of gut or leather, each with a ball of metal at the end. To win this game, the contestants had to wind the ropes of their *ayllus* as many times as possible around a rod thrown into the air. The king himself played this game with a ritual instrument in the shape of a serpent made of wool.[81] The Incas played both games in specific festive contexts in order to predict the outcome of an enterprise or the success of the coming agricultural season. The *ayllus* may have also provided the ruler with a means to acquire more subjects to work on consecrated fields, and to reassign agricultural plots among his kin of mixed Inca blood.[82] As Albornoz reports, the king did indeed gamble the herds, arable lands, and service of labourers pertaining to provincial *wakas* in order to redistribute their resources. Most importantly, evidence in primary sources reveals that such competitions also took place during, or just after, the initiation ceremony

---

76 Anónimo [c. 1575]: 165-166. My translation.
77 Gose 1994: 117-118; Gentile 1998; Salomon 2002; Arellano Hoffmann 2003.
78 Arriaga: 66. See also González Holguín, *Vocabulario*: 284; Santo Tomás, *Lexicón*: 283.
79 Isbell 1978: 6-7.
80 Guaman Poma: 243 [245], 217.
81 Albornoz: 174-175.
82 Zuidema 1989a: 256-272; Ziólkowski 1996: 257-285.

of young Incas, suggesting that both were played twice a year at the time of the transition between the two Andean seasons. The first indication is provided by the association of the *ayllus* with the cosmological serpent that appears in the night sky at the beginning of the rainy season. In Cuzco, the image of this sacred snake, like the instrument used in *ayllus* contests, was made of multicoloured wool and taken out of its temple only once a year, in January, following the completion of the *chaqma* labours. Providing further support to the association of ritual games with this particular period, an Inca narrative relates how Tupa Yupanqui lost five *pueblos* at an *ayllus* contest against one of his sons. This episode would have taken place during the *warachiku* of this young boy whose mother was a native of the Huayro nation and the Inca's favourite wife. The story tells us that the ruler's feelings for her were such that he decided to give her name to the ace of the *pichqa* die, which became known as *wayru*.[83] This element, as well as the symbolic number of the five prizes won by the youth, demonstrates that the Inca's dice and the *ayllus* were closely related games occupying the attention of the nobility between December and January, as well as after the harvest in May. Finally, Guaman Poma associates the time of the *warachiku* celebration with the festival of the dead, whose presence among the living during *puquy mit'a* was believed to influence the sprouting of seeds (Figure 5).[84] The concurrence of these two ceremonies would therefore confirm that the *pichqa* die was a game associated with the spirit of the dead lingering and not yet departed for the afterworld.

Salomon recently recorded similar events taking place at the start of the rainy season in the district of Huarochirí. He observed there that around New Year's Day all of the communities organize a plenary meeting that includes a *pichqa* contest placing the outgoing political authorities in opposition to their successors. In addition to taking place in a *kancha* divided into an upper and lower *barrio*, the game is composed of two gender-specified sequences. On this occasion, the dice throws are believed to express the will of a divine pair called *dueños del agua* (rulers of the water), which are interpreted as an augury for the coming agro-pastoral year. Each year, these two deities are invoked a second time, also towards the end of the rainy season, prompting Salomon to analyse this ritual as a celebration of two traditional structures: one is "the division of the space between opposed halves and [the other] is the division of time in alternating intervals"

83 Cobo: T. 2, Bk. 12, Ch. 15, 86-87.

84 On this issue, see Gose 1994: 114-123; Harris 2000; Stobart 2006: 199.

Figure 5. November: The Festival of the Dead. In Guaman Poma 1980: 256 [258].

corresponding to the two seasons.[85] In light of his argument, but also thanks to the descriptions provided by Guaman Poma and the anonymous chronicler (c. 1575), it is possible to relate the time setting of the pre-Hispanic *pichqa* and *ayllus* with the ritual games of today. Primary sources indicate that Inca noblemen were occupied with these recreations at the time of agricultural labours, which were clearly identified in one instance with the harvesting activities held between April and May.[86] Given the evidence associating the *pichqa* and the *ayllus* with the ceremonies conducted in December and January, it is most likely that they also concurred with the *chaqma* of the rainy season. Molina situates these activities after the mock battle that closed the *warachiku* festival and just before the celebration of the snake-like figure in early to mid-January. These indications set the ritual games on the very first days of the Gregorian year, just like the

[85] Salomon 2002: 17.
[86] Guaman Poma: 243 [245], 217.

contests conducted in the present-day district of Huarochirí. It seems, therefore, that the intermediary period between December and January was regarded as a crucial transition in both the Inca and the Christian calendars, which would explain in part the persistence of the *pichqa* as an augury device for predicting the New Year in contemporary communities.

Molina provides a last piece of information that completes this reconstruction. Interrupting his account of the initiatory rite, the chronicler reveals that another celebration, separate from the youths' ordeals, was held outside Cuzco precisely twenty-three days after the start of *warachiku*. On that occasion, a procession carrying the image of Wayna P'unchaw left Cuzco and made its way to Puquín hill beyond Cayaucache, on the outskirts of the city. Leading the cortège were the *suntur pawqar* insignia and the two precious llamas figures that had also opened the march of the celebrants at *inti rayme*.[87] The procession reached a temple dedicated to the Sun in *pukyu kancha* (the enclosure of the water spring, Cu. 10:2), the name of which was reminiscent of the water source from which the solar star arises.[88] There, they offered their oblations for the prosperity of the people and their land before commemorating the regeneration of the young sun. The significance of this celebration in reference to the sun's mythical birth and its resemblance to the festivities of the June solstice suggest that it was conducted on the day of the summer solstice (December 21-22), when the sun is at its southernmost point at noon.[89] If such an assumption were correct, it would bring the opening of the initiation ceremony to exactly twenty-three days before this date, on November 30 (GC), which today coincides with the Christian festivity of San Andrés marking the beginning of *puquy mit'a*. These elements provide corroboration of Urton and Zuidema's hypothesis (1976) that the combination of the heliacal rise of α and β Centaurii—the eyes of the dark Llama cloud—and the zenith sun on October 30 served to fix the date for the Inca initiation rite one month later, in late November. Stemming directly from this conclusion, the mock battle that closed the *warachiku* after the solstice celebration and the re-entrance of the foreigners into Cuzco would have taken place on the last week of December (GC). This point eventually

---

87 The *suntur pawqar* has been given different descriptions. Depending on the chroniclers, it was a "bulto", a "rich and coloured house" or a "royal insignia". See Cobo: T. 2, Bk. 13, Ch. 27: 214; González Holguín, *Vocabulario*: 332; Molina: 102. Guaman Poma: (329-330 [331-332], 302) mentioned the royal palace of *suntur wasi* where the figure would have been kept.

88 *Pukyu*: spring of water. See Bauer 1998: 124-125, 129-130.

89 Zuidema & Urton 1976: 89-94.

establishes a correlation in time between the rituals of Inca communion held in Cuzco after these festivities and the confession rite at *puquy mit'a* in the central highlands, which occurred "around Christmas" as reported by Arriaga.

In sum, the Inca festivities conducted from late April to early June constituted an indivisible series of celebrations revolving around the change of seasons and the reciprocity of labour. This group of rituals, coinciding with the disappearance of the Pleiades, may be described as a transition period starting with the sun's passage through the nadir and ending with the redistribution of wealth and the sharing of sacrificial meat, all of which were orchestrated by the Inca elite. During this interval, the antagonisms placing the ruling deities of *chiraw* and *puquy mit'a* in opposition disrupted the cosmological order and affected the pre-existent interrelations that structured human societies. Even today, this interval of time is conducive to the formulation of new kin ties between peoples of different ecological levels, leading to the temporary alleviation of social rules condemning sexual intercourse between the *puna* and valley residents. The next transition period, which took place in November, replicated the same structure starting with the sun's passage through the local zenith on October 30. On that date, the Inca novices assembled in order to celebrate the *itu* festival, and for the next thirty days the nobility prepared the ceremonial food and garments that they would later offer to their youths in honour of their coming of age. On this occasion, the reversal of the cosmological order presented a pattern opposite to its earlier counterpart; this time around, P'unchaw was preparing to defeat Wiraqucha and the Pleiades were close to their culmination period. In this context, human activities also had a different significance, as the collection of labour services was no longer intended for redistribution, but was strictly devoted to the Inca nobility and their divinities. Eventually, the start of *warachiku* on November 30 ended this intermediary period on the same day that the communities of the central sierra conducted a ritual hunt of wild camelids. This last event, held at San Andrés, could be seen as a mirror image of the celebration of *unquy mit'a* held on June 3 when young Inca males enacted the capture of four vicuñas on the main plaza (Figure 6).

This overall picture is concordant with the chroniclers' testimonies giving two different dates, June or December, for the start of the Inca year. Each of these dates can be seen as initiating a new cycle. Much like the two narrative variants of the Incas' settlement in the Cuzco valley, this discordance can be attributed to moiety membership so that Hurin members

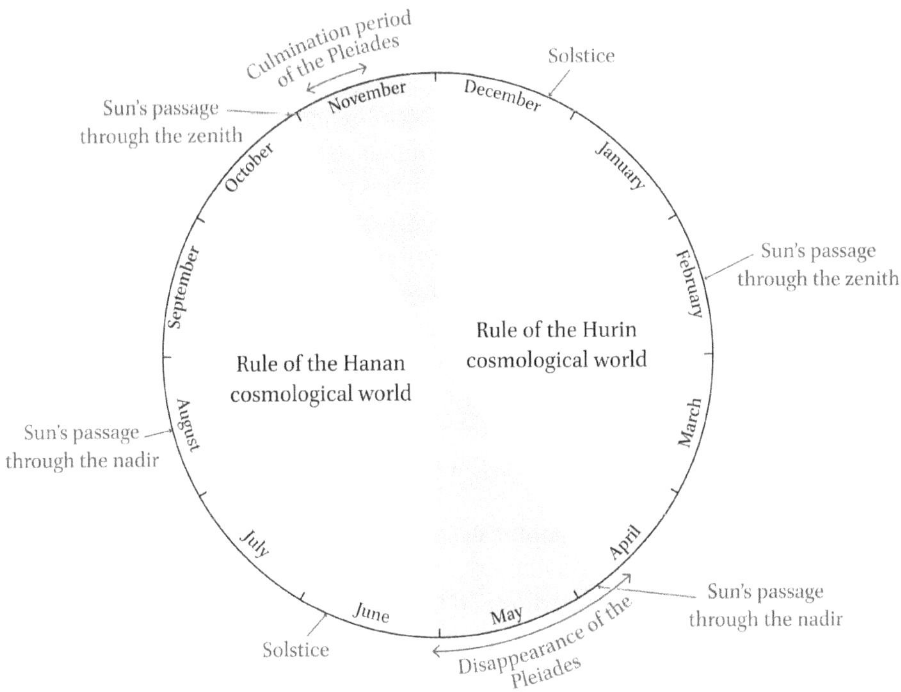

Figure 6. The ritual calendar of the Inca year, with its transition periods.

started the year with the celebration of their tutelary deity in December, whereas Hanan members situated it in June, with the advent of Wiraqucha. This construction shows that the bipartition of the metropolitan calendar had progressively detached itself from strict ecological manifestations to be more directly informed by astronomical events and social interactions. It also gives a similar length to each partition of the year, that is, 147-148 days for *puquy mit'a* and 149-150 days for *chiraw mit'a*, which are ideally divisible into five months of 29.4 to 30 days.[90] The resulting total of ten months corresponds in turn to the number of royal lineages that arguably rotated to assume the responsibilities of the ritual calendar. To further evaluate the reliability of this model and to investigate the possible link between the panacas and the conduct of the festivities, let us now turn to the study of the celebrations that comprised each temporal division.

90 The one-day variation for each *mit'a* depends on the fluctuation of astronomical events, especially the passage of the sun through the local zenith and nadir.

The Rule of Wiraqucha

The activities of the new agricultural year started in June when the prediction for the coming cultivation cycle was announced. Still performed today, this augury consists of observing the luminance of the Pleiades on June 24, the day of San Juan, two to three days after the winter solstice.[91] This time of year coincides with the Inca celebration of *inti rayme*, which commemorated for more than twenty-five days Wiraqucha's mythical path from Lake Titicaca to the sea.[92] This journey also marked the dawn of a new humanity as the deity and his attendants walked across the land and ordered the nations who had been following the course of the subterranean river to emerge from their *paqarina* and populate the earth. It was this particular tradition, tracing the origin of the Inca ancestors back to the *qullasuyu* region, that the people of Hanan Cuzco claimed as their own. Accordingly, it was the Sapa Inca, as head of the Upper moiety, who conducted the festivities on Mantucalla hill while a group of celebrants from the Tarpuntay non-noble ayllu, also members of Hanan Cuzco, proceeded towards Vilcanota to pay tribute to the *wakas* at which Wiraqucha had stopped on his way to the sea.[93] From this celebration rose the rule of the subterranean Sun, who then received the central devotions of the dry season festivals centred on cleansing, irrigation, and the early maturation of crops.

Pachayachachiq and the Earth Regeneration

After the closing rite of *inti rayme*, the Cuzqueños and villagers of the high *puna* irrigated their fields in preparation for sowing the early crops and tubers also known as *maway* potatoes.[94] This activity usually lasted from July to mid-August, a period that Betanzos calls *qarpay killa*, "the month of irrigation."[95] In most documents, however, the problematic word

[91] Urton 1981: 118-199; Orlove et al. 2002.

[92] The duration of *inti rayme* is difficult to estimate. Molina indicates that the first two days of this festivity took place in Cuzco and Huanacauri before the Tarpuntay officiants left the city for a pilgrimage of twenty-two days. Meanwhile or after this pilgrimage, the Inca and the nobility went to Mantucalla for an undetermined number of days.

[93] Molina's description of the *sitwa* festival and Sarmiento's account of the Inca ancestors' migration both associate the Tarpuntay Ayllu to the Hanan moiety. See Molina: 75; Sarmiento: Ch. 11, 53.

[94] "Mahuay o uripapa: the very early potatoes" in González Holguín, *Vocabulario*: 235; Guaman Poma: 248 [250]-249 [251], 223.

[95] Betanzos: Pt. 1, Ch. 15, 72; González Holguín, *Vocabulario*: 137; Beyersdorff 1984: 82.

"chahuaruay" designates the month of July.[96] Three possibilities are more likely to translate its meaning, all constructed around the verb *rura-*, "to make, realize."[97] *Ch'awar* is the native name for maguey (*agave americana*), a plant commonly used in Andean housing, the leaves of which yield fibres "the Indians use [to make] bags for coca, sandals and arquebus' fuses."[98] Although no sources record these activities as taking place around July in Inca times, house rethatching in modern Andean communities occurs approximately during that period, from August to mid-September.[99] Gose observes that this activity coincides with a seasonal shift from domestic appropriation to the collective production initiated by the growing season.[100] "Chahua" could alternatively be associated with *chawa*, "the fruit still to ripen" or "something raw" in reference to the preparation of the seeds in July,[101] but a third option of translation deriving from the Aymara appears to be more plausible. In Bertonio's lexicon, the "cchahua" is "a club used to break up the clods (*desterronar*) in the chakra, it is a stone attached to a stick."[102] This implement was originally used on dry soil after it had been ploughed and before sowing began, three conditions that characterize the fields in July. Gutiérrez also indicates that "chahuaruay" meant "month of labour", when several offerings of chicha were poured into the irrigation canals and the rivers to induce a prosperous year with abundant rain and plentiful yields. In light of these descriptions, it is more likely that the Inca month corresponding to July was named after the ploughing tool of the dry season. During this period, the Incas sacrificed a hundred head of brown llamas and dedicated two of them to the *waka* of Tocoripuquiu (*Tuquri pukyu*), a spring of water located in the *chinchaysuyu* division and "from which issues a stream that passes through the city."[103] Traditions recount that Inca Roca generated it by pressing his newly pierced and bloodstained ear onto the ground of Chacca hill, where water suddenly burst forth from the soil to become Cuzco's main water supply. Each year in July, members of the ruler's panaca immolated two llamas in memory

---

[96] Betanzos: Pt. 1, Ch. 15, 72; Polo, "Tratado": 22; Anónimo [c. 1570]: 160; Molina: 71; Fernández: Pt. 1, Bk. 1, Ch. 10, 86; Gutiérrez: Ch. 65, 255; Cabello: 351; Cobo: T. 2; Bk. 13; Ch. 28, 216.

[97] González Holguín, *Vocabulario*: 322; Santo Tomás, *Lexicón*: 349.

[98] Cobo: T.1, Bk. 6, Ch. 125, 283; *Ritos y Tradiciones de Huarochirí*: 335, 411.

[99] Isbell 1978: 168; Gose 1994: 74-90.

[100] Gose 1994: 74.

[101] González Holguín, *Vocabulario*: 92; Santo Tomás, *Lexicón*: 259.

[102] Bertonio: 74.

[103] Polo, "Relación de los adoratorios": 42. The location of this shrine is uncertain, see Bauer 1998: 71-72.

of this wonder, one where the irrigation of the valley began and the other where it ended.[104] With this act, the descendants of the first Hanan ruler not only commemorated their command of this fountainhead but also placed the Upper moiety in control of the city's irrigation canals.[105] It also initiated the ritual cleansing of the water channels (*yarqa aspiy* or *yarqa faena*), which was performed at the beginning of every agricultural season and sometimes lasted several months (Figure 7).[106]

In the Cuzco valley today, the early sowing of maize starts immediately after the feast of the Virgin's Assumption (August 15) and usually lasts up to a month. However, in many parts of the Andes, this activity ends in late December because it is not only contingent on climatic and ecological conditions, but also varies according to the nature of the produce. In Inca times, as in present-day Cuzco, the official sowing season started in the same period—when the pillars erected outside the city indicated for the second time the passage of the sun through the local nadir on August 16—and it would have lasted several weeks. When the moment had come to sow the *chakras*, the Incas conducted several sacrifices of llamas and guinea pigs in and around Cuzco for the preservation of their crops. As they started to work on the fields, they intoned a chant of victory called *haylli* that celebrated triumphal warfare. As they did so, they laboured "as if they triumphed over the earth, ploughing it up and penetrating the soil so it will bear fruit."[107] Bauer (1996) notes that the violent connotation of this agricultural task was overtly paralleled with warlike activities, for both possessed fertilizing qualities.

While sowing was under way, an important ritual was held in the designated *chakra* of Sausero, where several *akllakuna* dedicated to the Sun poured chicha around a white llama attached in the centre of the field. This offering occurred at a time of year when the vital power of Pacha Mama (Mother Earth) is depleted by the last agricultural cycle and needs renourishment. The ceremonial beverage, spilled on the soil during sowing, would appease her thirst and impregnate her womb, ready to be fertilized. This

---

[104] Cobo: T. 2, Bk. 13, Ch. 28, 216.

[105] For a more thorough analysis of the Cuzco irrigation system, see Sherbondy 1987; Sherbondy 1994.

[106] Depending on the altitude and the type of cultivation, today *yarqa aspiy* is performed from July to October. See Isbell 1978: 138-145; Urton 1986: 52; Zuidema 1986b: 184; Paerregaard 1994: 196-197; Ossio 2002: 482-492.

[107] Garcilaso: Bk. 5, Ch. 2, 228. "Hayllini, gui: to capture someone" in Santo Tomás, *Lexicón*: 290; "haylli: song joyously performed in warfare, or on the chakras well finished and vanquished" in González Holguín, *Vocabulario*: 157; Pachacuti Yamqui: f. 25 v., 232.

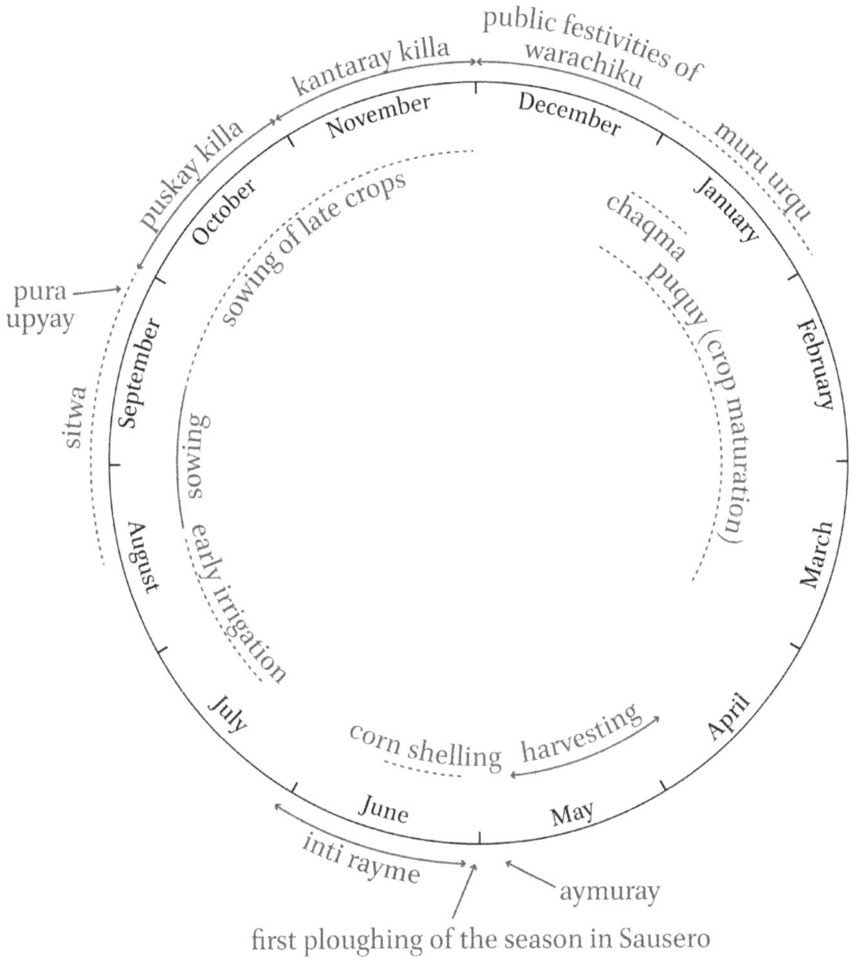

Figure 7. Identifiable festivals of the Inca calendar in parallel with the agricultural cycle (The dashed lines indicate moveable feasts and extendible periods).

practice is concordant with contemporary beliefs prevalent in Peru as well as Bolivia, where locals maintain that the earth is enraged in August and February before "resurrecting" and "opening up" to receive ritual offerings.[108] Molina reports that in pre-Hispanic times, the priests of the Tarpuntay Ayllu, who had already served Wiraqucha at *inti rayme*, fasted from the time the maize was planted until it reached one finger high.

---

108 Isbell 1978: 138, 151-163; Hopkins 1982: 178; Tomoeda 1993: 290; Rozás Álvarez and Calderón García 2000: 317; Ossio 2002: 482.

Throughout this critical period of germination, while the seed grew underground and benefited from the warmth of the subterranean sun, the *hatun runa* danced in honour of Wiraqucha and asked for his benevolence in the year to come. Meanwhile, the sole drink consumed by the Tarpuntay celebrants was cloudy chicha, also known as *qunchu* or *aqha mama*, which is the deposit that collects at the bottom of the fermented liquor and is kept aside as a fermenting agent for the preparation of the next batch. Robert Randall argues that this sediment is comparable to a fertilizing component indispensable for the completion of the ceremonial drink. By drinking only this substance during the crops' early growth, the Tarpuntay ministers accompanied and replicated the yields' maturation process in their own bodies.[109] This symbolic act appears all the more significant in that the same ayllu officiated at the service of Wiraqucha several times during the annual cycle in order to regenerate the god's fertility: it fed the subterranean Sun during the deity's terrestrial journey along the Vilcamayu River at *inti rayme*, it fasted during the early phase of the maize maturation in July, and it finally closed the agricultural year by sacrificing a llama to Wiraqucha to "ask him to always send them good years."[110] The close association of this non-royal ayllu of the Hanan moiety with the ancestor-god is perhaps enclosed in the etymology of its name, because *tarpuy* means "to sow" (any type of seed, whether tubers or legumes).[111] The sacerdotal duties of the Tarpuntay Ayllu, then, would have been directed at the preservation of the agricultural cycle, and thereby to a particular aspect of Wiraqucha, the *Pachayachachiq* who animated the earth and the crops.

While sowing was under way, large offerings of chicha were made to the rivers and irrigation canals of the valley between July and August.[112] These oblations, which are still performed today in several localities after the

---

109 Randall 1993: 82.

110 Molina: 72, 118. It is peculiar that Molina regularly mentions the role of this non-noble ayllu in the course of the Inca ritual calendar, while he does not identify the lineage affiliations of the rest of the Inca priesthood. One of his informants may have been affiliated to the Tarpuntay Ayllu. Like Sarmiento, Molina gathered the material necessary for his chronicle during the inquests ordered by viceroy Francisco de Toledo. Sarmiento and Molina are the only Spanish chroniclers to mention the presence of non-noble ayllus, including the Tarpuntay Ayllu, next to the panacas at *sitwa* and in the narrative of the Inca ancestors' migration.

111 Beyersdorff 1984: 108-109. In the Southern and northern Andes today, *tarpuy* refers more precisely to "the action of planting the tubers (seed) in a land already laboured" (Cerrón Palomino et al. 1992: 108).

112 Fernández: Pt. 1, Bk. 1, Ch. 10, 86; Gutiérrez: Ch. 65, 255.

communal cleaning of the waterways,[113] were the separate responsibility of each ayllu. They may have conveyed a symbolism similar to the Tarpuntay's consumption of *aqha mama* because streams also carry sediments (*qunchu unu*) that rise to the surface and cloud the water during *puquy mit'a*. These, however, are absent from the low and clear water that flows unhindered throughout the dry season. Randall argues convincingly that the fermented chicha poured into the rivers from July to August re-created the conditions brought about by the precipitations and regenerated the streams so that they could fertilize the crops.[114] This belief may also be paralleled with the rite of *yaku cambio* held today in several communities of Peru and Bolivia in order to attract the rainfall. This modern practice varies somewhat from the pre-Hispanic offering because it consists of mixing the waters of different sites, and sometimes of contrasting types (stagnant versus flowing waters). However, the opposing nature of the two ingredients and the aim of the Inca and present-day rituals concur, as both were meant to generate a beneficial encounter inducing clouds and precipitations.[115] After these offerings, the Inca celebrations that accompanied the early growth of the crops ended once the priests concluded their fast and everyone assembled in Limaqpampa dressed in colourful garments and ready to perform a *taki* called "yahuayra."[116] This dance put an end to a little more than two months of agrarian and ritual activities centred around the division of water resources, during which offerings to the rivers and springs were entrusted to separate lineages. The festivities of the following two months reversed this situation by exalting ritual communion and collective production in much the same way the growing season today inaugurates "inter-household cooperation" in the Andes.[117]

Tiqsi Wiraqucha and the Rites of Purification

The next important festivity of the year was *sitwa*, which began three days after the conjunction of the moon, in the month of *quya rayme*.[118] Despite detailed accounts of this event, the testimonies of early and late chroniclers are equally divided in situating the festival in either August or

---

113 Isbell 1978: 162; Paerregaard 1994: 195-197; Ossio 2002: 482-492.
114 Randall 1993: 80-82.
115 Van den Berg 1989: 68-70; Sikkink 1997.
116 Molina: 72; Polo, "Tratado": 22; Guaman Poma: 251 [253], 225; Cobo: T. 2, Bk. 13, Ch. 28, 216.
117 Gose 1994: 74-75.
118 Molina: 73.

September.[119] Reliable authors such as Fernández or Molina, for example, situate *sitwa* in August, whereas Garcilaso maintains that it began on "the first day of the September moon following the equinox,"[120] a proposition widely accepted among scholars today.[121] This last hypothesis, however, situates *sitwa* in the very last days of September, on a date that cannot be harmonized with the early data provided by Fernández and Molina, even with the addition of the lapse of ten days introduced by the Gregorian calendar. Moreover González Holguín, who published his lexicon twenty years after the Gegorian reform and only a few months before Garcilaso's *Comentarios*, still situates *sitwa* in August. Garcilaso's assumption appears tenuous, not least because he is the first chronicler, half a century after the conquest, to mention explicitly the use of the vernal equinox as fixing the start of this celebration. Yet, if September 23 did not determine the date of *sitwa*, another solar observation, harmonizing early and late primary sources, may have marked it.

The passage of the sun through the local nadir (August 16) constitutes a better point in time for fixing *sitwa* because the new moon following this event would have extended from August 16 to September 13.[122] This interpretation, which admittedly contrasts with most scholarly works, not only resolves the discrepancy observable in the sources but also places *sitwa* within a more appropriate time span. Indeed, the chronicles describe it as one of the major festivals of the Inca calendar, during which all living beings were purified of any illnesses and defects that could endanger the community. In Molina's words, it took place when "the first rainfalls started [because] the first rainfalls usually brought many diseases."[123] Complementing this statement, Guaman Poma provides an extensive list of illnesses that break out in the Andes at that time of year, affecting men,

---

119 Garcilaso: Bk. 7, Ch. 6: 383; Betanzos: Pt. 1, Ch. 15, 72; Polo, "Tratado": 23; Cabello: 351; Acosta: Bk. 5, Ch. 28, 269; Guaman Poma: 253 [255], 227; Molina: 73; Fernández: Pt. 1, Bk. 1, Ch. 10, 86; Gutiérrez: Ch. 65, 255.

120 The vernal equinox occurs on September 23.

121 Zuidema & Urton 1976: 90; Ziółkowski 1987-1988; Sallnow 1987: 38; Silverblatt 1987: 65.

122 Following Zuidema (1981), Bauer and Dearborn (1995) contend that the passage of the sun through the local nadir occurs around August 18 (variability is +/- 1 day). In 2006, a professional of the Collecte Localisation Satellites (Toulouse) kindly re-calculated this occurrence at my request and suggested that August 16 was a more accurate date. This date is also closer to the Virgin of Assumption day (15 August), a major festival in the Andes. I count August 16 as the first possible date for the *sitwa* celebration because the observation of the sun's passage through the local nadir occurred at daytime, and could therefore coincide with the new moon.

123 Molina: 73.

crops, and water supplies. It is a time when the irrigation waters rise up from their underground realm, bringing to the surface of the earth the dangers of the subterranean world in which the dead journey. Popular beliefs today associate the month of August with an unsettled period when a warm insalubrious wind (*wayra*), considered to originate from corpses and caves, can infect the villagers.[124] The characteristics of this *remolino* conjure up the attributes of the pre-Hispanic god Huari, the dark Sun of the underground, who manifested himself on earth as a powerful and great wind. The meaning of the word *sitwa* supports this connection because it refers to "the resplendence of the sun, of that which reflects [light]."[125] During August and throughout September, dense clouds gather progressively in the sky before producing any rain, while the *puna* vegetation completely disappears after the llamas have grazed. For Hiroyasu Tomoeda, Andeans perceive this transition as "a period of crisis, when the breath of the vital force is hardly perceptible."[126] The earth, animals, and men have few supplies left during this period of privation, so the world needs to regenerate by expelling the ailments and misfortunes that could affect the *kallpa* of living beings.

Sitwa, as I described earlier, began with the ritual expulsion of diseases by several Inca lineages and *mitmaq*. After this cleansing, a group of celebrants would light balls of dry grass bound with cords and, dancing with these torches in the streets, loudly beseech the illnesses to leave the city. Everyone, including the sovereign and his wife, re-entered their respective homes, where they rubbed their faces with a sort of maize mush called *sanku* before spreading it over the main parts of their habitation, in the belief that it protected its occupants and their food supplies from afflictions. In the different sanctuaries of Cuzco, the mummy bundles and the statues of Inca gods were also "warmed" with the same mush and, this done, both the deities and the people were able to enjoy rich food in the privacy of their residence.

A glance at the religious practices performed today in the same month confirms that *sitwa* was a ritual of regeneration. Every year, in several communities of the high plateaux, the pastoralists conduct the *Herranza* ritual, also known as *Agustukuy*, during which the participants renew the vital

---

124 Kato 1989; Rozás Álvarez & Calderón García 2000: 321. This wind bears the name of "agosto wayra" in the Calca Province (Cuzco). For specific studies on ethnomedicine, see Carey 1993; Larme 1998: 1008-1010.

125 González Holguín, *Vocabulario*: 85.

126 Tomoeda 1993: 290. See also Flannery et al. 1989: 177.

force of the herds and crops.[127] Tomoeda (1993) has shown that this celebration takes on the same patterns as the Inca festival, for both involve "warming" the beings chilled by their loss of *kamaq*, and both lead to the decontamination of the area defiled by ills. The contemporary rite also echoes its pre-Hispanic antecedent in that only a stranger to the family, like the *mitmaq* in Inca times, was able to carry away the impurities from the household and pour them into a distant watercourse. Zuidema (1986b), building upon Sherbondy's work, suggests that the ceremonial space of *sitwa* reflected the organization of Cuzco's irrigation system, in the way that the four squads of nobles and *mitmaq*, passing through the non-Inca communities, would have controlled the water channels around the valley. This argument is particularly interesting, for it concurs with other aspects of this festival that were related to the mythical deeds of the subterranean sun. *Sitwa*'s spatial organization focuses on the distribution of the arable lands, the diversity of peoples, and the opening of irrigation canals, all of which were the work of Wiraqucha, to whom this ritual was apparently dedicated. Primary sources report that He, "who lives at the edge of this world", was believed to carry away the illnesses from the rivers into the sea and to protect the Incas from any affliction.[128] The orations that the celebrants declaimed on that occasion were dedicated to the Hanan ancestor-god, together with thirty immolated figurines made of *kishwar*, similar to those he received for *inti rayme*.[129]

In the following days, the nobility danced the *arawi sitwa* in honour of Wiraqucha. Several thousands of llamas, claimed from the four *suyus* and pledged to the Inca divinities, were then brought to the main plaza. Four immaculate animals were chosen from the flock to be sacrificed. Their lungs were inspected to foretell the coming year, and their blood was mixed with the *sanku* mush. A share of this mixture, called *yawar sanku*, was then distributed among the participants with a slice of raw llama meat, which they ate as a sign of confederation with the Inca and the divinities present. Two goddesses, the Moon and Pacha Mama, were next to be honoured, and on the following day both received similar oblations while the officiants uttered prayers. Gender roles seem to have played a central part in this ceremony, for in addition to the expressions of devotion bestowed upon these two female deities, *sitwa* took place during the Inca month of *quya*

---

[127] Quispe 1969; Isbell 1978: 151-164; Flannery et al. 1989: 143-182; Rozas Alvarez & Calderon Garcia 2000; Vivanco Guerra 2001; Rivera Andía 2003.

[128] Molina: 81-95.

[129] Cobo: T. 2, Bk. 13, Ch. 29, 218.

rayme, commonly translated as the "festival of the queen."[130] Julien recently challenged this interpretation and demonstrated that the word *quya* extended to a class of noblewomen related to the living ruler and from whom his wife was elected.[131] It was also from this elite group that the female victim, sacrificed as the Sun's spouse during *sitwa*, was chosen. Indications that gender duality was present in August rituals can be found from colonial times through to the present day. For David Cahill (2002), the festival of Our Lady of Loreto, conducted on 22 August 1692 by indigenous officials, may have echoed the pre-Hispanic cleansing ritual. At the heart of this feast was the veneration of a particular representation of the Virgin who had become the female patron of the colonial Inca nobility. Today, the Assumption feast in August is often equated with the celebration of Mother Earth, believed to be renewed and therefore "virgin" when she (i.e., the earth) opens during this period of the year. In several communities, this characteristic would have favoured the merging of indigenous lore with the Christian feast, where a certain aspect of the Virgin Mary is frequently associated with the Pacha Mama and the early growth of crops.[132] The festivities of *sitwa* ended after the re-entry of foreign delegates and the distribution of goods and wives. The ashes of all of the sacrifices conducted during that month were dispersed over the pastures of the *puna*, an act that helped regenerate the vegetation of the high altitudes after the general grazing of the llamas.

Pura Upyay

Two weeks after the beginning of *sitwa*, at the time of the full moon, the Incas carried out another festivity called *pura upyay*, which was dedicated to the first phase of the crops' maturation process.[133] This ritual was held on the edge of the city during a period of four days, while in Haucaypata the common people shared the early fruits and crops that had just ripened. This festival shows strong similarities with the dispersion of the ashes performed in early January, which Molina called *mayu qatiy*, or "following the

---

130 Guaman Poma: 252 [254], 227; Zuidema 1986b: 180; Sallnow 1987: 38.

131 Julien 2000: 35-37.

132 Mariscotti 1978a; Harris 1999: 201-219; Albó 2002: 406-410. This parallel has its limitations, because certain characteristics attributed to the Virgin Mary, such as chastity and mercy, cannot be equated with those of Pachamama. See Salles-Reese 1997: 30-39.

133 Primary sources report that this celebration started fourteen days after the beginning of *sitwa*, that is the interval separating the first crescent that followed the conjunction of the moon to the full moon.

river."[134] Gutiérrez confirms this parallel and gives the same name (*pura upyay*) to both events, suggesting that it was performed at least twice a year, upon the opening and closing of the heavy rains.[135] On this occasion, the ruling elite gathered at the junction of the two rivers in Pumap Chupan and cast several offerings of llama blood and precious garments into the watercourse. They incinerated a great number of llamas in a fire lit *in situ* and scattered the ashes of the sacrifices at the rivers' junction, together with ground coca and flowers they had brought "from all the plants of the fields afar."[136] Eventually, every *orejón*, having poured a tumbler of chicha into the streams, drank from one "as if [they were] toasting with the waters."[137] While these oblations were being made, two lords, one affiliated with Hanan Cuzco and the other with Hurin Cuzco, posted themselves on opposite sides of the river, each one with a group of ten or so assistants carrying long poles. The offerings accomplished, both teams followed the current downstream for approximately a hundred and fifty kilometres, using their staffs to remove the items stuck in the backwaters. They finally ended their journey in the town of Ollantaytambo, which stood as the farthest settlement area of the Incas-by-Privilege northwest of Cuzco,[138] hence circumscribing the *pura upyay* festival, like that of *sitwa*, within a ritual space associated with the Incas' allies with whom the ruling elite had contracted matrimonial alliances. Molina, who carefully described this event as being held in January, reports that these offerings were intended for Wiraqucha, who received them once the river met the waters of Mama Qucha that surround the universe.

The name of this ritual, as recorded by Betanzos and Gutiérrez, and its focus on dual organization may cast some light on the meaning of these events. To begin with, González Holguín notes several uses of the term *pura* in reference to the opposition between two contrasting sides. Followed by *killa*, it refers to the full moon and its waning crescent, which provided Zuidema with the substance for his translation of the ceremony: "to drink after the full moon."[139] This rite effectively took place at this specific time,

134 Molina: 114-117.

135 Gutiérrez: Ch. 65, 256.

136 Betanzos: Pt. 1, Ch. 15, 72-73.

137 Ibid.

138 Zuidema 1985: 223. See also Bauer 1992: 18-35.

139 Zuidema (2002: 22) also considers *pura upiay* to not only be the name of this ritual, but also that of this specific month. However, the testimonies of Fernández and Gutiérrez indicate that this ceremony was performed at least twice a year so that *pura upiay* could not have designated a month, but rather the celebration in itself.

but the gloss *pura* possessed a broader meaning in Inca times. In several entries of González Holguín's lexicon, it refers to an intermediate position in the middle of two factions that accommodates or disparages both parties.[140] Holguín records another meaning of *pura*: the complementary aspect of two opposing elements forming a couple, a definition that may have influenced its contemporary meaning of "pure". All of these variations in meaning ideally depict the concepts that characterized the moieties' relationships at the heart of the September festival. As a result, the combination of *pura* and *upyay* (to drink) describe the act of "drinking to/with the division that opposes and brings together [the moieties]", this division being, namely, the watercourse. This ceremony was addressed to the Vilcamayu River, which proceeded from the union in Pumap Chupan of the four tributaries supplying the Hanan and Hurin divisions with irrigation water. Upstream, the Upper part supervised two major watercourses called Tullumayu and Saphi, while the lower one controlled the Chunchulmayu and the Huancaru. Sherbondy shows that the headsprings supplying these rivers were all located in a different *suyu*. She suggests that each of them was supervised by a panaca associated with the respective division.[141] In this way, *pura upyay* commemorated the complementary role of each moiety in the maintenance of these watercourses and honoured the salutary confluence of these forces essential for the irrigation of the valley and the growth of the early cultivations. It also mirrored and counterbalanced the early rituals conducted by the Wikakiraw Panaca (Hanan Cuzco) in the primary feeder canal of the city, because the second stage of *pura upyay* took place in the *kuntisuyu* region, a location associated with Hurin Cuzco. While the offerings conducted after the cleansing of the water ditches were entrusted to each lineage on the specific water channel of which it was in charge, *pura upyay* engaged the members of both moieties as a conciliation of the two divisions, which was only made possible once the city had been ritually purified and the foreigners reintegrated into its centre.

140 "Pura: entre si uno con otro, o uno y otro"; " Purap man sayani: hazer entrambos vandos contrarios, o arrimareados"; "Purapman sonco: Ynclinado a rreboluer y embarrar, o ayudar a entrambas partes con dezir a unos mal de otros"; "Purap man sayak: El de dos bandos, o embarrador rebolvedor"; "Purap man, o purap simiyoc: El que rebuelue y lleua chismes entre dos", González Holguín, *Vocabulario*: 296-297. The semantic field of this word recalls the concepts of *taypi* or *chawpi*, see Bouysse-Cassagne 1986; Harris 1985.

141 Sherbondy 1987: 121-122, 127-128.

Preparations for the Rite of Passage

After *pura upyay*, another religious event soon occupied the attention of the nobility. For the next two months every household in Cuzco that included a male youth ready for initiation lived in expectation of the next *warachiku* and gathered the accessories necessary for the ceremony. To begin with, the male sponsors of high status provided a large quantity of black wool to their female relatives, who would spend the entire month spinning (*puskay*) and weaving (*away*) a garment destined for the probationer. In reference to this activity, the period in question was called *puskay killa*, or the "month of spinning."[142] If the rainfall were to be late in arriving during this crucial period for the growth of young crops, the Incas would tether a black llama in Haucaypata and pour chicha around it. They then would let it starve and bleat from hunger for several days (Figure 8). The animal would have remained tethered in the main plaza until Wiraqucha sent the rain. At the origin of this practice was the belief, still widespread today, that the llamas' plaintive cry attracts the rains.[143]

It is often assumed that *warachiku* commenced with the public festivities of *qhapaq rayme* when the novices were tonsured and began their pilgrimage outside the city. However, before this event occurred, the passage of the sun through the local zenith on October 30 marked an important date. This was when the young boys requested of their ancestor Huanacauri the privilege of being initiated. In order to do so, they first assembled in the main square to drink with the deities and the mummies of their past sovereigns before praying for courage and good fortune in the coming ordeals. Then, a hundred llamas chosen for their purity were brought to Haucaypata and distributed among thirty attendants. Each day of the month, one of these celebrants walked around the main square with three, sometimes four llamas and sacrificed them to Wiraqucha. This ritual lasted until no animal remained to be slaughtered, thus coming to an end thirty days later, on November 30, when the actual ordeals of the rite of passage started. The ashes of these sacrifices were then carried away to a house in Pumap Chupan where the Incas stored the remains of their past

142 "Puchcani puchcaccuni: to spin" in González Holguín, *Vocabulario*: 292. "Puchayquis" in Polo, "Tratado": 23; "puzquay quiz" in Gutiérrez: Ch. 65, 255; "puzcuay quiz" in Fernández: Pt. 1, Bk. 1, Ch. 10, 86; and "puchay quiz" in Cabello: 352; Acosta: Bk. 5, Ch. 28, 270; Murúa, *Historia general*: Bk. 2, Ch. 39, 440; Cobo: T. 2, Bk. 13, Ch. 30, 219.

143 Guaman Poma: 255 [257], 229; Cobo: T. 2, Bk. 13, Ch. 30, 219; Isbell 1978; Stobart 2006: 218.

Figure 8. October, *uma rayme*. Inserted in the drawing: "Carnero negro ayuda a llorar y a pedir agua a dios con la hambre que tiene". From Guaman Poma de Ayala: 254.

offerings.[144] Betanzos, when recording the details of Pachacuti Yupanqui's reforms, also mentions the observance of a lapse of thirty days between the decree of the *warachiku* and its enactment. According to him, the king called for his three loyal captains to gather in Cuzco, where he imparted to them the festivity's events that he had devised. This assembly, Betanzos specifies, occurred thirty days before the actual celebration of the

[144] Cobo: T. 2, Bk. 13, Ch. 25, 208-209; Molina: 97-98. Molina mentions briefly the sacrifices of three llamas every day of the month. However, Cobo records this information in detail for his description of the *qhapaq rayme* festival. Although he situates this celebration in December, his invaluable relation does not fit in the timeframe of a single month but rather embraces the whole festival, starting from October with the spinning activities. He therefore fixes the beginning of the *warachiku* on October 30, when the Inca nobility started to perform the *itu* celebration and the young men asked Huanacauri the licence to be initiated. However, Cobo's description becomes confusing later on, as he merges the first visit of the boys to Huanacauri with the second, which was held several weeks later at the start of the public celebrations of *warachiku*.

initiation ritual, in order to prepare the garments and offerings of chicha.[145] Altogether, this evidence confirms that *warachiku* started on November 30 (GC).

In the same month, and five days after the first sacrifice of llamas, the young boys left Cuzco with several adults and spent the night in Huanacauri hill. In Molina's words, the reason for this journey was to "reproduce the pilgrimage their forefathers had made in this place", in reference to the Ayar siblings' migration to the valley.[146] Arriving at the shrine, the celebrants gashed the legs of six old llamas, the blood of which they smeared on the boys' faces. These creatures were then walked around the hill, bleeding, until they dropped dead, and were then cremated.[147] Once they made these offerings, the youths were granted the licence to be initiated. Before returning to Cuzco, they collected another bundle of straw, which they later distributed among the women who had spent the previous few weeks fasting and preparing the ceremonial chicha. According to Betanzos, this drink was kept in the household during the festival and only consumed on the day of the ear piercing. Therefore, the offerings to the gods required an additional quantity of chicha, which the young maidens or the novices themselves would have prepared by chewing maize and mixing it with the water of Calispuquio (Ch 3:8).[148] The preparation of this ceremonial drink gave its name to the month, *kantaray killa*, which refers literally to this activity.[149]

November ended with the *itu* festival, when the probationers walked solemnly around Haucaypata playing on small drums, gravely rehearsing the daily preliminaries of the *warachiku*. This procession, however, not only announced the rite of passage to come, but was also directed at the protection of the recently sowed crops. The Incas believed that the sound of drums incited abundant downpours and, importantly, prevented the appearance of hail and frost that could damage the yields at this crucial time of year.[150] Drums were closely associated with Illapa, the warlike deity of the high plateaux who ruled these atmospheric phenomena. Such an association also explains why the same instruments accompanied other activities of the *puna*, such as the llamas' ritual mating held later in the

145 Betanzos: Pt. 1, Ch. 14, 70.
146 Molina: 98.
147 Cobo: T. 2, Bk. 13, Ch. 2, 209.
148 Betanzos: Pt. 1, Ch. 14, 69; Gutiérrez: Ch. 65, 255-256; Cobo: T. 2, Bk. 13, Ch. 30, 220.
149 *Kantaray* referred to a specific type of chicha, see Molina: 97.
150 Murra 1983: 43; Zuidema 1985: 201; Gruszczynska-Ziólkowska 1995: 116-117.

rainy season, and were played before warfare encounters.[151] The sound of drums at *itu* similarly heralded the warlike ordeals of *warachiku,* whose time span coincided with the herds' breeding and early birthing periods. While these events unfolded, the dark cloud of the Llama watched over the terrestrial activities that marked the start of the ceremonial cycle ruled by the Sun god P'unchaw.

Prior to the inauguration of this new cycle, the official celebrations of the metropolitan calendar constituted an indivisible body of traditions. From early June to September, these public representations placed Wiraqucha at the head of the hierarchy of Inca gods because it was He who regulated the underground waters supplying the irrigation systems. At the end of the dry season, this function was ultimately crucial because the fields required artificial watering in anticipation of the rains. It was also Wiraqucha who bore the illnesses away from the sacred centre of Cuzco when the first rainfalls fell, drenching the city. He was the ancestor-god who received the Incas' offerings in the sea bordering the edge of the world, and He who animated the different ethnic groups at the dawn of the new era, bestowing them with their own territory, customs, and language. Accordingly, his rule over *chiraw mit'a* opened with the re-enactment of his journey from Lake Titicaca to the sea during the June festival conducted by the Sapa Inca. It proceeded with the celebration of his deeds at *sitwa* when the communities' boundaries were reasserted and foreign delegates performed their traditional dances at the close of the revelry. The last festivities of the Hanan cycle focused on regulating the precipitations brought about by Illapa, the divinity with whom Wiraqucha was closely associated. With the apparition of the first rains at *quya rayme*, artificial irrigation had become obsolete, and Tiqsi Wiraqucha was now living on the edge of the world, where he had carried away the illness. From this time forth, Illapa ruled over the atmospheric elements and was assisted in this task by the mummies of late Inca lords, also known as *illapa*.[152] These corpses were bundled within many layers of rich fabrics and had their heads covered similarly to the thunder god's figure, Ch'uqi Illa, whose "face was not visible."[153] Like the divinity of lightning and thunderbolt, they were believed to regulate the precipitations, the reason for which Sinchi Roca's mummy was carried throughout the fields when the annual rains had not

151 Zuidema 1985: 196-199; Niles 1999: 40, 61-62; Tomoeda 2004: 180; Gudemos 2005: 44.
152 Albornoz: 167; Guaman Poma: 288 [290], 263.
153 Cobo: T. 2, Bk. 14, Ch. 19, 275; Molina: 67.

been abundant enough to nourish the land.[154] Their attendance at public ceremonies also largely coincided with the rule of Illapa on earth: they received chicha and maize mash at *sitwa* and later attended the Incas' initiation rite, as Manco Capac's embalmed body was carried to Huanacauri hill.[155] Their benevolence during this crucial moment of the agrarian cycle was later remembered at the Sun festival marking the start of the harvesting season, when they received large offerings of food.[156] Guaman Poma is alone in suggesting that November (GC) corresponded as well to the annual celebration of the dead, when the mummies received food, garments, and livestock.[157] Even today, the dead ancestors are believed to look after the early maturation of crops, and for this purpose they receive rich offerings around All Saints and All Souls Days (November 1 and 2).[158] In Inca time as well, the mummies assisted Illapa in regulating the precipitations, but their public appearance extended beyond the cycle ruled by the tutelary deities of Hanan Cuzco: the royal mummies attended the *qhapaq rayme* festivities that inaugurated the reign of P'unchaw. This may be because their cult was more strictly associated with the meteorological season of *puquy mit'a*, regardless of the cultural construct that divided the ceremonial calendar between the two moieties, and perhaps because together, the embalmed bodies of past rulers represented the continuity of the Inca dynasty, "the key reference points in genealogical reckoning", in Salomon's words.[159] Be that as it may, the Hurin cycle started with the public celebrations of *warachiku*, during which the celebrants re-enacted the Ayar siblings' journey across the valley, as had been stored in the memory of the Lower moiety.

The Advent of P'unchaw

As with the narratives of the Inca migration to the Cuzco valley, the chroniclers' rendition of the *warachiku*'s foundation synthesizes two different

---

[154] Polo, "Tratado": 10; Cobo: T. 2; Bk. 12, Ch. 9, 73. See also Ziólkowski 1996: 132-133.

[155] Molina: 76, 78, 107, 112, 114; Sarmiento: Ch. 14, 64.

[156] Segovia: 81-83; Santillán: 34. In colonial times, communities of the central Andes conducted similar oblations at the beginning of the harvesting season. In San Pedro de Hacas (Cajatambo), for instance, the mummies were offered the first produce of the year together with the new chicha prepared from half-ripened corn. See Duviols 1986: 148-149.

[157] Guaman Poma: 257 [259], 231.

[158] *Ritos y Tradiciones de Huarochirí*: Ch. 28, 365-371; Allen 1988: 56-57, 164-165; Gose 1994: 141-146; Harris 2000.

[159] Salomon 1995: 339.

traditions. To begin with, the stories of the Ayars' emergence from Paqariq Tampu attribute the institutionalization of this ritual to Manco Capac following the injunction of his brother Ayar Uchu in Huanacauri. Unlike this variant, the epic of Pachacuti Yupanqui assigns to this king the entire conception of *warachiku*. Given that the initiation festival was a celebration for the entire Inca nobility, these discrepant accounts would have corresponded to the two divergent Inca modes of recording the past. The first followed the conventions of the genealogical accounts that trace the origin of every Inca institution back to the apical ancestor. The second conformed to the life story genre that credits Hanan lords with the reorganization of the established order. This explains why Wiraqucha, the head deity of the Upper moiety, is sometimes said to have bestowed part of the *warachiku* customs upon Manco Capac. Still, the itinerary of the *warachiku* celebrants within the topography of the Cuzco region, Huanacauri's central role as P'unchaw's servant in these events, and the link of several ceremonial shrines with the pre-Deluge era, suggest that *warachiku* inaugurated the annual cycle of the Lower cosmological half supervised by P'unchaw, the Hurin moiety, and the first generation of settlers. In addition, the concurrent commemoration of the Deluge and the appearance of the dark Llama cloud in the sky, together with the procreation cycle of this animal, give the *warachiku* pilgrimage another level of significance. Several myths of the central sierra recount how legendary llamas had warned humans about an imminent flood.[160] This gift of prophecy was related to the belief that the *yana phuyu* of the llama, who walks in the middle of a river at night (the *Mayu*, or Milky Way), drinks the water of the sea when no one watches, because "if it does not do so, the sea would immediately inundate the earth."[161] The potentially destructive and heavy rainfalls of the rainy season were therefore believed to originate from the water of the celestial river, which only "the *kamaq* of llamas" could prevent from becoming cataclysmic. It was in order to attain an equilibrium in the vital force of the herds that the *warachiku* celebrants performed the *wari* dance and honoured the emergence of the first llama together with their ancestors. The fulfilment of these ritual requirements ensured that a new and stable era opened before them in much the same way that the rainbow arching over Huanacauri at the Ayars' entrance in the valley indicated that "the world would never again be destroyed by water."[162]

---

160 Zuidema & Urton 1976: 82-83; Molina: 57; *Ritos y tradiciones de Huarochirí*: Ch. 3: 33-37; *Carta Annua* 1614 ARSI Perú 19: ff. 258-262, in Polia Meconi 1999: 358-359.

161 *Ritos y tradiciones de Huarochirí*: Ch. 29: 377.

162 Sarmiento: Ch. 12, 56; Murúa, *Historia general*: Bk. 1, Ch. 2, 41; Cabello: 262.

After having accomplished their pilgrimage outside Cuzco, the novices and their families congregated in Haucaypata with the deities and mummy bundles of the kings and queens. Around the plaza, a large golden rope strung on gold and silver posts enclosed the entire space where for the next six days the Hanan and Hurin nobility of Cuzco performed a *taki* called "caua" or "aucayo" twice a day.[163] The dancers wore a red tunic cloaked in puma hides, while each dancer's face was covered with the head of a puma adorned with golden teeth, earrings, and a headband. Four *Indios principales*—two from each moiety—played the tune on large drums. Zuidema (1985) has thoroughly discussed the theme of the puma in Andean mythology and posits that it represented the "body politic" in periods of transition and uncertainty. Puma skins were worn in rites of seasonal changeover: at *qhapaq rayme* and during harvesting festivals. It also ensured victory to those who sought to overcome the established order: at times of royal succession or at the turn of a new era.[164] Hence, the young Inca Yupanqui, who had not yet received the *maska paycha*, was said to have placed a puma hide on his head before confronting and defeating the Chancas. A Huarochirí tradition also recounts how a poor Yauyo man won a dancing contest against a wealthy Yunca local by wearing a puma skin, whereupon a rainbow appeared above his head, announcing the imminent rule of his protector, Pariaqaqa.[165] As for the drums, they were associated with military invasion. The Incas played them when the stability of the nations making up Tawantinsuyu was threatened.[166] In this way, the four moieties' drums that accompanied the "caua" or "aucayu" dance would have indicated that the stability of the Inca polity was endangered. Zuidema suggests that this *taki* served to reassert Inca political borders at a transitional time when the youths underwent a transformation to become warriors.[167] An alternative, and perhaps complementary, perspective can be articulated here. We observed earlier that the Inca nobility played the *ayllu* and *wayru* (also *pichqa*) games while the boys underwent their *warachiku*. The stake of these contests was in part to redistribute cultivable parcels among the ruler's kin. Polo indicates as well that the lands, *akllas*, and livestock

[163] Garcilaso evokes a golden rope that Huayna Capac would have ordered to be made when his firstborn, Huascar, was born. See Garcilaso: Bk. 6, Ch. 2, 293; Bk. 9, Ch. 1, 505-506.

[164] Cieza: Pt. 1, Ch. 117, 306.

[165] Cieza: Pt. 2, Ch. 45, 133; *Ritos y tradiciones de Huarochirí*: Ch. 5, 65-67. In the central sierra traditions, the Yauyo people are highlanders commonly opposed to the Yuncas who resided in the lowlands. See Rostworowski 1978.

[166] See this Chapter, n. 151.

[167] Zuidema 1985: 242.

allocated to the empire were redistributed annually at the *rayme* festival, which he situates in February.[168] These lands were not owned in the literal sense, but rather exploited on a rotational basis, and the fruit of their farming was granted to the royal lineages as tribute.[169] Like the allotment of communal lands today, this procedure would have generated conflicts and rivalry among the panacas, and potential disputes would have been averted through the entertainment of the *ayllu* and *wayru* competitions. In modern Pacariqtambo, the ayllus' headmen anticipate these hostilities by engaging in a ritual battle, each one grouped according to his moiety affiliation. In Urton's words, this encounter serves "to re-establish hierarchical relations of authority."[170] A similar rationale may have prevailed in the "caua" or "aucayo" dance, also set prior to the redistribution of lands and other goods. In this way, the four drum players evoked the asymmetric relations that shaped the social organization of Cuzco (e.g., moieties and moieties' subdivisions) and echoed the interlineage rivalries that could endanger the Inca polity. The presence of the puma-dancers not only confirms that a period of transition and instability was taking place; it also transcended this volatile situation to give way to a unified and stable polity.[171]

The novices eventually had their ears pierced in the privacy of their homes, after which the foreign delegates re-entered the ceremonial space. They all assembled in the main plaza, where a group of Inca celebrants, descendants of Lluque Yupanqui affiliated with Hurin Cuzco, presented them with the maize mush.[172] Although priesthood membership is largely undocumented for the rest of the month, this last piece of information

---

168 Polo, "Tratado": 69-70; Polo, *El Mundo*: 56, 80. This date would have corresponded to the end of the *qhapaq rayme* in the Gregorian calendar.

169 Murra [1995] 2002; Rostworowski 1993a; Ramírez 1996: 42-61.

170 Urton 1993: 129.

171 This dance is either called "cayo" (Molina: 108), or "aucayo" (Cobo: Pt. 2, Bk. 13, Ch. 25, 212). For Cabello it was performed in June, which corresponded to the harvesting season. In different sources, June (GC) is referred to as "aucay cuzqui" or "cauay" (see Tables 11 and 12). The spelling "cauay" is particularly interesting because Pachacuti Yamqui (f. 13v., 208) refers to the puma as "caua or chuqi chinchay" in his drawing of the Coricancha altar. Guaman Poma (321 [323], 296) also describes "caua" as a dance. This information suggests that the dances with puma skins performed in January and June were closely related and may have originally bore the same name, which transformed by metathesis.

172 Acosta: Bk. 5, Ch. 23, 256. Acosta is the only chronicler to mention the role of this particular panaca at the end of this celebration. This unusual information is however too specific to be disregarded and appears coherent with the general picture of a festival commemorating the advent of P'unchaw, ruler of the lower half of the cosmos and tutelary god of the Hurin moiety.

suggests that the sacerdotal duties of *qhapaq rayme* were entrusted to the Lower moiety so that their members would have presided over the celebration marking the advent of their tutelary god, P'unchaw, on earth.

Rituals of Fertility and Impregnation

As the *warachiku* reached an end, the Incas inaugurated another festival for the multiplication of the llamas. For this event, officials were sent throughout the land to count the livestock allocated to their ruler and divinities, while in Cuzco the celebrants sprinkled the animals with chicha. During their visits, the royal inspectors dispensed presents to the pastoralists who had taken the best care of the herds and punished those who had neglected their task. Today, a similar custom is still perpetuated in the high plateaux, which follows a scenario of purification, reward, and occasional punishment.[173] Yet, besides expressing the elite's control over the empire's goods and supplies, this event was marked by the llamas reaching sexual maturity, a stage that can be paralleled with the transformation undergone by the novices, who were now facing their first public appearance since their ears had been pierced. The novices were now reborn as mature beings that had acquired sexual potential after they had been ritually killed. This new aptitude was at the centre of a new ordeal during which the blood they shed was destined to fertilize the earth. On the new moon of the following month, called *kamay killa* (the month of the vital energy), the recently initiated males gathered in Haucaypata and lined up in two factions according to their moiety affiliation. In this fashion, the Hanan and Hurin youths fought against each other in a mock—albeit excruciating—battle by hurling thorny fruits with the slings they had recently acquired. During the hostilities, the participants could suffer deadly injuries and often engaged in a hand-to-hand confrontation before the Sapa Inca separated the two camps.[174]

The Incas performed mock battles during at least three ceremonial occasions: at the end of the *warachiku*, for the royal funerals of *purukaya*, and to commemorate the triumphal return of their leaders from far-off campaigns.[175] These confrontations dramatized a passage between two temporal and social states, from childhood to adulthood and from life to

---

173 Tomoeda 2004: 175.

174 Molina: 111; Gutiérrez: Ch. 65, 256.

175 Betanzos: Pt. 1, Ch. 31, 145-148; Pachacuti Yamqui: f. 25 v., 232; Zuidema 1991. An early seventeenth-century document reports a ritual battle, opposing the Cañares to the Canas, performed in Cuzco on the Sunday following Asumption. It commemorated the

death.[176] The timing of the young men's last ordeal also corresponded to another transition, situated between the end of the sowing season and the maturation of the crops. The mock battle performed by the novices was closely linked to the agricultural cycle. It had this characteristic in common with the ritual battles (*tinku*) enacted in present-day Peru. Andean communities conduct these fights at different times of the year, principally on January 1 and 20 (San Sebastian), and at pre-Lenten time, around Carnival.[177] In the Bolivian highlands, the Macha also engage in a ritual fight at Corpus Christi, while Olivia Harris observed this practice among the Laymi at All Saints Day.[178] In the former cases, the outcome of the ritual battles serves to predict the forthcoming agricultural cycle because the winning side is believed to be the fortunate beneficiary of a prosperous year. The blood spilt by the participants is also seen as an offering to the earth that fertilizes its soil[179] at a time of year when young adults often display erotic familiarity in rituals.[180] In P'isaq, for instance, the young males engage in a fierce *tinku* on the Sunday of Carnival. Earlier the same day, the unmarried men and women who had danced separately over the past six weeks joined each other to perform sexual games.[181] Likewise, in Pacariqtambo, young fighters hurl unripened fruit at their opponents during the Mardi Gras *tinku*.[182] Muelle, who witnessed this ritual in 1945, reports that the unmarried men and women of this town celebrated this day by throwing green fruit and chasing one another with thorny branches. These events, he argues, were decisive for the constitution of matrimonial unions.[183] In many respects, the modern *tinkus* are comparable to the ritual battle of the recently initiated—and unmarried—Incas who also hurled fruit, called *pitahayas*, at their opponents. There exist two main species of these *vine cacti*, yet only one yields prickly fruit—the yellow *pitahayas*, which can

---

victory of the Cañares, "the Inca's guard", against the Canas. See "Relación de la Fiesta, 1610" in Bradley & Cahill 2000: 156-161.

176 Molinié 1988: 55.

177 Bastien 1978: 57-58; Hopkins 1982: 171; Molinié 1988: 50; Allen 1988: 183-187, 205-207; Urton 1993.

178 Platt 1986: 239-240; Harris 1999: 38. Both these authors emphasize that mock battles were performed during different festivities throughout the year. Among them, the *tinkus* held at Corpus Christi and All Saints Day seem to be the most important and the best documented of all.

179 Zorn 2002: 140.

180 Hopkins 1982: 174-176; Allen 1988: 206; Stobart 2006: 248-255.

181 Hopkins 1982: 179-180; Sallnow 1987: 137, 139.

182 Urton 1993: 126-128. In the community of Kaata (Bolivia), the fighters sling ripe fruit (Bastien 1978: 57).

183 Muelle, in Urton 1993: 129-130.

grow thorns up to two centimetres but loses them upon ripening. This means that the young Inca fighters at *kamay killa* also threw unripe fruit as projectiles, which denoted their early and vigorous sexuality. The youths' fertility concurred in turn with that of the livestock that had just begun breeding. Today, many *tinkus* conducted in the Cuzco district on the day of Comadres (the Thursday before Ash Wednesday) exploit the vitality of both humans and herds. Alpacas and llamas are symbolically or actively present among the fighters.[184] About this festival, Michael Sallnow reports that the residents of Qamawara cover the animals' fleece with the red soil in which *tinku* victims had been buried, before deploying the herd into the path of their enemy.[185] In this way, men and llamas join to become the sacrificial victims whose fertile blood nourishes the earth. A similar symbolism pervaded the Inca ritual battle, which closed a cycle during which the probationers and the llamas shared a common destiny dramatized by the *warachiku* ordeals. During this time, the youths were ritually slaughtered before being reborn under a new name, while the same ceremony celebrated the emergence of the prototypical llama in the presence of the *napa*. Both animals and men initiated at *qhapaq rayme* a new cycle that opened with the re-enactment of their respective ancestors' journey through the land and ended with a celebration of their sexual potential. During this ritual process, the nature of the young novices mingled with that of the wilderness, whereas the llamas assumed human traits. Thus, the white *napa* whose ears were pierced and adorned with the large pendants as a sign of noble distinction led the *warachiku* cortège. Meanwhile, the youths' physical substance underwent a transformation that resulted in their own flesh being likened to the raw meat of the llama slaughtered in Yauira.[186] As they were given this meat, the novices were reminded of their enemies' cannibalistic behaviour and symbolically consumed their own bodies, which had at this liminal moment become the incarnation of sacrificial meat. Finally, the *wari* dances performed several times daily at *warachiku* were a variation of the *wariqsa* dance conducted in April, which reproduced the vocalizations of the herds.[187] Guaman Poma reports that the *wariqsa* consisted of the celebrants imitating the plaintive cry of the llama to the sound of a small *pinkillu* flute. He offers an illustration of this practice in which the Inca is depicted singing with a *puka llama* (red llama)

---

[184] Hopkins 1982: 174-175.
[185] Sallnow 1987: 138-140.
[186] Cieza: Pt. 2, Ch. 7, 19.
[187] Gudemos 2005: 30.

(Figure 9).[188] Stobart observes that in present-day Macha (Bolivia), the animal's yammering sound is also closely associated with the *pinkillu* flute, which is an instrument mostly played during the rainy season. Both are believed to influence the maturation of the crops, so that on All Souls Day humans imitate the llamas' mating habits by suggestively putting the flute in between their legs before playing on it, after which "men and women have sex, copulate like llamas."[189] In pre-Hispanic times, a similar symbolism, associating the animal's vocalizations with fertility, would have connoted the *wari* dance.

The ritual assimilation of the novices with the livestock reinforces the thesis, put forward by Zuidema and Urton (1976), that the heliacal rise of the Llama dark cloud and the start of the breeding season together determined the beginning of the Inca initiation rite. Urton indicates that the superior culmination of this constellation at midnight occurs in late April, concurrently with the end of the llamas' breeding and birthing season.[190] In the Inca calendar, this date coincides with the Sun festival opening the harvesting season when, Betanzos indicates, the newly initiated youths ended their fast, but also terminated the sacrifices they had started in November. On that day, they dressed in "tunics of woven gold, silver and iridescent feathers" and assembled in Limaqpampa to honour the Sun before harvesting.[191] These data suggest that the novices' probation period extended beyond the scope of the *warachiku* ordeals and lasted for the entire length of the cycle ruled by P'unchaw, alongside the llamas' mating season and production of offspring. Thus, the April festival closed a period of a little less than six months, which began with the sun's passage through the zenith on October 30, when the novices requested of Huanacauri the licence to be initiated. The song *puka llama* performed that month would have closed this cycle by echoing the *wari taki* with which the novices opened it.

---

188 Guaman Poma: 243 [245], 217; 318 [320], 292; 319 [321], 293. This author details the dance called *warisqa arawi*, while Pachacuti Yamqui retold that the Ayar ancestors sang the *chamay warisqa* when they conquered the Cuzco valley. See also Mannheim 1986; Zuidema 1986c.

189 Stobart 1996: 478. For similar festivities in other parts of the Andes, see Isbell 1978: 201, Hopkins 1982: 179-180.

190 Urton 1981: 187.

191 Betanzos: Pt. 1, Ch. 15; translation by Hamilton & Buchanan 1996: 66; Molina: 118.

Figure 9. Drawing of the *wariqsa arawi*. From Guaman Poma de Ayala: 318.

The Celebrations of the Daylight God

The Inca nobility celebrated the December solstice at the turn of *qhapaq rayme* whilst the *warachiku* trials were still under way. On that day, a procession carrying Wayna P'unchaw's figure left Cuzco, together with the *suntur pawqar*, the *napa,* and the two llama figures that had composed the cortège of the June solstice festival at *inti rayme*. Three kilometres southwest of the city, the celebrants joined the temple of *pukyu kancha*, on the height of Puquín hill, where they burnt garments, conch shells, and thirty white llamas to honour the regeneration of the young Daylight deity who was believed to have been reborn from a *pukyu* (spring). With these offerings and the procession of the solar emblems, this festival echoed the composition of the celebration conducted in Mantucalla at *inti rayme*. Puquín hill, however, was located in *kuntisuyu*, which belonged to the Hurin section of the Cuzco valley, while Mantucalla's structure was situated in *antisuyu*, associated with the Hanan division of the valley.[192] Accordingly, both ceremonies were held in the geographical division related to the cosmological half ruling the annual cycle at the time of the respective solstices. This characteristic also extended to the two ritual dispersions of sacrificial ashes that followed the January festival of the snake-cord and *inti rayme*, respectively. The first ceremony took place on the nineteenth day of the lunar month following the December solstice, as the Incas assembled at the junction of the Saphi and Tullumayu rivers in Pumap Chupan (*kuntisuyu*) and poured the remains of their past offerings into the flowing water.[193] The second dispersion of sacrificial ashes was held in the opposite quarter of *antisuyu* in the second week of July, when the nobility "collected all the charcoal and bones of the burnt offerings and strewed them over a plain nearby the [Mantucalla] hill, where nobody could enter other than those who had brought them."[194] Each of these ritual scatterings of ashes followed the first festival of the two respective divisions of the metropolitan calendar, supporting the assumption that they each closed a season of ritual offerings and inaugurated a new cycle that began with the solstices. Within this picture, the Inca month of *qhapaq rayme* initiated the reign of P'unchaw as ruler of the Lower half of the cosmos, whose celebrations the members of the Hurin moiety sponsored. Accordingly, the next festivity of the calendar honoured the *amaru*, a supernatural figure closely associated with P'unchaw.

On the day following the full moon of *kamay killa*, a group of female and male dancers dressed in black attire adorned with white fringe and feathers walked to a house located near the Coricancha, from where they took a large woollen cord. They carried this image to Haucaypata, the men on one side and the women on the other, dancing along the streets and around the plaza before dropping the cord on the ground, leaving it in the shape of a coiled snake. This figure was the *muru urqu*, literally "blotched male snake."[195] Its ritual appearance corresponded to the precise time of the year when the *yana phuyu* of the Serpent rules the night sky. Called *mach'aqway* today, this constellation becomes apparent in the Milky Way with the first rainfalls and gradually disappears by early February. Its presence in the firmament also replicates the active period of the reptile on earth, when it emerges from its hibernation with the coming of the rains and the warm

192 Sherbondy 1987; Zuidema 1986b: 190.
193 Molina: 114-117.
194 Cobo: T.2, Bk. 13, Ch. 28, 216.
195 *Urqu* refers to any male animal, see González Holguín, *Vocabulario*: 357.

weather. As I indicated earlier, the constellation of the Snake is commonly associated with the rainbow, which is deemed a giant multicoloured *amaru* rising from springs of water. This rainbow-snake was closely associated with Huanacauri, P'unchaw's servant and messenger, in the account of the Ayar siblings' original migration. It was above this hill that the founders of the Inca dynasty spotted a rainbow they interpreted as a favourable omen. This apparition, they believed, was a sign that the rising waters of the subterranean sea will not annihilate the earth, and that a new sun will emergence in the midst of the downpours to rule the chaos. This new sun, whose rays on the raindrops created the spectral colours of the rainbow, can be identified with P'unchaw. He was the Daylight deity in his vigorous form (Wayna P'unchaw) that had established his authority from the December solstice, as the sun reached its southernmost point, causing the days to last longer. He thus prevented the deluge by overcoming Wiraqucha, the god who ruled the underground water. Very much like the presence of the rainbow above Huanacauri hill, the dance of the snake-cord can be seen as announcing the advent of the Daylight god on earth. The imagery of the *muru urqu* dance suggests that this Inca ritual celebrated the supernatural manifestation of the multicoloured cosmic snake, as if it were moving against the starry night sky evoked by the black and white costumes of the dancers.

The End of the Maize Maturation

Following this change of power, a time of respite and anticipation set in as every man and woman returned to his or her home and waited for the crops to be fully grown before the reaping. Partly for this reason, the period stretching from the end of *warachiku* to the harvesting of the crops in late April is poorly documented. Molina forthrightly records that the Incas did not hold any specific festival during those months, but only worked on their *chakras*. Betanzos and Polo are just as laconic, and other sporadic sources only provide data on agricultural labour. An overview of these last accounts reveals that this period was dominated by the maturation of the crops (*puquy*), during which the labourers weeded their *chakras* and broke the clumps of soil softened by the heavy rains (*chaqma*). In the month corresponding to February (GC), the priests immolated one hundred chestnut-coloured llamas in order to attract the gods' benevolence. They also gathered around in a field and sacrificed a score of guinea pigs (*quwi*), which they burnt over a wood fire. The women of the Sun assisted in this ceremony and were eventually offered rich food, which they set aside to serve their godly spouse.

In the following month, one hundred black-coated llamas were sacrificed at a time when the maize started to dry and turn yellow. In several sources, this period is called *pawqar waray killa*, that is, the month of "a diversity of flowers, when they all blossom", with no further explanation.[196] Guaman Poma uses this designation in reference to February when he suggests that the newly initiated boys put on their loincloths (*wara*). Although he, like other chroniclers, situates *warachiku* in November, Guaman Poma may have referred here to a private stage of the rite of passage, when the youths finally donned the *wara* they had been given in Yauira hill. Finally, the rituals held in April included the sacrifice of one hundred mottled llamas known in Quechua as *muru muru*.[197] In addition to these, the daily offering of a ceremonial beverage was made to the earth in P'unchaw's name. To accomplish this, the two insignia of the Sun god were placed in the middle of the main plaza, the *suntur pawqar* next to the white *napa*. Every day, two attendants placed a jar filled to the brim at the feet of the sacred llama, which would spill the contents on the ground by kicking it over.[198] With this offering, it was believed that the thirst of the Pacha Mama was quenched and the maize would dry appropriately. Every man could now head to the fields to start harvesting. At this time, the young men who had been recently initiated received a young wife to marry and parcels of land to cultivate,[199] bringing to an end to both their period of trial and the rule of the Sun god P'unchaw over the annual cycle.

Politics of the Calendar

In Inca times, the calendrical regulation of social activities only remotely coincided with the alternation of the meteorological seasons. Although the sequential structure of *puquy mit'a* and *chiraw mit'a* was crucial to the fulfilment of organic life, crops and livestock included, these periods did not serve to divide social practice *per se*. By the time Pizarro's troops marched into Cuzco, the Incas had considerably expanded their astronomical knowledge and implemented various observation devices to fix the setting

---

196 González Holguín, *Vocabulario*: 282; Molina: 117; Fernández: Pt. 1, Bk. 1, Ch. 10, 86; Gutiérrez: Ch. 65, 256.

197 In his ethnographical work on the classification of camelids, Flores Ochoa (1986) noted the persistence of this type of denomination among present-day pastoralists.

198 Polo, "Tratado": 20; Cobo: T. 2, Bk. 13, Ch. 27, 214.

199 Anónimo [c.1575]: 156-157.

of their festivities. This knowledge enabled the metropolitan elite to divide time around two ideologies: one that reflected the ruling nobility's dual identity, and another that served its imperialist values. The cornerstone of this time construction was the *rayme* months, held at separate times of the year and fixed by the sun's passage through pairs of pillars erected in different locations around Cuzco. The primary purpose of the grandiose gatherings held during these months was to commemorate the supremacy of the elite's solar religion before the Incas' subjects. It was on these specific occasions that foreign delegates brought their share of the tribute to the imperial capital. Through these requirements set by the king and more generally through the implementation of a determined timeframe for the completion of political and economic activities, the ruling elite ensured an effective means of social control over every constituent population of Tawantinsuyu. For the Inca nobility, as for many civilizations throughout the world, control over time was a strategy of hierarchic dominion.[200]

Besides its function as an imperialist means of socio-economic regulation, the Inca calendar also reinforced hierarchy by contrasting the identity of the ruling elite with that of its subjects, a distinction that opposed insiders versus outsiders. The festivities of the Inca metropolis emphasize the symbolic value of maize not only as the favoured staple of ritual offerings to the gods, but also as the crop imported by the dynasty's founders. It was around this produce that the Inca nobility conducted its periodic rites of propitiation. Although the Spanish chroniclers may have partly overlooked the activities related to tuber agriculture in the course of Inca yearly occupations, maize was undeniably a prestige good associated with imperial power.[201] By contrast, the calendar of Altiplano communities reflected a different social time directed at the cultivation of high-altitude produce and the maintenance of pastoral activities. By imposing an additional timeframe over the annual activities of these populations and by requesting that these people perform their traditional dances at *quya rayme*, the Incas intended to maintain ethnic distinctions, and through them, hierarchical differences.[202]

Yet, the Inca calendar not only operated as an identity marker and a subjection tool, it also mirrored the dual social division of the ethnic group

---

200 There is a substantial literature on the function of calendars as tools of social control and identity making. See notably Burman 1981; Zerubavel 1981; Aveni 1989: 334-339; Munn 1992: 109-112; Gell 1992: 306-313.

201 Cavero Carrasco 1986; Hastorf & Johannessen 1993.

202 Bouysse-Cassagne 1987: 290-291.

in power. In this way, the calendar in use in Cuzco before the Spanish invasion reflected the order of the Inca cosmos and rested upon the alternate rules of the moieties' tutelary deities, Wiraqucha and P'unchaw, both of whom shared solar aspects. It was probably for this reason that the observation of the sun's activity determined the beginning and ending of these two cycles. An intermediary period corresponding to a changeover of power between the lower and upper cosmological halves separated each cycle. This argument is based on an elementary rereading of primary sources, which reveals that the commemorations of the June solstice were dedicated to the tutelary divinity of the Upper moiety, Wiraqucha, and celebrated the ethnic groups' subterranean migration from Lake Titicaca to the site of their final settlement. During the next few months, every performance was directed at assisting the circulation of water from the cosmic river to the underground canals supplying the valley's irrigation systems. Halfway through this cycle, at *quya rayme*, the Thunder god replaced Wiraqucha on earth following the latter's disappearance in the waters of Mama Tiqsi Qucha, situated on the edge of the existing world. Illapa now regulated the rains by pouring the water of the Milky Way down to the earth until the December solstice announced P'unchaw's rise. To celebrate the start of this new cycle, the Inca novices re-enacted their ancestors' journey in the Cuzco valley as part of their initiation ordeals. Their pilgrimage echoed the Hurin narratives that confined the Ayars' migration to the Inca heartland, whereas the *inti rayme* procession to the *qullasuyu* had recalled the Hanan traditions locating the Incas' origin in the Lake Titicaca Basin. Within this balanced partition of ritual time, it remains difficult to grasp the involvement of the panacas and other ayllus in the conduct of the annual festivities; whether there existed particular classes of ministers; and what their mode of recruitment would have been. Despite the scarcity of data on this issue, the evidence suggests that Hanan and Hurin alternated in officiating the metropolitan festivities: Upper Cuzco and the Sapa Inca were responsible for seeing to the ceremonies held between June and November, while Lower Cuzco attended to the rituals stretching from December to April. Each divinity would have received a service of priests who were affiliated with him/her by moiety membership. Such an organization appears very similar to the way present-day Andean communities continue to divide their civico-religious duties.

CONCLUSION

A BIDIMENSIONAL PERSPECTIVE ON INCA HISTORICAL NARRATIVES

To gain a renewed insight into the historical lore and cosmology of the Cuzco ruling elite, I have examined a miscellaneous corpus of primary data ranging from translations of Inca kingship narratives, to ceremonial descriptions, reports of kinship practice, as well as published reports on the Extirpation of idolatry. In exploring this material, cognisance of the agendas and biases of their respective authors was fundamental, but in addition to this historiographical dimension, I have attempted to delineate the preconquest cultural framework that provided a fertile ground for hybrid concepts to develop, such as Wiraqucha's "creative" power or the evangelist tone of his journey on earth. This emic approach, coupled with a reinterpretation of the Inca moiety system, was instrumental in identifying, and *ergo* reconciling the narrative inconsistencies that could not be explained merely through Spanish misinterpretation or the chroniclers' intentional deformation. This conclusion draws attention to the value of literary analysis in the reading of Spanish chronicles. This line of investigation, for which Catherine Julien strongly advocated, provides keys to distinguishing between different native historical genres and various narrative traditions within a vast and diverse body of primary sources. The Incas' recollections of their dead were as tainted with cultural conventions and political interests as any European historical discourse because these narratives had been produced to legitimate the status and claims of individuals vis-à-vis the prestige hierarchy. Their content was necessarily heterogeneous and remained malleable to the contemporary circumstances that surrounded their enunciation.

These observations highlight the value of including in our field of analysis several chronicles whose agendas and biases were long assumed to have distorted the oral traditions they record, in particular Gutiérrez's controversial *Quinquenarios* and Las Casas' *Apologética Historia*. By putting Gutiérrez's narrative of migration into perspective with other variants of the same myth, this particular excerpt of his work no longer appears to be an isolated fable proceeding from the chronicler's fantasy. Instead, it may be seen as part of a larger bulk of traditions documented by several authors, which reflected the perspective of Hanan Cuzcos on their own mythical origin. Many other fragments of the *Quinquenarios*, such as the elaborated narrative of Yahuar Huacac's abduction or the account of Tupa Inca Yupanqui's division of Cuzco into asymmetrical moieties, also hint that Gutiérrez had access to the detailed life stories of Hanan kings. Examination of his material proved that whether or not Gutiérrez extensively copied passages from other sources, and whether or not he had seen with his own eyes the collapse of the Inca empire, his text, like those of other unheeded chroniclers, contains fragments of indigenous traditions that can be uncovered by isolating and comparing the narrative patterns it shares with other colonial sources. I believe that this approach, which should be complementary to the existing historiographical tradition, refocuses Inca history onto the peoples it describes. It demonstrates that despite the layers of interpretation and recomposition that shaped the colonial transcription of pre-Hispanic narratives, new insights into Inca lore are at hand.

Building on this methodological framework, this book offers a study of the different features that characterized the Inca royal descent group, examined together with expressions of dual antagonisms in historical discourse and ritual action. This double focus expands our understanding of Hanan/Hurin membership, Inca noble identity, and gender dynamics by revealing that the characteristics attached to each opposite half of Cuzco varied according to the narrative conventions of the historical genre in which they appear. Although the Inca dual organization was primarily a system of social stratification by kinship whereby insiders to the dynastic line opposed relatives by marriage, its many expressions in dynastic accounts did not confine it to a static model of social reproduction. Hornborg had once described dual social organizations in Lowland South America as all "alternative expressions of kin/affine duality."[1] Despite sharing this feature, the Qhapaq Ayllu was probably not composed of exogamous clans required to marry into the opposing division. No primary source suggests that Hanan and Hurin were distinct descent groups functioning as matrimonial classes. In fact, the term moiety may appear to be misleading to kinship purists when applied to the binary divisions of pre-Hispanic Cuzco. Rather, the Inca dual organization borrowed the idiom of kinship to articulate a political discourse on dynastic perpetuation and kingship legitimacy. In addition to resorting to the first settlers/newcomers opposition commonly found in other Andean communities, it contrasted primogeniture (innate status) versus personal aptitudes (acquired

[1] Hornborg 1988: 275.

position), women of *qhapaq* extraction versus women from foreign ayllus, cognatic relations versus affinal networks, localized politics versus imperialist expansion, and established order and continuity versus disruption and reforms, all of which fit two different reflections on the past and on state ideology explaining the success of Inca state building, its longevity, and its endurance. The former conceived history as a narrative of social reproduction in which the extension of linear time proved the viability of the rules prescribed by the apical ancestor. It emphasized continuity in the dynastic line and situated the reigns of Hurin lords before those of Hanan leaders in a chronological sequence. The latter regarded history as a locus of dynamic contingencies. It gave preference to the law of the most able and exalted change as a means to legitimate power. Following these conventions, the life stories of Hanan kings retell their incredible feats in battle and exult in their combative qualities as opposed to their elder brothers' irresolution and indolence. In other words, Inca kings vindicated their position through two different perspectives on past events, which were both deemed as equally true/legitimate. This reading of native history, in redefining rules of royal descent, integrates a time perspective within the cultural dynamics that mediated dual antagonisms. It also emphasizes a crucial point of methodology: while the anthropological perspective has long warned us to not read primary sources as faithful records of "what really happened", the historical perspective reminds us that selecting from these descriptions the sociological data that are *a priori* more consistent with a plausible reconstruction of the Inca kinship or political system is equally misguided. By overlooking the existence of a preferred rule of primogeniture and close-relative marriage among the Inca nobility, past studies have failed to identify fundamental aspects of the Cuzco dual organization.

Wiraqucha was the divine ancestor who sponsored the illegitimate upstarts in their rise to power. Like them, who were outsiders to the main dynastic line, he was the Sun of the high plateaux who had migrated from his remote land to animate his valiant descendants with an incredible strength (*kallpa*). As the story of his emergence over Lake Titicaca tells, he was the second Sun who destroyed the primordial, weak humanity to establish his rule. He succeeded the older Sun, much like Pachacuti Yupanqui is said to have placed Wiraqucha at the head of the Inca religious system after he had understood the superiority of the "Maker" over the sun whose daylight could be overshadowed by clouds. Although most chroniclers record this change of divine authority with reference to the theological argument of the First Cause, the Inca cosmological model reveals a hierarchical structure conducive to the elaboration of a hybrid narration. In addition to Wiraqucha's and P'unchaw's supervision over different modes of kingship legitimacy (personal merit or birthright), the rationale that underpinned these deities' association with either moiety was not based on an abstract analogy that identified their domains (subterranean and celestial) with the spatial partition of the city (Lower and Upper). Rather, Wiraqucha's and P'unchaw's affiliations with Hanan and Hurin depended on the influence they exerted over the environmental phenomena and the agricultural activities of the two divisions of the Inca calendar. In turn, each moiety maintained a particular account of its ancestors' origin and a separate listing of Inca rulers that contradicted in many ways those elaborated by the opposite division. At the centre of Inca cosmology was the concept that divinities alternated twice a year in regulating water through the different planes of the cosmos. Their descent on earth also rotated in conducting the annual festivals of the metropolitan calendar so that moiety membership determined the ritual services that each individual was entitled to accomplish in honour of the divinities of his/her division.

Overall, Inca historical representation was bidimensional, and its two facets reflected divergent interests and preoccupations. The first was outward oriented and aimed to create a homogenous identity that served the imperialist ideology. It emphasized a coalition of opposites, thereby forging a linear history in which continuity characterized descent, and durability legitimized leadership. By the same token, it depicted the Incas as a unified and unwavering people whose authority was supported by a single divine hand: Inti. The official religion spread by the ruling elite revolved around the cult of one great denomination of the Sun that all nations of Tawantinsuyu were compelled to revere, and which embraced the two solar devotions of the panacas. The other facet of Inca historical representation was inward oriented and expressed the particular perspectives of the royal lineages. It focused on conflicts, rupture, and personal achievements, but also stressed the hierarchical pre-eminence of Hanan over Hurin. Accordingly, the narratives pertaining to this tradition emphasized the superiority of Wiraqucha over P'unchaw and accredited to the rulers of Upper Cuzco the articulation of a pseudo-metaphysical discourse that evinced the ascendency of the subterranean god over Daylight. A dynamic view of the past mediated these two pictures, fluctuating between the paradigms of duration and change, depending on the contexts of enunciation. This perspective on dual opposition thus distinguishes between the social praxis that moieties articulate in relation to the totality they form, on the one hand, and that which they elaborate in their rapport with each other, on the other. In the former context, Inca ritual and historical representations accentuate identity and coalition, whereas, in the latter they play up contrast and asymmetry. Such a framework, which derives from Dumont's formal model of binary oppositions, successfully decentres conceptions of time consciousness in Inca Cuzco, resulting in a more subtle understanding of the narratives elaborated by the pre-Hispanic elite.

An acknowledgement that the Incas devised a historical consciousness of their own is therefore the first step towards a sharper and more inflected approach to Inca studies, one that decolonizes the Andean past by uncovering native voices inscribed in Spanish texts, and enriches our wider comprehension of pre-colonial societies. It also opens promising routes into the thinking and culture of ancient Andean polities. Many contradictions in primary sources, which are often disparaged as Spanish constructs, can thus be investigated through the light of dualism and enhance our understanding of Inca civilization. The Qhapaq Ayllu's relations with local lords and, notably, the role of foreign wives and concubines in political decision-making would particularly benefit from this approach. Fragmentary evidence in the chronicles alludes to the influence these women played during the wars of succession and conquest, showing that the stability of the Inca imperial edifice was heavily reliant on bride exchange and on the mutual obligations in goods and services attached to this practice. The study of Inca kinship, and particularly the examination of royal descent rules, transmission of names, and marriage preferences, has been instrumental in arriving at this renewed perspective on historical representations. The outlook gained through this analysis calls for an approach to Inca historiography wherein history and anthropology are interrelated and intersecting domains of research, whose aim is not to reconstruct past events or fixed social structures, but is rather to unravel how the Incas related to their past as well as to delineate the contours of alternative forms of historical representation.

Two major points arise from these considerations. The first relates to our handling of primary material to establish a chronological history of the spread of Inca sovereignty over an estimated nine to sixteen million people. The second, which we shall deal with subsequently, has implications for anthropological theories of dual organizations. To begin with, let us examine the issue of the historicity of Inca narratives. For centuries, scholars have debated over the reliability of these accounts for the chronological reconstruction of Inca state formation. While some have denied any temporal depth to these traditions, others have envisaged the Inca ruler list as a sufficiently accurate record to suggest that imperial expansion began only a century before Pizarro landed in Peru. The rise of Pachacuti Yupanqui in the late 1430s would have initiated an extraordinarily rapid enforcement of Inca jurisdiction over one of the largest empires of this period. Before Pachacuti's reign, the founders of the Inca dynasty would have settled in the valley around AD 1200, where they engaged exclusively in local politics for the two centuries that followed. A processual approach to archaeological data, based on the distribution of the K'illke ceramic style, has recently contradicted these conclusions.[2] Bauer and Covey, in particular, point out that settlement patterns and material remains do not substantiate any population shift in the Cuzco region after the downfall of Wari dominion in the Late Intermediate Period. In other words, there is no processual evidence to the Ayar siblings' migration. The same archaeologists also argue that evidence of state expansion was found outside the valley around AD 1300, which would suggest that Inca imperialist politics began earlier than Rowe expected.[3] Altogether, these data appear to be more consistent with the enduring practicalities involved in setting up a state administration with extended infrastructures like those of Tawantinsuyu.

Yet, we should not invalidate the information enclosed in Inca historical narratives because their apparent chronology does not fit a realistic time span. The key to reading this corpus lies in a better understanding of the aims and conventions of historical production among the Cuzco elite. Until now, however, scholars have enquired as to whether pre-Hispanic traditions were mythical constructs or whether they formed an accurate record of sequential facts. They thus questioned the nature of this documentary material in ways that overlooked the very aim of historical constructions. Nevertheless, a different outlook on Inca historical narratives is possible—that which takes into account native modes of representing the kingship institution. This renewed perspective is intrinsically linked to the definition of the royal descent group. Despite the evident structural

---

[2] Processual archaeology is an anthropologically orientated approach to the interpretation of archaeological data. Influenced by the systems theory, it is concerned with culture process and posits that environmental variables, whether natural or anthropogenic, condition social behaviour and cultural evolution whose traits are manifest in material remains.

[3] Bauer & Covey 2002; Covey 2006a.

features of the Qhapaq Ayllu, this system also provides broad temporal sequences for the Inca dynasty because genealogical distance was not exclusively metonymic. For a panaca to aspire to a significant role in the political arena, it must have held sufficient prestige to create extended affinal relations with powerful foreign ayllus. The royal lineages that did not enjoy large matrimonial networks (or maintained only localized ones) would once have assumed important offices, but gradually lost their political precedence. In this way, the Qhapaq Ayllu's genealogical organization partly reflected a historical reality: the lower in rank the founder of a panaca was, the longer the time since he and his descent group had exerted leadership. Of course, this picture remains approximate and does not provide a precise dynastic chronology, but it is necessary to acknowledge here the limitations of our historical sources for this modern enquiry. To offset these constraints, the intensification of archaeological excavations and the systematization of dating analyses constitute a promising avenue of research. In this investigation, we must also consider the imprint of positional inheritance on Inca oral transmission. The life stories of distinguished characters, whether they were kings, queens, or army commanders, were compiled from the deeds of namesakes who had inherited their common appellation from an eminent ancestor. In this way, the events recalled in Inca historical traditions were not merely invented, but instead took place over several generations. This social practice explains adequately why chroniclers like Murúa and Sarmiento provide reign lengths for Inca kings exceeding a hundred years. Altogether, these observations extend the chronology of Inca state expansion and offer an alternative strategy for looking at the archaeological record.

To conclude, the argument of this book also challenges the view elaborated in earlier anthropological studies that dual organization is a system for discerning order that always tends towards equilibrium. Although this argument may apply to a number of Lowland South American societies, a similar quest for harmony cannot be ascribed to the Inca ruling elite. The Qhapaq Ayllu was a highly stratified descent group with imperialist designs whose organization was based on profoundly asymmetric values. In this system, the rotation of ritual duties and ceremonial cycles did not counteract the prominence of Hanan over Hurin Cuzco, nor did it place P'unchaw and Wiraqucha on the same cosmological pedestal. Through this hierarchical structure, which sublimated the deeds of lords who defied the established order, the Inca dual organization exalted its openness to historical contingencies. It recognised the role of individual actions in reinforcing the foundations of their imperialist agenda against the structural order of the elite's kinship relations. Therefore, the Inca dual divisions cannot be said to have striven towards reciprocity, but instead mediated upward social mobility to allow the most able individuals to enter the prestige hierarchy. Attaining a better social status consisted of positioning oneself closer to the senior line and converging towards the apical ancestor. This concentric movement operated through time according to two principles that effected transmission of statuses: marriage and personal merit. The former was because a man's ability to secure a union with a high-ranking woman increased the rank of their progeny, thereby moving the children of this marriage closer to the senior line; the latter because personal merit was rewarded with prestige names and related politico-ritual offices. Although, theoretically, time condemned the descent group to statutory decline, in practice it mediated upward mobility and allowed the Inca authority structure to reproduce itself. The assimilation of outsiders was crucial to this dynamic. Like the three-generation model of alliance that concluded the narrative of Yahuar Huacac's abduction, the provincial ethnic groups that entered into a symmetric alliance with the ruling elite gradually converged towards the core of the Inca body politics: from "outsider" nations, they first became *wakcha* Incas (poor, without ancestry), and then *kaka* Cuzcos (maternal uncle, wife-giver to the king). In this way, the Inca royal descent group and its historical mode of self-portrayal shared several features in common, albeit intertwined in different ways, with other kingship systems organized in conical clans, as in Hawaii, or even Tonga, where a form of diarchy operated. This observation highlights the relevance of cross-cultural comparison in the study of kingship and dual social systems while stressing the variability that exists within these models. As Hornborg observed about dual organizations of Lowland South America, "there can be no master narrative. Yet, there are serial congruities between various representations."[4] In the hands of the Cuzqueñan nobility, the dual organization model became a vehicle of imperial authority, allowing the Inca prestige hierarchy to reproduce itself through name/rank transmission and marriage strategies, while at the same time extending the elite's affinal network throughout the various provinces of Tawantinsuyu. It simultaneously preserved the divine essence of the dynasty while adapting its mechanisms to the imperialist need for expansion. These adjustments were made possible because the Inca dual organization held

[4] Hornborg 1998: 182.

intrinsic hierarchical principles, like many examples of dualistic social structure from the pre-Hispanic authoritarian states to the Amazonian non-centralized polities.[5] This formal model, once regarded as a mere instrument of equity and social reproduction, enshrines asymmetric principles that are conducive to creating its own transformations, thus enabling a regional polity like Inca Cuzco to become an imperial power. The Qhapaq Ayllu's historical traditions are here to attest to it: they celebrated the paradigm of individual-generated changes versus social inertia.

5 Turner 1996.

GLOSSARY

Aklla: Chosen women selected among the Inca royal descent group and the provincial governing elites. They were sent to state facilities (*aklla wasi*) and served in religious activities or distributed as wives of the elites.

Ayllu: Before the conquest, a non-localized segmentary organization of preferred unilineal descent, most commonly in its patrilineal variety, in which individuals are ranked according to closeness to the apical ancestor. The term also referred to a weapon, otherwise called *riwi*, composed of three cords of gut or leather, each with a ball of stone or metal at the end. It was thrown around the legs of people or animals.

Chakra: Cultivated field.

Chicha: Word of Caribbean origin (*aqha* in Quechua), fermented beverage, generally of maize.

Hanan: The Upper division of Cuzco. It also designates objects/people located outside or above.

Hurin: The Lower division of Cuzco. It also designates objects or people located inside or below.

Kallpa: The physical or spiritual might of powerful humans, infused by an ancestor.

Kamaq: The vital force that ancestors transmitted to animate living beings.

Kancha: Enclosure. The word applied to monumental architecture, temporary ritual precincts, or livestock corrals.

Khipu: Set of knotted, dyed strings on which the Incas recorded their narratives and a variety of quantitative data.

Kuraka: Native leader. In Spanish writings, this title is often replaced by the word *cacique* of Caribbean origin.

Llawtu: Headband with pendant fringe. Its Spanish word is *borla*.

Mama Qucha: The subterranean sea on which floated the earth.

Maska paycha: Insignia of Inca kingship.

Mit'a: Unilateral tributary system; also any event held periodically, such as the seasons.

Mitmaq (also *mitimaes*): Ethnic groups, allies of the Incas, which were permanently displaced from their land of origin and assigned land in other parts of the empire, where they resettled to ensure political stability.

Panaca: Inca royal lineage, ramification of the main dynastic line.

Paqarina: Birthplace of an ayllu founder.

Puna: An ecological zone situated between 3200m and 3500m high, natural habitat of camelids.

Purun awqas: Literally, the stone warriors who assisted the young Pachacuti Inca in winning victory against the Chancas.
Qhapaq ñan: Royal road.
Qhapaq: Royal, rich because of having jurisdiction over many people and resources.
Qullqa: Storehouses located in all provinces of the empire, used to dry and store foodstuff as part of the tribute.
Quya: Women of *qhapaq* descent from both the paternal and maternal lines.
Rutuchiku: Ceremony of the first hair cutting, during which the infant received his first name.
Sapa Inca: "Sole, unique Inca" in reference to the living king.
Seqe: Imaginary line that radiated from a central point in Cuzco to cross several shrines in its surroundings. It may have denoted a limit, but it also described arable land divided into several cultivation parcels.
Suyu: A division of land with its people, which came to describe any of the four quarters of the Inca realm.
Taki: Ritual dance with singing.
Tampu: State lodging of varying sizes with storage facilities, constructed along the Inca road system. Some may have been used as administrative centres.
Tiyana: Low stool on which the Inca king and Andean officials sat.
Ushnu: Primary sources give a variety of meanings to this word. It applies to stone platforms erected in the centre of some Inca cities for ceremonial purposes. *Ushnus* would have served as seats for officials during important festivities and as places for sacrificial libations.
Waka: Shrine linked to the mythical past of an ancestor.
Warachiku: "The donning of the loincloth", the initiation ritual during which the Inca novices went through several ordeals, had their hair shaved, had their ears pierced, and received a new name.
Yana phuyu: Black cloud constellation.
Yanakuna: Retainers of various ethnic backgrounds attached to the ruling elite. They were permanently removed from their community of origin and exempt from tribute.

BIBLIOGRAPHY

Manuscript Source

Archivo Departamental del Cuzco, Perú (ADC), Real Audiencia, Causas Ordinarias, Leg. 42 (1670 [1589]), Expediente sobre el Cacigazgo de Chumbivilcas.

Primary Sources

Acosta, José de. *Historia natural y moral de las Indias* [1590]. México: Fondo de Cultura Económica, 1940.

Albornoz, Cristóbal de. "Instrucción para descubrir todas las guacas del Perú y sus camayos y haciendas." [1582] In *Fabulas y mitos de los Incas*, edited by Henrique Urbano and Pierre Duviols, 163-198. Madrid: Historia 16, 1989.

Anónimo. "Carta sobre el gobierno, las costumbres y los antiguos ritos del Perú." [c. 1551] In *Zeitschrift für ethnologie* 66 (1934): 402-416.

———. "Discurso de la sucesión y gobierno de los Yngas." [c. 1570] In *Juicio de límites entre el Perú y Bolivia, prueba peruana presentada al gobierno de la república argentina*, vol. 8, edited by Víctor M. Maúrtua, 149-165. Madrid: Imprenta de los hijos de M.G. Hernández, 1906.

———. "Aviso de el modo que havia en el gobierno de los indios en tiempo del inga y como se repartían las tierras y tributos." [c. 1575] In *Revista Española de Antropología Española* 5 (1970): 163-173.

———. "Relación de señores indios que servieron a Tupac Yupanqui y Huayna Capac." [c. 1580] In *Informaciones sobre el antiguo Perú*, edited by Horacio Urteaga, Colección de Libros y Documentos Referentes a la Historia del Perú, 2nd ser. Lima: Sanmartí y Ca., 1920.

Arriaga, Pablo José de. *La extirpación de la idolatría en el Perú* [1621]. Cusco: Centro de Estudios Regionales Andinos Bartolomé de Las Casas, 1999.

Atienza, Lope de. "Compendio historial del estado de los indios del Perú." [c. 1575] In *La religión del imperio de los Incas*, Apéndice I, edited by Jacinto Jijon y Caamaño, 1-235. Quito: Escuela Tipográfica Salesiana, 1931.

Augustinos. "Relación de idolatrías en Huamachuco por los primeres Augustinos." [1560] In *Informaciones acerca de la religión y gobierno de los Incas*. Lima: Sanmartí y Ca, 1918.

Avila, Francisco de. *The Huarochirí manuscript: a testament of ancient and colonial Andean religion* [1598]. Translated by Frank Salomon and George L. Urioste. Austin: University of Texas Press, 1991.

Bandera, Damían de la. "Relación del origen y gobierno que los Ingas tuvieron y del que había antes de ellos." [1557] In *Informaciones sobre el Perú antiguo*, edited by Horacio H. Urteaga, Colección de Libros y Documentos Referentes a la Historia del Perú, 2nd ser. Lima: Sanmartí y Ca., 1921.

Bertonio, Ludovico. *Vocabulario de la lengua Aymara* [1612]. Cochabamba: Centro de Estudios de la Realidad Económica y Social, 1984.

Betanzos, Juan de. *Suma y narración de los Incas* [1551]. Madrid: Editorial Atlas, 1987.

———. *Narratives of the Incas*. Translated and edited by Roland Hamilton & Dana Buchanan. Austin: University of Texas Press, 1996.

Cabello Balboa, Miguel. *Miscelánea Antártica, una historia del Perú antiguo* [1586]. Lima: Universidad Nacional Mayor de San Marcos, Instituto de Etnología, 1951.

Castro, Cristóbal de, and Ortega Morejón, Diego de. "Relación y declaración del modo que este valle de Chincha y sus comarcanos se gobernaban antes que oviese Yngas y despues que los vuo hasta que los cristianos entraron en esta tierra."[1558] In *Quellen zur Kulturgeschichte des präkolumbischen Amerika*, edited by Hermann Trimborn. Stuttgart: Strecker und Schröder, Verlag, 1968.

Cieza de León, Pedro. *Primera parte de la Crónica del Perú* [1553]. Lima: Pontificia Universidad Católica del Perú, 1996.

———. *Segunda parte de la crónica del Perú: El señorío de los Incas* [1554]. Lima: Pontificia Universidad Católica del Perú, 1996.

Cobo, Bernabé. *Historia del Nuevo Mundo* [1653]. Biblioteca de Autores Españoles 91-92. Madrid: Editorial Atlas, 1964.

Discurso sobre la descendencia y gobierno de los Incas [1543-1608]. Edited by Horacio Urteaga, Colección de Libros y Documentos Referentes a la Historia del Perú, 3. Lima: Sanmartí y Ca.

Estete, Miguel de. *Noticia del Perú* [1535]. Colección de Libros y Documentos Referentes a la Historia del Perú, 2nd ser. Lima: Sanmartí y Ca., 1924.

Fernández, Diego. *Historia del Perú* [1571]. Biblioteca de Autores Españoles 165. Madrid: Editorial Atlas, 1963.

Garcilaso de la Vega. *Comentarios Reales de los Incas* [1609]. Lima: Librería Internacional de Perú, 1959.

———. *Royal commentaries of the Incas, and general history of Peru*. Austin: University of Texas Press, 1989.

González Holguín, Diego. *Vocabulario de la lengua general de todo el Perú* [1608]. Lima: Imprenta Santa Maria, 1952.

———. *Gramática y arte nueva de la lengua general de todo el Perú* [1552]. Genève: Pagano, 1842.

Guaman Poma De Ayala, Felipe. *El primer nueva corónica y buen gobierno* [c. 1615], 3 vols. México: Siglo Veintiuno Editores, 1980.

Gutiérrez de Santa Clara, Pedro. *Quinquenarios o Historia de las guerras civiles del Perú* [c. 1590]. Biblioteca de Autores Españoles 166. Madrid: Editorial Atlas, 1963-64.

Herrera y Tordesillas, Antonio de. *Historia general de los hechos de los Castellanos en las islas i tierra firme del mar océano* [1601-1615]. Madrid: Universidad Complutense de Madrid, 1991.

"Informaciones al Virrey Toledo." [1571-1572] In *Informaciones sobre el antiguo Perú*, edited by Horacio Urteaga, Colección de Libros y Documentos Referentes a la Historia del Perú, 2nd ser. Lima: Sanmartí y Ca., 1920.

Las Casas, Bartolomé de. *Apologética historia de las Indias* [1562-1564], Biblioteca de Autores Españoles 105-106. Madrid: Editorial Atlas, 1909.

Lizárraga, Reginaldo de. *Descripción breve de toda la tierra del Perú, Tucumán, Rio de la Plata y Chile* [c. 1603-1609], Biblioteca de Autores Españoles 216. Madrid: Editorial Atlas, 1968.

López de Gómara, Francisco. *Historia General de las Indias y conquista de México* [1552]. Madrid: Calpe, 1932.

Mateos, Francisco (ed.). *Historia general de la compania de Jesús en la provincia del Perú* [c. 1600]. Madrid: Consejo Superior de Investigaciones Científicas, 1944.

Matienzo, Juan de. *Gobierno del Perú* [1567]. Travaux de l'Institut d'Etudes Andines 11. Paris, Lima: Institut Français d'Etudes Andines, 1967.

Molina, Cristóbal de. "Relación de las fábulas y ritos de los Incas." [1575] In *Fábulas y mitos de los Incas*, edited by Henrique Urbano and Pierre Duviols, 48-134. Madrid: Historia 16, 1989.

Murúa, Martín de. *Historia general del Perú* [Wellington ms., 1590-1602]. Madrid: Dastin Historia, 2001.

———. *Historia del origen y genealogía real de los reyes Incas del Perú* [Loyola ms., 1590]. Madrid: Consejo Superior de Investigaciones Científicas, Instituto Santo Toribio de Mogrovejo.

Pachacuti Yamqui Salcamaygua, Juan de Santa Cruz. *Relación de antigüedades deste reyno del Pirú* [c. 1615]. Lima/Cuzco: Institut Français d'Etudes Andines/Centro de Estudios Regionales Andinos Bartolomé de Las Casas, 1993.

Pizarro, Pedro. *Relación del descubrimiento y conquista de los reinos del Perú* [1571]. Colección de Libros y Documentos Referentes a la Historia del Perú. Lima: Sanmartí y Ca., 1917.

Polo Ondegardo, Juan. "Tratado sobre los errores y supersticiones de los Indios sacados del tratado y averiguación que hizo licenciado Polo." [1585] In *Informaciones acerca de la religión y gobierno de los Incas*. Colección de Libros y Documentos Referentes a la Historia del Perú. Lima: Sanmartí y Ca., 1916.

———. (attributed to). "Relación de los adoratorios de los indios en los quatro caminos (ceques) que salían del Cuzco." [c. 1560] In *Informaciones acerca de la religión y gobierno de los Incas*. Colección de Libros y Documentos Referentes a la Historia del Perú. Lima: Sanmartí y Ca., 1917.

———. *El mundo de los Incas* [c. 1560-1570]. Madrid: Historia 16, 1990.

Ramos Gavilán, Alfonso. *Historia de Nuestra Señora de Copacabana* [1621]. La Paz: Cámara de comercio, 1976.

Relaciones geográficas de Indias, Perú [c. 1570-1586], edited by Marcos Jiménez de la Espada, Biblioteca de Autores Españoles 183-184. Madrid: Editorial Atlas, 1965.

Ricardo, Antonio. *Vocabulario y phrasis en la lengua general de los indios del Perú, llamada Quichua y en la lengua española* [1586]. Lima: Instituto de Historia de la Facultad de Letras, Universidad Nacional Mayor de San Marcos, 1951.

Ritos y tradiciones de Huarochirí [1598]. Translated and edited by Gerald Taylor. Travaux de l'Institut Français d'Etudes Andines 116. Lima: Banco Central de Reserva del Perú, Universidad Particular Ricardo Palma, 1999.

Román y Zamora, Jéronimo. *Repúblicas de Indias, idolatrias y gobierno antes de la conquista* [1575]. Colección de Libros Españoles Raros o Curiosos que Tratan de América XIV-XV. Madrid: Victoriano Suarez Editor, 1897.

San Pedro, Fray Juan de. *La persecución del demonio. Crónica de los primeros Agustinos en el norte del Perú* [1560]. Málaga: Algazara, 1992.

Santillán, Hernando de. *Relación del origen, descendencia, politica y gobierno de los Incas* [1563]. Biblioteca de Autores Españoles 209. Madrid: Editorial Atlas, 1968.

Santo Tomás, Domingo de. *Lexicón o vocabulario de la lengua general del Perú* [1560]. Lima: Universidad Nacional Mayor de San Marcos, 1951.

———. *Grammatica o arte de la lengua general de los Indios de los Reynos del Peru* [1560]. Lima: Universidad Nacional Mayor de San Marcos, 1951.

Sarmiento de Gamboa, Pedro. *Historia de los Incas* [1572]. Madrid: Miraguano/Polifemo Ediciones, 1988.

Segovia, Bartolomé. *Relación de muchas cosas acaescidas en el Perú* [1553]. Biblioteca de Autores Españoles 209. Madrid: Editorial Atlas, 1916.

Valera, Blas (attributed to). "Relación de las costumbres antiguas de los naturales del Pirú." [c. 1595] In *Tres relaciones de antigüedades peruanas*, 133-203. Lima: Sanmartí y Ca., 1927.

Vázquez de Espinosa, Antonio. *Compendio y descripción de las Indias Occidentales* [1629], Biblioteca de Autores Españoles 231. Madrid: Editorial Atlas, 1969.

Yupanqui, Titu Cusi. *History of how the Spaniards arrived in Peru/Relación de como los españoles entraron en el Perú* [1570]. Translated by Catherine Julien. Idianapolis/Cambridge: Hackett Publishing Company, 2006.

Zárate, Augustín de. *Historia del descubrimiento y conquista del Perú* [1555]. Buenos Aires: Publicaciones del Instituto de Historia Argentina y Americana, Universidad de Buenos Aires, 1965.

Secondary Sources

Abercrombie, Thomas A. *Pathways of memory and power, ethnography and history among an Andean people.* Madison: The University of Wisconsin Press, 1998.
Adorno, Rolena. *Guaman Poma, writing and resistance in colonial Peru.* Austin: University of Texas Press, 1988.
Adorno, Rolena, and Boserup, Ivan. *New Studies of the autograph manuscript of Felipe Guaman Poma de Ayala's Nueva corónica y buen gobierno.* Copenhagen: Museum Tusculanum Press, 2003.
Alberti Manzanares, Pilar. "La influencia económica y política de las Acllacuna en el Incanato." *Revista de Indias* 45, n° 176 (1985): 557-585.
———. "una institución exclusivamente feminina en la época incaica: las acllacuna." *Revista Española de Antropología Americana* 16 (1986): 153-190.
Albó, Xavier. "Preguntas a los historiadores desde los rituales andinos actuales." In *Incas e indios cristianos. Elites indígenas e identidades cristianas en los Andes coloniales,* edited by Jean-Jacques Decoster, 395-438. Travaux de l'Institut Français d'Etudes Andines 149. Lima, Cuzco: Centro de Estudios Regionales Andinos Bartolomé de Las Casas, Institut Français d'Etudes Andines, Asociación Kuraka, 2002.
Alconini, Sonia. "The southeastern Inka frontier against the Chiriguanos: structure and dynamics of the Inka Imperial borderlands." *Latin American Antiquity* 15, n°4 (2004): 389-418.
Allen, Catherine J. *The hold life has. Coca and cultural identity in an Andean community.* Washington, London: Smithsonian Institution Press, 1988.
———. "Time, place and narrative in an Andean community." *Journal de la Société Suisse des Américanistes* 57-58 (1993-1994): 89-95.
Alonso Sagaseta, Alicia. "Las momias de los Incas: su función y realidad social." *Revista Española de Antropología Americana* XIX (1989): 109-135.
Amado Gonzales, Donato. "El alférez real de los Incas: resistencia, cambios, y continuidad de la identidad indígena." In *Incas e Indios cristianos: elites indígenas e identidades cristianas en los Andes coloniales,* edited by Jean-Jacques Decoster, 221-249. Travaux de l'Institut Français d'Etudes Andines 149. Lima, Cuzco: Centro de Estudios Regionales Andinos Bartolomé de Las Casas, Institut Français d'Etudes Andines, Asociación Kuraka, 2002.
Andersson, Hans. "One aspect of Wedin's work." *The Americas,* 25 n°2 (1968): 180-190.
Appadurai, Arjun. "The past is a scarce resource." *Man* 16, n°2 (1981): 201-219.
Arellano Hoffmann, Carmen. "Hanan/Urin: reflexiones acerca de un viejo concepto dual Inka y su aplicación en el Chinchaysuyu." In *50 años de Estudios Americanistas en la Universidad de Bonn: nuevas contribuciones a la arqueología, ethnohistoria, ethnolingüística y ethnografía de las Américas,* B.A.S. 30, edited by Sabine Dedenbach-Salazar Sáenz et. al, 473-493. Bonn Verlag: Anton Saurwein, 1998.
———. "Quipu y tocapu: sistemas de comunicación inca." In *Los Incas: Arte y símbolos,* edited by Franklin Pease et al., 215-261. Lima: Banco de Crédito del Perú, Colección Arte y Tesoros del Perú, 1999.
———. "El juego de Chuncana entre los Chimú un tablero de madera que prueba la hipótesis de Erland Nordenskiöld." *Bulletin de l'Institut Français d'Etudes Andines* 32, n°2 (2003): 317-345.
Armas Asín, Fernando. "Wiracocha, pastoral católica y mitología del Titicaca, consideraciones desde la mitografía y la andinística." *Anuario de Historia de la Iglesia* 11 (2002): 191-213.
Arnold, Denise and Espejo Ayca, Elvira. "On drinking cups and constellations: some relations between Aymara astronomical and textual practices in Qaqachaka ayllu (Bolivia)." *Journal of the Latin American Cultural Studies* 15, n°2 (2006): 183-213.

Ascher, Marcia and Ascher, Robert. *Code of the quipu. Study in media, mathematics, and culture.* Ann Arbor: University of Michigan Press, 1981.

Ávarez-Calderón, Annalyda. "La crónica de Fray Martín de Murúa: mentiras y legados de un mercedario vasco en los Andes." *Revista Andina* 45 (2007): 159-186.

Aveni, Anthony F., ed. *Native American Astronomy*. Austin: University of Texas Press, 1977.

———. "The Incas and their orientation calendar." In *Empires of times, calendars, clocks and cultures*, edited by Anthony F. Aveni, 278-304. New York: BasicBooks, 1989.

Bastien, Joseph W. *Mountain of the condor: metaphor and ritual in an Andean ayllu.* Saint Paul: West Publishing Co., 1978.

Bataillon, Marcel. "Gutiérrez de Santa Clara, escritor Mexicano." *Nueva Revista de Filología Hispánica* 15 (1961): 405-440.

Bauer, Brian S. *The development of the Inca state*. Austin: University of Texas Press, 1992.

———. "Legitimization of the state in inca myth and ritual." *American Anthropologist* 98, n°2 (1996): 327-337.

———. *The sacred landscape of the Inca: the Cusco ceque system*. Austin: University of Texas Press, 1998.

———. *Ancient Cuzco, heartland of the Inca*. Austin: University of Texas Press, 2004.

Bauer, Brian S., and Dearborn, David S. P. *Astronomy and empire in the Ancient Andes*. Austin: University of Texas Press, 1995.

Bauer, Brian S., and Stanish, Charles. *Ritual and pilgrimage in the Ancient Andes, the Islands of the Sun and the Moon*. Austin: University of Texas Press, 2001.

Bauer, Brian S., and Covey, Alan. "Processes of state formation in the Inca heartland (Cuzco, Peru)." *American Anthropologist* 104, n°3 (2002): 846-864.

Baumann, Max P. "Andean music, symbolic dualism and cosmology." In *Cosmología and Música en los Andes*, edited by Max Peter Baumann, 15-66. Frankfurt am Main, Madrid: Vervuert Iberoamericana, 1996.

Beyersdorff, Margot. *Léxico Agropecuario Quechua*. Cuzco: Centro de estudios Rurales Andinos Bartolomé de las Casas, 1984.

Bietenholz, Peter G. *Historia and fabula: myths and legend in historical thought from antiquity to the modern age*. Leiden, New York: Brill, 1994.

Bixler, Ray H. "Sibling incest in the royal families of Egypt, Peru, and Hawaii." *Journal of Sex Research* 18, n°3 (1982a): 264-281.

———. "Comment on the incidence and purpose of royal sibling incest." *American Ethnologist* 9, n°3 (1982b): 580-582.

Bolin, Inge. *Rituals of respect. The secret of survival in the High Peruvian Andes*. Austin: University of Texas Press, 1998.

Bolton, Ralph. "Guinea pigs, protein and ritual." *Ethnology* 18, n°3 (1979): 229-252.

Bonilla, Heraclio, and Fonseca Martel, César. *Tradición y conservadurismo en la área cultural del Lago Titicaca: Jesús de Machaca, una communidad aymara del altiplano andino*. Lima: Instituto de Estudios Peruanos, 1967.

Bouysse-Cassagne, Thérèse. "Urco and uma: Aymara concepts of space." In *Anthropological history of Andean polities*, edited by John V. Murra, Nathan Wachtel, and Jacques Revel, 201-227. Cambridge/Paris: Cambridge University Press/Maison des Sciences de l'Homme, 1986.

———. *La identidad Aymara, aproximación histórica (siglo XV-XVI)*. La Paz: HISBOL, Institut Français d'Etudes Andines, 1987.

———. *Lluvias y cenizas: dos Pachacuti en la historia*. La Paz: HISBOL, Institut Français d'Etudes Andines, 1988.

———. "Le lac Titicaca: Histoire perdue d'une mer intérieure." *Bulletin de l'Institut Français d'Etudes Andines* 21, n°1 (1992): 89-159.

———. "'El rayo que no cesa' d'un dieu pukina à un dieu inca." In *Religions des Andes et Langues indigènes. Equateur-Pérou-Bolivie, avant et après la Conquête espagnole*, edited by Pierre Duviols, 165-180. Aix en Provence: Publication de l'Université de Provence, 1993.

———. "De Empédocles a Tunupa: Evangelización, hagiografía y mitos." In *Saberes y Memorias en los Andes. In Memoriam Thierry Saignes*, edited by Thérèse Bouysse-Cassagne, 157-212. Paris, Lima: Institut des Hautes Etudes de l'Amérique latine, Institut Français d'Etudes Andines, 1997.

Bouysse-Cassagne, Thérèse, and Bouysse, Philippe. "Volcan Indien, volcan Chrétien. A propos de l'éruption de Huaynaputina en l'an 1600." *Journal de la Société des Américanistes* 70 (1984): 43-68.

Bradby, Barbara. "Symmetry around a centre: music of an Andean community." *Popular music* 6, n°2 (1987): 197-218.

Bradley, Peter T., and Cahill, David. *Habsburg Peru, images, imagination and memory*. Liverpool Latin American Studies, New Series 2. Liverpool: Liverpool University Press, 2000.

Bravo Guerreira, Concepción. "Del poder dual a la diarquía en el estado Inca." *Revista Complutense de Historia de América* 18 (1992): 11-62.

Bray, Tamara L., Minc, Leah D., Ceruti, María Constanza, Chávez, José Antonio, Perea, Ruddy, and Reinhrad, Johan. "A compositional analysis of pottery vessels associated with the Inca ritual of *capacocha*." *Journal of Anthropological Archaeology* 24 (2005): 82-100.

Brokaw, Galen. "The poetics of khipu historiography: Felipe Guaman Poma de Ayala's *Nueva corónica* and the *Relación de los quipucamayos*." *Latin American Research Review* 38, n°3 (2003): 111-147.

Browman, David L. "Titicaca basin archaeolinguistics: Uru, Pukina and Aymara AD 750-1450." *World Archaeology* 26, n°2 (1994): 235-251.

Brundage, Burr Cartwright. *Empire of the Inca*. Norman: University of Oklahoma Press, 1963.

Burke, Peter. *Varieties of cultural history*. Cambridge: Polity Press, 1997.

Burman, Rickie. "Time and socioeconomic change in Simbo, Solomon Islands." *Man* 16, n° 2 (1981): 251-267.

Cáceres Chalco, Efraín. "La muerte como sanción y compensación: vision de equilibrio y reciprocidad en Cusco." *Chungará* 33, n°2 (2001): 187-200.

Cahill, David. "After the fall: constructing Inca identity in Late Colonial Cuzco." In *Constructing collective identities and shaping public spheres: Latin American paths*, edited by Luis Roniger and Mario Sznajder, 65-99. Brighton: Sussex Academic Press, 1998.

———. "The Inca and Inca symbolism in popular festive culture: the religious processions of seventeenth-century Cuzco." In *Habsburg Peru: images, imagination and memory*, Peter T. Bradley and David Cahill, Part II, 87-150. Liverpool: Liverpool University Press, 2000.

———. "The Virgin and the Inca: An Incaic procession in the city of Cuzco in 1692." *Ethnohistory* 49, n°3 (2002): 611-649.

Carey, James W. "Distribution of culture bound illnesses in the Southern Peruvian Andes." *Medical Anthropology Quaterly* 7, n°3 (1993): 281-300.

Carpenter, Lawrence K. "Inside/Outside, which side counts? Duality-of-self and bipartization in Quechua." In *Andean cosmologies through time, persistence and emergence*, edited by Robert V. H. Dover, Katharine E. Seibold, John H. McDowell, 115-136. Bloomington: Indiana University Press, 1992.

Carr, Christopher. "The question of ranking in Havana Hopewellian societies, a retrospective in the light of multi-cemetery ceremonial organization." In *Gathering Hopwell: society, ritual and ritual interaction*, edited by Christopher Carr and D. Troy Case, 238-257. New York: Kluwer Academic/Plenum, 2005.

Cavero Carrasco, Ranulfo. *Maíz, chicha y religiosidad Andina*. Ayacucho: Universidad Nacional de San Cristóbal de Huamanga, 1986.

Cereceda, Verónica. "The semiology of Andean textiles: the talegas of Isluga." In *Anthropological history of Andean polities*, edited by John V. Murra, Nathan Wachtel, and Jacques

Revel, 149-173. Cambridge/Paris: Cambridge University Press/Maison des Sciences de l'Homme, 1986.

Cerrón Palomino, Rodolfo. *Gramática Quechua: Junín-Huanca*. Lima: Instituto de Estudios Peruanos, 1976.

———. "El cantar del Inca Yupanqui y la lengua secreta de los Incas." *Revista Andina* 32 n°2 (1998): 417-458.

———. "Hurin: un espejismo léxico opuesto a hanan." In *El Hombre y los Andes: Homenaje a Franklin Pease G. Y.*, edited by Javier Flores Espinoza and Rafael Varón Gabai, Vol. I, 219-235. Lima: Pontificia Universidad Católica del Perú, 2002.

———. *Lingüística Quechua* [1987]. Cuzco: Centro de Estudios Rurales Andinos Bartolomé de las Casas, 2003.

Cerrón Palomino, Rodolfo, Ballón Aguirre, Enrique, and Chambi, Emilio. *Vocabulario razonado de la actividad agraria andina*. Cuzco: Centro de Estudio Regionales Bartolomé de las Casas, 1992.

Ceruti, María Constanza. "Human bodies as objects of dedication at Inca mountaintop shrines (northwest Argentina)." *World Archaeology* 36, n°1 (2004): 103-122.

Charles, John. "Unreliable confessions: *khipus* in the colonial parish." *The Americas* 64, n°1 (2007): 11-33.

Chávez Hualpa, Fabiola. "Embarazo y parto en los cronistas de los siglos XVI-XVII." *Anthropologica* 15 (1997): 97-134.

Christinat, Jean Louis. *Des parrains pour la vie. Parenté rituelle dans une communauté des Andes Péruviennes*. Paris: Maison des Sciences de l'Homme, Recherches et Travaux de l'Institut d'ethnologie 9, 1989.

Classen, Constance. *Inca cosmology and the human body*. Salt Lake City: University of Utah Press, 1993.

Cock C., Guillermo, and Doyle, Mary Eileen. "Del culto solar a la clandestinidad de Inti y Punchao." *Historia y Cultura* (Lima) 12 (1979): 51-71.

Colás, Alejandro. *Empire*. Cambridge, Madlen: Polity Press, 2007.

Conrad, Geoffrey W. and Demarest, Arthur A. *Religion and empire, the dynamics of Aztec and Inca expansionism*. Cambridge: Cambridge University Press, 1984.

Covey, Alan R. *How the Incas built their heartland: State formation and the innovation of imperial strategies in the sacred valley, Peru*. Ann Arbor: University of Michigan Press, 2006a.

———. "Chronology, succession, and sovereignty: the politics of Inka historiography and its modern interpretation." *Comparative Study of Society and History* 48, n°1 (2006b): 169-199.

Crocker, William H. *The Canela (Eastern Timbira), I: an ethnographic introduction*. Smithsonian Contributions to Anthropology 33. Washington DC: Smithsonian Institution Press, 1990.

Cummins, Tom. *Toasts with the Incas: Andean abstraction and colonial images in quero vessels*. Ann Arbor: University of Michigan Press, 2002a.

———. "Los Quilcakamayoq y los dibujos de Guaman Poma." In *Libros y escritura de tradición indígena, ensayos sobre los codices prehispanicos y coloniales de México*, edited by Carmen Arellano Hoffman, Peer Schmidt, Xavier Noguez, 187-215. Toluca, México: El colegio Mexiquense, 2002b.

Cunow, Heinrich. *El sistema de parentesco peruano y las comunidades gentilicias de los Incas* [1891]. Paris: le "Livre Libre", 1929.

———. *La organización social del imperio de los Incas* [1896], Lima: Editorial Librería Peruana, 1933.

D'Altroy, Terence N. *The Incas*. Malden, MA: Blackwell Publishers, 2002.

D'Altroy, Terence N., and Bishop, Ronald. L. "The provincial organization of the Inka ceramic production." *American Antiquity* 55 (1990): 120-138.

D'Altroy, Terence N., and Earle, Timothy K. "Staple finance, wealth finance, and storage in the Inka political economy." *Current Anthropology* 28, n°2 (1985): 187-206.
Da Matta, Roberto. *A divided world: Apinayé social structure*. Harvard Studies in Cultural Anthropology 6. Cambridge, MA: Harvard University Press, 1982.
Davenport, William H. *Pi'o: an inquiry into the marriage of brothers and sisters and other close relatives in old Hawai'i*. Lanham, MD: University Press of America, 1994.
Dean, Carolyn. *Inka bodies and the body of Christ: Corpus Christi in colonial Cuzco, Peru*. Durham: Duke University Press, 1999.
———. "Andean androgyny and the making of men." In *Gender in Pre-Hispanic America*, edited by Cecelia F. Klein, 143-182. Washington DC: Dumbarton Oaks, 2001.
Dearborn, David S. P., Schreiber, Katharina J. "Houses of the Rising Sun." In *Time and calendars in the inca empire*, edited by Mariusz Ziólkowski and Robert Sadowski, 49-74. Oxford: British Archaeological Reports, 1989.
Dedenbach-Salazar Saenz, Sabine. *Un aporte a la reconstrucción del vocabulario agrícola de la época incaica*. Bonner Amerikanistische Studien 14. Bonn: Bonn University, 1985.
Demarest, Arthur A. *Viracocha, the nature and antiquity of the Andean god*. Peabody museum monographs 6. Cambridge, MA: Harvard University, 1981.
Domínguez Faura, Nicanor. "Betanzos y los Quipucamayos en la época de Vaca de Castro (Cuzco, 1543)." *Revista Andina* 46 (2008): 155-192.
———. "El licenciado Polo: ¿fuente del 'Discurso sobre la descendencia y govierno de los ingas' (ms. 1602/1603-1608)?" *Historica* XXXIV n°1 (2010): 131-144.
Doyle, Mary Eileen, "Ancestor cult and burial ritual in seventeenth and eighteenth century central Peru." PhD diss., University of California, 1988.
Dumont, Louis. *Homo Hierarchicus, the caste system and its implications*. Chicago, London: University of Chicago Press, 1980.
———. *Essays on individualism, modern ideology in anthropological perspective*. Chicago: University of Chicago Press, 1986.
Duviols, Pierre. *La lutte contre les religions autochtones dans le Pérou colonial: l'extirpation de l'idolatrie entre 1532 et 1600*. Lima: Institut Français d'Etudes Andines, 1971.
———. "Huari y Llacuaz, agricultores y pastores. Un dualismo prehispánico de oposición y complementaridad." *Revista del Museo Nacional* (Lima) XXXIX (1973): 153-191.
———. "Punchao, idolo mayor del Coricancha, Historia y Tipología." *Revista Antropología Andina* 1-2 (1976a): 156-182.
———. "La Capacocha." *Allpanchis* 9 (1976b): 11-58.
———. *La destrucción de las religiones andinas (durante la conquista y la colonia)*. México: Universidad Nacional Autonoma de México, 1977a.
———. "Los nombres Quechua de Viracocha, supuesto 'Dios Creador' de los evangelizadores." *Allpanchis* 10 (1977b): 53-63.
———. "Datation, paternité et idéologie de la 'Declaración de los quipucamayos a Vaca de Castro' (Discurso de la descendencia y gobierno de los Ingas)." In *Les cultures ibériques en devenir: essais publiés en hommage à la mémoire de Marcel Bataillon (1895–1977)*, 583-591. Paris: La Fondation Singer-Polignac, 1979a.
———. "Un symbolisme de l'occupation, de l'aménagement et de l'exploitation de l'espace. Le monolithe 'huanca' et sa fonction dans les Andes préhispaniques." *L'Homme* XIX n°2 (1979b): 7-31.
———. "La dinastía de los Incas : monarquia o diarquía ? Argumentos heurísticos a favor de una tesis estructuralista." *Journal de la Société des Américanistes* LXVI (1979c): 67-83.
———. *Cultura Andina y Represión*. Cuzco: Centro Bartolomé de las Casas, 1986.
———. ed. *Religions des Andes et langues indigènes: Equateur, Pérou, Bolivie avant et après la conquète espagnole*, Actes du Colloque III d'Etudes Andines. Aix-en-Provence: CARLAM, Université de Provence, 1993.

———. "Del discurso escrito colonial al discurso prehispánico: hacia el sistema sociocosmológico inca de oposición y complementaridad." *Bulletin de l'Institut Français d'Etudes Andines* 26, n°3 (1997): 279- 303.

Escribens, Augusto. "Debilitamiento de vocables y metatesis en el Quechua de Ferreñafe." In *Tercero Congreso peruanos del hombre y de la cultura andina*, edited by Ramiro Matos Mendieta, Vol. V, 994-1002. Lima: Editores Lasontay, 1980.

Farrington, Ian S. "Ritual geography, settlement patterns and the characterization of the provinces of the Inka heartland." *World Archaeology* 23, n°3 (1992): 368-385.

Fernández Juárez, Gerardo. *Yatiris y Ch'amakanis del Altiplano: sueños, testimonios y prácticas ceremoniales.* Quito: Abya Yala, 2004.

Firth, Raymond. "A note on descent groups in Polynesia." *Man* 57 (1957): 4-8.

Flannery, Kent V., Marcus, Joyce, Reynolds, Robert G. *The flocks of the Wamani. A study of llama herders on the punas of Ayacucho, Peru.* San Diego: Academic Press, Inc., 1989.

Flores Ochoa, Jorge. *Pastoralists of the Andes, the alpaca herders of Paratía.* Philadelphia: ISHI, 1979.

———. "The classification and naming of South American camelids." In *Anthropological History of Andean polities*, edited by John V. Murra, Nathan Wachtel, and Jacques Revel, 137-148. Cambridge/Paris: Cambridge University Press/Maison des Sciences de l'Homme, 1986.

———. and Núñez Del Prado, Juan, eds. *Q'ero, el útimo ayllu Inka: Homenaje a Oscar Núñez del Prado Castro.* Cusco: Centro de Estudios Andinos, 1983.

Fossa, Lydia. "Two khipu, one narrative: Answering Urton's questions." *Ethnohistory* 47, n°2 (2000): 453-458.

Fujii, Tatsuhiko, and Tomoeda, Hiroyasu. "Chacra, laime y auquénidos." In *Estudios etnógraficos del Perú meridional*, edited by Shozo Masuda, 33-63. Tokyo: Tokyo University Press, 1981.

García Escudero, María del Carmen. "El Arco Iris en la tradición prehispánica centroandina." *Gazeta de Antropología* 23 (2007). Accessed 10 April 2008. http://www.ugr.es/~pwlac/G23_15Carmen_Garcia_Escudero.html.

Gareis, Iris. "Las religiones Andinas en la documentación de la extirpación de las idolatrías." In *Religiones Andinas*, edited by Manuel M. Marzal, Enciclopedia Iberoamericana de Religiones 04, 115-141. Madrid: Editorial Trotta, 2005.

Geertz, Clifford. *Negara: the theatre state in nineteenth-century Bali.* Princeton, NJ: Princeton University Press, 1980.

Gell, Alfred. *The anthropology of time, cultural constructions of temporal maps and images.* Oxford, Providence: Berg, 1992.

Gelles, Paul H. "Equilibrium and extraction: dual organization in the Andes." *American Ethnologist* 22, n°4 (1995): 710-742.

Gentile, Margarita E. "La pichca: oráculo y juego de fortuna (su persistencia en el espacio y tiempo andinos)." *Bulletin de l'Institut Français d'Etudes Andines* 27, n°1 (1998): 75-131.

Godoy, Ricardo A. "The fiscal role of the Andean ayllu." *Man* 21, n°4 (1986): 723-741.

Goldman, Irving. *Ancient Polynesian society.* Chicago: University of Chicago Press, 1970.

Gose, Peter. *Deathly waters and hungry mountains, agrarian ritual and class formation in an Andean town.* Toronto: University of Toronto Press, 1994.

———. "The past is a lower moiety." *History and Anthropology* 9, n°4 (1996a): 383-414.

———. "Oracles, divine kingship, and political representation in the Inka state." *Ethnohistory* 43, n°1 (1996b): 1-32.

———. "The state as a chosen woman: brideservice and the feeding of tributaries in the Inka empire." *American Anthropologist* 102, n°1 (2000): 84-97.

Gow, David, Gow, Rosalina. "La alpaca en el mito y el ritual." *Allpanchis* 8 (1975): 141-165.

Gow, Peter. *Of mixed blood: Kinship and History in Peruvian Amazonia.* Oxford Studies in Social and Cultural Anthropology. Oxford: Clarendon Press, 1991.

Graubart, Karen B. *With our labor and sweat: Indigenous women and the formation of colonial society in Peru, 1550-1700*. Stanford: Stanford University Press, 2007.

Griffiths, Nicholas. *The Cross and the Serpent, religious repression and resurgence in colonial Peru*. Norman: University of Oklahoma Press, 1996.

Grimes, Ronald L. "Performance." In *Theorizing rituals: issues, topics, apporaches, concepts*, edited by Jens Kreinath, Jan Snoek and Michael Stausberg, 379-394. Leiden, Boston: Brill, 2006.

Grisel, Etienne. "The beginnings of International Law and general public law doctrine: Francisco de Vitoria's *De Indiis prior*." In *First Images of America: The Impact of the New World on the Old*, edited by Fredi Chiapelli, Michael J.B. Allen, and Robert L. Bension, 305-325. Berkeley, London: University of California Press, 1976.

Gruszczynska Ziólkowska, Anna. *El poder del sonido: el papel de las crónicas españolas en la etnomusicología andina*. Quito: Abya-Yala, 1995.

Gudemos, Mónica. "*Capac, Camac, Yacana*. El *Capac Raymi* y la música como emblema de poder." *Anales del Museo de América* 13 (2005): 9-52.

Hage, Per, and Harary, Frank. *Island networks. Communication, kinship, and classification structures in Oceania*. Cambridge: Cambridge University Press, 1996.

Harris, Olivia. "Complementarity and conflict: an Andean view of women and men." In *Sex and age as principles of social differentiation*, edited by Jean S. La Fontaine, 21-40. London, New York: Academic Press, 1978.

———. "Ecological duality and the role of the centre in Northern Potosí." In *Andean Ecology and Civilization, an interdisciplinary perspective on Andean ecological complementarity*, edited by Shozo Masuda, Izumi Shimada, and Craig Morris, 331-355. Tokyo: University of Tokyo Press, 1985.

———. "From asymmetry to triangle: symbolic transformations in northern Potosí." In *Anthropological history of Andean polities*, edited by John V. Murra, Nathan Wachtel, and Jacques Revel, 260-279. Cambridge/Paris: Cambridge University Press/Maison des Sciences de l'Homme, 1986.

———. "The Dead and the Devils among the Bolivian Laymi." [1982] In *To make the earth bear fruit: essays on fertility, work and gender in highland Bolivia*. Londres: Institute of Latin American Studies, 2000.

Hastorf, Christine A., Johannessen, Sissel. "Pre-Hispanic political change and the role of maize in the central Andes of Peru." *American Anthropologist* 95, n°1 (1993): 115-138.

Heggarty, Paul. "Enigmas en el origen de las lenguas andina: aplicando nuevas técnicas en las incógnitas por resolver." *Revista Andina* 40 (2005): 9-57, 70-80.

Hemming, John. *The conquest of the Incas*. London: Macmillan, 1993.

Hendon, Julia A. "Status and power in Classic Maya society: An archaeological study." *American anthropologist* 93, n°4 (1991): 894-918.

Hernández Astete, Francisco. "Roles sexuales en la organización incaica." *Histórica* 22, n°1 (1998): 93-134.

———. *La mujer en el Tahuantinsuyo*. Lima: Pontificia Universidad Católica del Perú Fondo Editorial, 2002a.

———. "El poder incaico: una aproximación a la figura del Inca hurin." In *El Hombre y los Andes: Homenaje a Franklin Pease G. Y.*, edited by Javier Flores Espinoza and Rafael Varón Gabai, Vol. II, 589-596. Lima: Instituto de Estudios Peruanos, 2002b.

Hill, Jonathan David. *Rethinking history and myth: indigenous South American perspectives on the past*. Urbana: University of Illinois Press, 1988.

Hiltunen, Juha J. *Ancient kings of Peru, the reliability of the chronicle of Fernando Montesinos*. Helsinski: Suomen Historiallinen Seura, 1999.

Hopkins, Diane. "Juegos de enemigos." *Allpanchis* 20 (1982): 167-188.

Hornborg, Alf. *Dualism and hierarchy in lowland South America, Trajectories of indigenous social organization*. Acta Universitatis Upsaliensis 9. Uppsala: Almqvist & Wiksell International, 1988.

———. "Serial redundancy in Amazonian social structure: is there a method for poststructuralist comparison?" In *Transformations of kinship*, edited by M. Godelier, T. Trautmann, F. E Tjon Sie Fat, 168-186. Washington: Smithsonian Institution Press, 1998.

Houseman, Michael, and Severi, Carlo. *Naven or the Other Self: a relational approach to ritual action*. Translated by Michael Fineberg. Leiden: Brill, 1998.

Hyslop, John. *Inka settlement planning*. Austin: University of Texas Press, 1990.

Ibarra Grasso, Dick Edgar. *Ciencia en Tiahuanacu y el incario*. La Paz: Los Amigos del Libro, 1982.

Inamura, Tetsuya. "Adaptación ambiental de los pastores altoandinos en el Sur del Perú." In *Estudios etnógraficos del Perú meridional*, edited by Shozo Masuda, 65-83. Tokyo: Tokyo University Press, 1981.

Isbell, Billie Jean. *To defend ourselves. Ecology and ritual in an Andean village*. Austin: University of Texas Press, 1978.

Isbell, William Harris. *Mummies and mortuary monuments: a postprocessual prehistory of Central Andean social organization*. Austin: University of Texas Press, 1997.

Itier, César. "La tradición oral Quechua antigua en los procesos de idolatrías de Cajatambo." *Bulletin de l'Institut Français d'Etudes Andines* 21, n°3 (1992): 1009-1051.

———. "Estudio etnohistórico y lingüístico." In *Relación de antigüedades deste reyno del Pirú*, by Juan de Santa Cruz Pachacuti Yamqui Salcamaygua. Lima/Cuzco: Institut Français d'Etudes Andines/Centro de Estudios Regionales Andinos Bartolomé de Las Casas, 1993.

———. "Le jeune homme et l'étoile ou le voyage au pays de la multiplication des grains. Un mythe andin sur les dangers de l'exogamie." *Journal de la Société des Américanistes* 82 (1996): 157-191.

Jenkins, David. "Temporal processes and synchronic relations: age and rank in the central Andes." *Ethnohistory* 42, n°1 (1995): 91-131.

———. "The Inka conical clan." *Journal of Anthropological Research* 57, n°2 (2001): 167-195.

Julien, Catherine. *Hatunqolla: a view of Inca rule from the Lake Titicaca region*. Berkeley, Los Angeles: University of California Press, 1983.

———. "History and art in translation: the paños and other objects collected by Francisco de Toledo." *Colonial Latin American Review* 8, n°1 (1999): 61-89.

———. *Reading Inca history*. Iowa City: University of Iowa Press, 2000.

———. "Punchao en España." In *El Hombre y los Andes: Homenaje a Franklin Pease G. Y.*, edited by Javier Flores Espinoza and Rafael Varón Gabai, Vol. II, 709-715. Lima: Instituto de Estudios Peruanos, 2002.

———. "Polo de Ondegardo y el 'Discurso sobre la descendencia y gouierno de los ingas'." *Histórica* XXXIII, n°2 (2009): 7-28.

Kirchhoff, Paul. "The social and political organization of the Andean peoples." In *Handbook of South American Indians*, edited by J. Steward, vol. 5, 293-311. Washington: Government Printing Office, 1949.

———. "The principles of clanship in human society." *Davidson Journal of Anthropology* 1 (1955): 1-10.

La Lone, Mary B., La Lone, Darrell E. "The Inka state in the southern highlands: state administrative and production enclaves." *Ethnohistory* 34, n°1 (1987): 47-62.

Larme, Anne C. "Environment, vulnerability and gender in Andean ethnomedicine." *Social Sciences and Medicine* 47, n°8 (1998): 1005-1015.

Lave, Jean Carter. "Eastern Timbira moiety systems in time and space: a complex structure." In *Actes du XLIIe Congrès International des Américanistes*, vol. 2, 309–321. Paris: Société des Américanistes, 1977.

Leach, Edmund R. *Rethinking anthropology*. London: Athlone Press, 1961.
———. *Political systems of highland Burma: A study of Kachin social structure* [1954]. Monographs on Social Anthropology 44. London: Athlon Press, 1970.
Levillier, Roberto. *Gobernantes del Perú. Cartas y papeles. Siglo XVI*, vol. IV. Buenos Aires: Biblioteca del Congreso de Argentina, 1924.
Levine, Terry Yarov. "Inka labor service at the regional level: the functional reality." *Ethnohistory* 43, n°1 (1987): 14-46.
Levi-Strauss, Claude. "On dual organization in South America." *América Indígena* 4, n°1 (1944): 37-47.
———. "Do dual organization exist?" *Structural Anthropology*, vol. 2, 132-163. New York: Basic Books, 1963.
———. *The elementary structures of kinship* [1949]. Boston: Beacon Press, 1969.
Lohmann Villena, Guillermo. "Unas notas acerca de curiosos paralelismos y correspondencias entre cuatro documentos históricos sobre la época incaica". *Fenix* 16 (1966): 174-197.
Lounsbury, Floyd G. "Some aspects of the Inka kinship system." In *Anthropological history of Andean polities*, edited by John V. Murra, Nathan Wachtel and Jacques Revel, 121-136. Cambridge/Paris: Cambridge University Press/Maison des Sciences de l'Homme, 1986.
Lynch, Thomas F. "Current research." *American Antiquity* 49, n°2 (1984): 416-420.
MacCormack, Sabine. "Pachacuti: miracles, punishments, and last judgment: visionary past and prophetic future in early colonial Peru." *The American Historical Review* 93, n°4 (1988): 960-1006.
———. *Religion in the Andes, vision and imagination in early colonial Peru.* Princeton: Princeton University Press, 1991a.
———. "Demons, imagination, and the Incas." *Representations* 33 (1991b): 121-146.
———. "Time, space and ritual action: the Inka and Christian calendars in Early colonial Peru." In *Native traditions in the Postconquest world*, edited by Elizabeth Hill Boone and Tom Cummins, 295-343. Washington DC: Dumbarton Oaks, 1998.
———. "Cuzco, another Rome?" In *Empires, Perspectives from Archaeology and History*, edited by Susan E. Alcock, Terence N. D'Altroy, Kathleen D. Morrison and Carla M. Sinopoli, 419-435. Cambridge: Cambridge University Press, 2001.
———. *On the wings of time. Rome, the Incas, Spain and Peru.* Princeton University Press, 2007.
Mackey, Carol J., Pereya, Hugo, Radicati Di Primeglio, Carlos, Rodríguez, Humberto, and Valverde oscar, eds. *Quipu y yupana: colección de escritos*. Lima: Consejo Nacional de Cienca y Tecnología, 1990.
Mannheim, Bruce. "Poetic form in Guaman Poma's *wariqsa arawi*." *Amerindia* 11 (1986): 41-67.
———. *The language of the Inkas since the European invasion.* Austin: University of Texas Press, 1991.
Mariscotti De Gorlitz, Ana María. "Pacha Mama Santa Tierra. Contribución al estudio de la religión autóctona en los Andes centro-meridionales." In Supplement of *Indiana* 8, 364-411. Berlin: Ibero-Amerikanisches Institut Preussischer Kulturbesitz, 1978a.
———. "Los curi y el rayo" In *Actes du XVLLe Congrès International des Américanistes*, 365-376. Paris: Société des Américanistes, 1978b.
Marzal, Manuel M. "Persistencia y transformación de ritos y sacerdocio andinos en el Perú." In *Mitos y ritual en América*, edited by Manuel Gutiérrez, 263-285. Madrid: Alhambra, 1988.
Masuda, Shozo, Shimada Izumi, and Morris Craig, eds. *Andean Ecology and Civilization, an interdisciplinary perspective on Andean ecological complementarity*. Tokyo: University of Tokyo Press, 1985.

Maybury-Lewis, David. "Cultural categories of the Central Gê." In *Dialectical societies, the Gê and Bororo of Central Brazil*, edited by David Maybury-Lewis, 218-248. Cambridge: Harvard University Press, 1979.
Maybury-Lewis, David, and Almagor, Uri, eds. *The attraction of opposites: thought and society in the dualistic mode.* Ann Arbor: University of Michigan Press, 1989.
Mazzotti, José Antonio. *Incan Insights. El Inca Garcilaso's Hints to Andean Readers.* Frankfurt: Iberoamericana Editorial Vervuert, 2008.
McEwan, Colin, and Van de Guchte, Maarten. "Ancestral time and sacred space in Inca state ritual." In *The ancient Americas: art from the sacred landscapes,* edited by Richard F. Townsend, 359-371. Chicago, Munich: The Art Institut/Prestel Verl, 1992.
McEwan, Colin, Chatfield, Melissa, and Gibaja, Arminda. "The archaeology of Inca origins, excavations at Chokepukio, Cuzco, Peru." In *Andean archaeology (volume 1): variations in sociopolitical organization,* edited by William H. Isbell and Helaine Silverman, 287-301. New York: Kluwer Academic/Plenum Publishers, 2002.
Meddens, Frank M. "Function and meaning of the Usnu in Late Horizon Peru." *Tawantinsuyu* 3 (1997): 5-14.
Meddens, Frank M., Branch, Nicolas P., Vicanco Pomocanchari, Cirilo, Riddiford, Noami, and Kemp, Rob. "High altitude *ushnu* platforms in the department of Ayacucho, Peru: Structure, ancestors, and animating essence." In *Pre-Columbian landscapes of creation and origin,* edited by John Edward Staller, 315-355. New York: Springer, 2008.
Meyers, Albert. "Algunos problemas en la clasificación del estilo incaico." *Pumapunku* 8 (1975): 7-25.
———. "Los Incas ¿bárbaros advenedizos o herederos de Tiahuanaco?" In *El Hombre y los Andes: Homenaje a Franklin Pease G. Y.,* Vol. II, 525-535. Lima: Pontificia Universidad Católica del Perú, 2002.
Mills, Kenneth. *An evil lost in view? An investigation of post-evangelisation Andean religion in mid-colonial Peru.* Liverpool: University of Liverpool, Institute of Latin American Studies, 1994.
———. *Idolatry and its enemies, colonial Andean religion and extirpation, 1640-1750.* Princeton: Princeton University Press, 1997.
Mishkin, Bernard. "The contemporary Quechua." In *Handbook of South American Indians,* edited by Julian H. Steward, vol. 2, 441-470. Washington DC: Smithsonian Institution, Bureau of American ethnology, 1946.
Molinié, Antoinette. "Instruments symboliques aux frontières andines." *Techniques et Cultures* 7 (1986): 145-179.
———. "El regreso de Viracocha." *Bulletin de l'Institut Français d'Etudes Andines* 16, n°3-4: (1987): 71-83.
———. "Sanglantes et fertiles frontières. A propos des batailles rituelles andines." *Journal de la Société des Américanistes* 74 (1988): 49-70.
———. "Héros mythique, dieu étatique: soleil aquatique." *Bulletin de la Société Suisse des Américanistes* 58-59 (1993-1994): 25-36.
———. "Buscando una historicidad andina: una propuesta antropológica y una memoria hecha rito." In *Arqueología, Antropología e Historia en los Andes: Homenaje a María Rostworowski,* edited by Rafael Varón Gabai, Javier Flores Espinoza, 691-708. Lima: Instituto de Estudios Peruanos, 1997.
Moore, Jerry D. "The archaeology of dual organization in Andean South America: a theoretical review and case study." *Latin American Antiquity* 6, n°2 (1995): 165-181.
Moore, Sally Falk. *Power and property in Inca Peru.* Westport: Greenwood Press, 1973.
Morales, Edmundo. *The guinea pig: healing, food and ritual in the Andes.* Tucson: University of Arizona Press, 1995.
Morlon, Pierre, ed. *Comprendre l'agriculture paysanne dans les Andes Centrales: Pérou-Bolivie.* Paris: Institut National de la Recherche Agronomique, 1992.

Morris, Craig. "Storage, supply and redistribution in the economy of the Inka state." In *Anthropological history of Andean polities*, edited by John V. Murra, Nathan Wachtel, and Jacques Revel, 59-68. Cambridge/Paris: Cambridge University Press/Maison des Sciences de l'Homme, 1986.

———. and Von Hagen Adriana. *The Inka empire and its Andean origin*. American Museum of Natural History. New York: Abbeville Press, 1993.

Munn, Nancy D. "The cultural anthropology of time: a critical essay." *Annual Review of Anthropology* 21 (1992): 93-123.

Muñoz-Bernand, Carmen. *Enfermedad, daño y ideología: Antropología médica de los renacientes de Pindilig*. Quito: Abya Yala, 1999.

Murra, John V. "El 'control vertical' de un máximo de pisos ecológicos en la economía de las sociedades andinas." In *Visita a la provincia de León de Huánuco* [1562], by Ortiz de Zúñiga, Vol. II, 427-476. Huánaco: Universidad Nacional Hermilio Valdizán, 1972.

———. "The mit'a obligations of ethnic groups to the Inka state." In *The Inca and Aztec States 1400-1800, Anthropology and History*, edited by George A. Collier, Renato I. Rosaldo, and John D. Wirth, 237-262. New York: Academic Press, 1982.

———. *La organización económica del estado inca*. Tercera edición. México: Siglo XXI, Instituto de Estudios Peruanos, 1983.

———. "Did tribute and markets prevail in the Andes before the European invasion?" In *Ethnicity, markets, and migration in the Andes, At the crossroads of History and Anthropology*, edited by Brooke Larson and Olivia Harris, 57-72. Durham, London: Duke University Press, 1995.

———. "Las etnocategorías de un *khipu* estatal." [1973] In *El mundo andino. Población, medio ambiente y economía*, by John V. Murra, 248-260. Lima: Instituto de Estudios Peruanos, Pontificia Universidad Católica del Perú, 2002.

———. "Derechos a tierras en el Tawantinsuyu." [1995] In *El mundo andino. Población, medio ambiente y economía*, by John V. Murra, 294-307. Lima: Instituto de Estudios Peruanos, Pontificia Universidad Católica del Perú, 2002.

Needham, Rodney. "Alliance and classification among the Lamet." *Sociologus* 10, n°2 (1960): 97-119.

———., ed. *Rethinking kinship and marriage*. Asssociation of Social Anthropologists of the Commonwealth Monographs 11. London: Tavistock Publications, 1971.

———., ed. *Right and left: essays on dual symbolic classifications*. Chicago, London: University of Chicago Press, 1973.

———. *Counterpoints*. Berkeley, Los Angeles: University of California Press, 1987.

Netherly, Patricia J. "Out of many, one. The organization of rule in the north coast polities." In *The Northern dynasties: kingship and statecraft in Chimor*, edited by Michael E. Moseley and Alana Cordy-Collins, 461-487. Washington DC: Dumbarton Oaks Research Library and Collection, 1990.

Niles, Susan A. *The shape of Inca history: narrative and architecture in an Andean empire*. Iowa City: University of Iowa Press, 1999.

Nimuendajú, Curt, and Lowie, Robert H. "The dual organizations of the Ramko'kamekra (Canella) of northern Brazil." *American Anthropologist* 39, n°4 (1937): 565-582.

Nowack, Kerstin. "Las intenciones del autor: Juan de Betanzos y la Suma y Narración de los Incas." *Revista Andina* 34 (2002): 47-64.

Ogburn, Dennis. "Power in stone: the long distant movement of building blocks in the Inca empire." *Ethnohistory* 51, n°1 (2004): 101-135.

Orlove, Benjamin S., Chiang, John C. H., and Cane, Mark A. "Ethnoclimatology in the Andes: a cross-disciplinary study uncovers a scientific basis for the scheme Andean potato farmers traditionally use to predict the coming rains." *American Scientist* 90, n°5 (2002): 428-435.

Ortiz, Alfonso. *Contrapunto Cubano del tabaco y el azúcar*. La Havana: Consejo Nacional de Cultura, 1940.

———. *The Tewa world: space, time, being and becoming in a pueblo society*. Chicago, London: University of Chicago Press, 1969.

Ossio A., Juan M. "La imagen de la unidad social en las fiestas andinas." In *Incas e Indios cristianos. Elites indígenas e identidades cristianas en los Andes coloniales*, edited by Jean-Jacques Decoster, Travaux de l'Institut Français d'Etudes Andines 149, 475-496. Cuzco: Centro de Estudios Rurales Andinos Bartolomé de Las Casas, Institut Français d'Etudes Andines, Asociación Kuraka, 2002.

Paerregaard, Karsten. "Complementarity and duality: oppositions between agriculturalists and herders in an Andean village." *Ethnology* 31, n°1 (1992): 15-26.

———. "Why fight over water? Power, conflict, and irrigation in an Andean village." In *Irrigation at high altitudes: the social organization of water control systems in the Andes*, edited by William P. Mitchell, David Guillet, and Inge Bolin, 189-202. Society For Latin American Anthropology Publications Series vol. 12. Washington: American Anthropological Association, 1994.

Pagden, Anthony. *The fall of natural man: the American Indian and the origins of comparative ethnology*. Cambridge: Cambridge University Press, 1986.

———. *Spanish Imperialism and the political imagination*. New Haven, London: Yale University Press, 1990.

Parkin, Robert. "Asymétrie dualiste ou opposition hiérarchique? Le legs de Robert Hertz dans le travail de Rodney Needham et de Louis Dumont." *Recherches sociologiques* 23, n°2 (1992): 43-68.

———. *The dark side of humanity: the work of Robert Hertz and its legacy*. Amsterdam: Harwood Academic Publishers, 1996.

———. *Louis Dumont and hierarchical opposition*. New York: Berghahn Books, 2003.

Pärssinen, Martti. *Tawantinsuyu: the Inca state and its political organization*, Studia Historica 43. Helsinki: Societas Historica Finlandiae, 1992.

Pärssinen, Martti, and Siiriäinen, Arii. "Inka style ceramics and their chronological relationship to the Inka expansion in the southern Lake Titicaca area (Bolivia)." *Latin American Antiquity* 8, n°3 (1997): 255-271.

Paulsen, Allison C. "The thorny oyster and the voice of god: *spondylus* and *strombus* in Andean Prehistory." *American Antiquity* 39, n°4 (1974): 597-607.

Pease, Franklin G. Y. "The Andean Creator God." *Numen* 17 (1970): 161-175.

———. *Del Tawantinsuyu a la historia del Perú*. Lima: Instituto de Estudios Peruanos, 1978.

———. *Los últimos incas del Cusco* [1972]. Madrid: Alianza Editorial, 1991.

———. *Los Incas: Una introducción*. Lima: Pontificia Universidad Católica del Perú, 1994.

———. *Las Crónicas y los Andes*. Lima: Pontificia Universidad Católica del Perú, 1995.

Petersen, Glenn. "Sociopolitical rank and conical clanship in the Caroline islands." *Journal of the Polynesian Society* 108, n°4 (1999): 367-410.

Platt, Tristan. "Mirrors and maize: the concept of yanantin among the Macha of Bolivia." In *Anthropological history of Andean polities*, edited by John V. Murra, Nathan Wachtel and Jacques Revel, 228-259. Cambridge/Paris: Cambridge University Press/Maison des Sciences de l'Homme, 1986.

Polia Meconi, Mario, ed. *La cosmovisión religiosa andina en los documentos inéditos del Archivo Romano de la Companía de Jésus (1581-1752)*. Lima: Pontificia Universidad Católica del Perú, 1999.

Poole, Deborah A. "Rituals of movement, rites of transformation: pilgrimage and dance in the Highlands of Cuzco, Peru." In *Pilgrimage in Latin America*, edited by N. Ross Crumrine and Alan Morinis, 307-338. Westport, CT: Greenwood Press, 1991.

Porras Barrenechea, Raúl. *Los Cronistas del Perú (1528-1650) y Otros Ensayos*. Biblioteca Clásicos del Perú 2. Lima: Banco de Crédito del Perú, 1986.

———. *Indagaciones peruanas: El legado Quechua.* Lima: Universidad Nacional Mayor de San Marcos, 1999.

Quilter, Jeffrey, and Urton, Gary. *Narrative threads: accounting and recounting in Andean khipu.* Austin: University of Texas Press, 2002.

Quispe M., Ulpiano. *La herranza en Choque Huarcaya y Huancasancos, Ayacucho.* Instituto Indigenista Peruano, monografía 20. Lima: Ministerio de Trabajo, 1969.

Ramírez, Susan Elizabeth. *The world upside down. Cross cultural contact and conflict in sixteenth-century Peru.* Stanford: Stanford University Press, 1996.

———. *To feed and to be fed, the cosmological bases of authority and identity in the Andes.* Stanford: Stanford University Press, 2005.

———. "Historia y memoria: la construcción de las tradiciones dinásticas andinas." *Revista de Indias* LXVI, 236 (2006a): 13-56.

———. "From people to place and back again: Back translation as decentering—an Andean case study." *Ethnohistory* 53, n°2 (2006b): 355-381.

———. "Lost—and found—in translation? The practice of translating, interpreting, and understanding the past." *The Americas* 62, n°3 (2006c): 305-312.

Randall, Robert. "Los dos vasos: cosmovisión y política de la embriaguez desde el inkanato hasta la colonia." In *Borrachera y memoria: la experiencia de lo sagrado en los Andes,* edited by Thierry Saignes, 73-112. La Paz, Lima: Hisbol/Institut Français d'Etudes Andines, 1993.

Regalado De Hurtado, Liliana. *La sucesión incaica. Aproximación al mando y poder entre los Incas a partir de la crónica de Betanzos.* Lima: Fondo Editorial Pontificia Universidad Católica del Perú, 1996.

Reinhard, Johan. "House of the Sun: the Inka temple of Vilcanota." *Latin American Antiquity* 6, n°4 (1995): 340-349.

Riva González, Palmira La. "Las representaciones del *animu* en los Andes del sur peruano." *Revista Andina* 41 (2005): 63-88.

Rivera Andía, Juan Javier. *La fiesta del ganado en el valle de Chancay (1962-2002). Religión, ritual y ganadería en los Andes: etnografía, documentos inéditos e interpretación.* Lima: Pontificia Universidad Católica del Perú, 2003.

Robin, Valérie. "El cura y sus hijos osos, o el recorrido civilizador de los hijos de un cura y una osa." *Bulletin de l'Institut Français d'Etudes Andines* 26, n°3 (1997): 369-420.

Rostworowski De Díez Canseco, María. "Succesion, cooption to kingship, and royal incest among the Inca." *Southwestern Journal of Anthropology* 16, n°4 (1960): 417-427.

———. *Los señoríos indígenas de Lima y Canta.* Lima: Instituto de Estudios Peruanos, 1978.

———. "La voz parcialidad en su contexto de los siglos XVI y XVII." In *Ethnohistoria y Antropología Andina,* edited by Marcia Koth de Paredes and Amalia Castelli, 35-48. Lima: Museo Nacional de Historia, 1981.

———. *Estructuras andinas del Poder, ideología religiosa y política.* Lima: Instituto de Estudios Peruanos, 1988a.

———. *Justicia 413: Conflicts over coca fields in XVIth century Peru (1558-1567).* Memoirs of the Museum of Anthropology 21. Ann Arbor: University of Michigan, 1988b.

———. "Nuevos datos sobre la tenencia de tierras reales en el Incario." In *Ensayos de Historia Andina. Elites, Etnías, Recursos,* by María Rostworowski, 105-146. Lima: Instituto de Estudios Peruanos, BCRP, 1993a.

———. "Los Ayarmaca." In *Ensayos de historia andina, elites, etnías, recursos,* by María Rostworowski, 241-290. Lima: Instituto de Estudios Peruanos, 1993b.

———. "La mujer en el Perú prehispánico." *Tawantinsuyu* 1 (1995): 4-10.

———. *History of the Inca realm* [1988]. Cambridge: Cambridge University Press, 1999.

———. "Adivinación y oráculos en el mundo andino antiguo." In *Peregrinaciones y procesiones rituales en los Andes,* edited by Marco Curatola Petrocchi & Mariusz Ziólkowski,

181-204. Lima: Institut Français d'Etudes Andines, Pontificia Universidad Católica del Perú.

Rostworowski De Díez Canseco, María, and Morris, Graig. "The fourfold domain: Inka power and its social foundations." In *The Cambridge history of the native peoples of the Americas,* edited by Frank Salomon and Stuart B. Schwartz, vol. 3, South America, Pt. 1, 769-863. Cambridge: Cambridge University Press, 1999.

Rowe, John H. *An introduction to the archaeology of Cuzco.* Papers of the Peabody Museum of American Archaeology and Ethnology, vol. 22. Cambridge, MA: Harvard University, 1944.

———. "Absolute chronology in the Andean area." *American Antiquity* 10, n°3 (1945): 265-284.

———. "Inca culture at the time of the Spanish conquest." In *Handbook of South American Indians,* edited by Julian H. Steward, vol. 2, 183-330. Washington, DC: Smithsonian Institution, Bureau of American ethnology, 1946.

———. "Inca policies and institutions relating to the cultural unification of the empire." In *The Inca and Aztec states 1400-1800: Anthropology and History,* edited by George A. Collier, Renato I. Rosaldo, and John D. Wirth, 93-118. New York: Academic Press, 1982.

———. "Probanza de los Incas nietos de conquistadores." *Histórica* 9, n°2 (1985): 35-73.

———. "Las tierras reales de los Incas." In *Arqueología, Antropología e Historia en los Andes: Homenaje a María Rostworowski,* edited by Varón Gabai Rafael, and Flores Espinoza Javier, 279-287. Lima, Instituto de Estudios Peruanos, 1997.

Rozas Alvarez, Jesús Washington, Calderon Garcia, María del Carmen. "La renovación: una interpretación de los rituales de agosto." In *Desde Afuera y desde adentro, ensayos de etnografía e historia del Cuzco y Apurímac,* edited by Luis Millones, Hiroyasu Tomoeda, and Tatsuhiko Fujii, 317-352. Osaka: National Museum of Ethnology, 2000.

Sadowski, Robert M. "A few remarks on the astronomy of R. T. Zuidema's "quipu-calendar." In *Time and calendars in the inca empire,* edited by Mariusz Ziólkowski and Robert Sadowski, 209-212. Oxford: British Archaeological Reports, 1989.

Sahlins, Marshall D. *Social stratification in Polynesia.* Seattle: University of Washington Press, 1958.

———. *Islands of History.* Chicago: University of Chicago Press, 1985.

Salles, Estela Cristina, and Noejovich, Héctor Omar. "La herencia femenina andina prehispánica y su transformación en el mundo colonial." *Bulletin de l'Institut Français d'Etudes Andines* 35, n°1 (2006): 37-53.

Salles-Reese, Verónica. *From Viracocha to the Virgin of Copacabana, representation of the Sacred at Lake Titicaca.* Austin: University of Texas Press, 1997.

Sallnow, Michael J. *Pilgrims of the Andes: regional cults in Cuzco.* Washington: Smithsonian Institution Press, 1987.

Salomon, Frank. *Native Lords of Quito in the Age of the Incas.* New York: Cambridge University Press, 1986.

———. "'The beautiful grandparents': Andean ancestors shrines and mortuary ritual as seen through colonial records." In *Tombs for the living: Andean mortuary practices,* edited by Tom D. Dillehay, 315-353. Washington, DC: Dumbarton Oaks, 1995.

———. "Testimonies: the making and reading of native South American historical sources." In *The Cambridge History of the Native Peoples of the Americas,* edited by Frank Salomon and Stuart B. Schwartz, vol. III, South America, Part 1, 19-95. Cambridge: Cambridge University Press, 1999.

———. "'¡Huayra huayra pichcamanta!': augurio, risa y regeneración en la política tradicional (Pacota, Huarochirí)." *Bulletin de l'Institut Français d'Etudes Andines* 31, n°1 (2002): 1-22.

———. *The cord keepers: Khipus and cultural life in a Peruvian village.* Durham, London: Duke University Press, 2004.

Sánchez-Concha Barrios, Rafael. "El licenciado Hernando de Santillán y sus observaciones en torno de las formas tiránicas de los curacas". *Histórica* 20, n°2 (1996): 285-302.
Schechner, Richard, and Appel, Willa, eds. *By means of performance: Intercultural studies of theatre and ritual.* Cambridge: Cambridge University Press, 1990.
Schneider, David M. *A critique of the study of kinship.* Ann Arbor: University of Michigan Press, 1984.
Seeger, Anthony. *Nature and society in Central Brazil: the Suya Indians of Mato Grosso.* Cambridge, MA: Harvard University Press, 1981.
Seibold, Katharine E. "Textiles and cosmology in Choquecancha, Cuzco, Peru." In *Andean cosmologies through time, persistence and emergence*, edited by Robert V. H Dover, Katharine E. Seibold, and John H. McDowell, 166-201. Bloomington: Indiana University Press, 1992.
Seler, Eduard G. "Über die sociale Stellung des Khapak Inca" [1894]. In *Compte rendu du Congrès International des Américanistes*, dixième session, 9-19. Stockholm: Ivar Hæggström, 1915.
Shapiro, Judith R. "Marriage rules, marriage exchange, and the definition of marriage in lowland South American societies." In *Marriage practices in lowland South America*, edited by Kenneth M. Kensinger, Illinois Studies in Anthropology 14, 1-30. Urbana, Chicago: University of Illinois Press, 1984.
Shaver Dwight, and Shaver, Gwynne. "Bosquejo de la metatesis en el Quechua de Incahuasi." In *Estudios ethno-lingüístico*, edited by Stephen Parker, 5-14. Lima: Instituto Lingüístico de Verano y el Ministerio de Educación, 1989.
Sherbondy, Jeanette. *Cuzco: aguas y poder.* Cuzco: Centro de Estudios Rurales Andinos Bartolomé de Las Casas, 1979.
———. "Organización hidráulica y poder en el Cuzco de los Incas." *Revista Española de Antropología Americana* 17 (1987): 117-151.
———. "Water Ideology in Inca Ethnogenesis." In *Andean cosmologies through time, persistence and emergence*, edited by Robert V. H Dover, Katharine E. Seibold, and John H. McDowell, 46-66. Bloomington: Indiana University Press, 1992.
———. "Panaca lands: re-invented communities." *Journal of the Steward Anthropological Society: Structure, knownledge and representation in the Andes* 24, n°1-2 (1996): 173-201.
Sikkink, Lynn. "Water and exchange: the ritual of *yaku cambio* as communal and competitive encounter." *American Ethnologist* 24, n°1 (1997): 170-189.
Sillar, Bill. "Caminando a través del tiempo: geografías sagradas en Cacha/Raqchi, departamento del Cuzco (Perú)." *Revista Andina* 35 (2002): 221-246.
Silverblatt, Irene. *Moon, sun and witches, gender ideologies and class in Inca and Colonial Perú.* Princeton: Princeton University Press, 1987.
———. "Imperial dilemnas, the politics of kinship, and Inca reconstructions of History", *Comparative Studies in Society and History* 30, n°1 (1988): 83-102.
Silverman, Gail P. *El tejido Andino, un libro de sabiduría.* Lima: Banco Central de Reserva del Perú, 1994.
Spalding, Karen. *De indio a campesino: Cambios en la estructura social del Perú colonial.* Lima: Instituto de Estudios Peruanos, 1974.
Sperber, Dan. *On anthropological knowledge.* Cambridge, Paris: Cambridge University Press, Maison de Sciences de l'Homme, 1985.
———. *Explaining culture: a naturalistic approach.* Oxford, Cambridge MA: Blackwell Publishers, 1996.
Stahl, Peter W. "Pre-Columbian Andean animal domesticates at the edge of empire." *World Archaeology* 34, n°3 (2003): 470-483.
Staller, John Edward. "Dimensions of place: the significance of centers to the development of Andean civilization: an exploration of the *ushnu* concept." In *Pre-Columbian land-*

scapes of creation and origin, edited by John Edward Staller, 269-313. New York: Springer, 2008.

Stanish, Charles. *Ancient Andean political economy*. Austin: University of Texas Press, 1992.

———. "Nonmarket imperialism in the Prehispanic Americas: the Inka occupation of the Titicaca basin." *Latin American Antiquity* 8, n°3 (1997): 195-216.

———. *Ancient Titicaca: the evolution of complex societies in southern Peru and northern Bolivia*. Berkeley: University of California Press, 2003.

Stern, Steve. *Peru's Indian peoples and the challenge of Spanish conquest. Huamanga to 1640*. Madison: University of Wisconsin Press, 1993.

Stirling, Stuart. *The last conquistador, Mansio Serra de Leguizamón and the conquest of the Incas*. Phoenix Mill: Sutton Publishing, 1999.

Stobart, Henry. "The llama's flute: musical misunderstandings in the Andes", *Early Music* 24, n°3 (1996): 470-482.

———. *Music and the poetics of production in the Bolivian Andes*, SOAS Musicology series. Aldershot: Ashgate, 2006.

Szemiński, Jan. "Wanakawri waka" in *El culto estatal del Imperio Inca*, edited by Mariusz Ziółkowski. Warsaw: University of Warsaw, 1991.

———. *Wira Quchan y sus obras, teología andina y lenguaje, 1550-1662*. Lima: Instituto de Estudios Peruanos, Banco Central de Reserva del Perú, 1997.

Taylor, Gerald. "Los dialectos Quechuas de Yauyos", *Bulletin de l'Institut Français d'Etudes Andines* 19, n°2 (1990): 293-325.

———. *Camac, camay y camasca y otros estudios sobre Huarochirí y Yauyos*. Lima, Cusco: Institut Français d'etudes Andines, Centro de Estudios Regionales Andinos Bartolomé de Las Casas, 2000.

———. *El Sol, la luna y las estrellas no son Dios...La evangelización en quechua (siglo XVI)*. Lima: Institut Français d'Etudes Andines, Fondo Editorial de la Pontificia Universidad Católica del Perú, 2003.

Thomas, Nicholas and Humphrey, Caroline. *Shamanism, History and the State*. Ann Arbor: University of Michigan Press, 1994.

Tomoeda, Hiroyasu. "Los ritos contempóraneos de camélidos y la ceremonia de la Citua." In *El mundo ceremonial andino*, edited by Luis Millones and Yoshio Onuki, Senri Ethnological Studies 37, 289-306. Osaka: Museo Nacional de Ethnología, 1993.

———. "The concept of vital energy among Andean pastoralists." In *Redefining Nature: ecology, culture and domestication*, edited by Roy F. Ellen and Katsu-Yoshi Fukui, 187-212. Oxford, Washington: Berg, 1996.

———. "Un poderoso espectáculo: el estado Inca visto desde un rincón de las punas." *Perspectivas Latinoamericanas* 1 (2004): 174-192.

Topic, John R., Topic, Theresa L., and Calva, Alfredo Melly. "Catequil: the archaeology, ethnohistory, and ethnography of a major provincial huaca." In *Andean archaeology: variations in sociopolitical organization*, edited by William H. Isbell and Helaine Silverman, 303-335. New York: Plenum Press, 2002.

Torero, Alfredo. *El Quechua y la historia social andina*. Lima: Universidad Ricardo Palma, 1974.

———. "Procesos lingüísticos e identificación de dioses en los Andes." *Revista Andina* 15, n°1 (1990): 237-263.

———. *Idiomas de los Andes. Lingüística y Historia*. Lima: Institut Français d'Etudes Andines, 2002.

Tschopik, Harry. "The Aymara." In *Handbook of South American Indians*, edited by Julian H. Steward, vol. 2, 501- 573. Washington, DC: Smithsonian Institution, Bureau of American Ethnology, 1946.

———. *The Aymara of Chucuito, Peru, I: Magic*. New York: Anthropological Papers of the American Museum of Natural History 44, n°2, 1951.

Turner, Terence. "Dual opposition, hierarchy, and value. Moiety structure and symbolic polarity in central Brazil and elsewhere." In *Différence, valeurs, hiérarchie. Textes offert à Louis Dumont*, edited by Jean-Claude Galey, 335-370. Paris: Ecole des Hautes Etudes en Sciences Sociales, 1984.

———. "Social complexity and recursive hierarchy in indigenous South American societies." *Journal of the Steward Anthropological Society* 24, n°1-2 (1996): 37-59.

Turner, Victor. *The ritual process: structure and anti-structure*. Ithaca: Cornell University Press, 1969.

———. *From ritual to theatre. The human seriousness of play*. New York: Performing Arts Journal Publications, 1982.

Urbano, Henrique. "Clases sacerdotales y sociedades andinas." In *Rituales y Fiestas de las Américas*, 45° Congreso Internacional de Americanistas, 90-102. Bogotá: Ediciones Uniandes, 1988.

Urton, Gary. "Orientation in Quechua and Incaic astronomy." *Ethnology* 17, n°2 (1978): 157-167.

———. "Celestial crosses: the cruciform in Quechua astronomy." *Journal of Latin American Lore* 6, n°1 (1980): 87-110.

———. *At the crossroads of the earth and the sky: an Andean cosmology*. Austin: University of Texas Press, 1981.

———. "Calendrical cycles and their projections in Pacariqtambo, Perú." *Journal of Latin American Lore* 12, n°1 (1986): 45-64.

———. *The history of a myth, Pacariqtambo and the origin of the Inkas*. Austin: University of Texas Press, 1990.

———. "Moieties and ceremonialism in the Andes: the ritual battles of the Carnival season in Southern Peru." In *El mundo ceremonial andino*, edited by Luis Millones and Yoshio Onuki, Senri Ethnological Studies 37, 117-142. Osaka: Museo Nacional de Etnología, 1993.

———. *The social life of numbers. A Quechua ontology of numbers and philosophy of arithmetic*. Austin: University of Texas Press, 1997.

———. "From knots to narratives: reconstructing the art of historical record keeping in the Andes from Spanish transcriptions of Inka *khipus*." *Ethnohistory* 45, n°3 (1998): 409-438.

———. *Signs of the Inka khipu, binary coding in the Andean knotted-string records*. Austin: University of Texas Press, 2003.

Valcárcel, Luis E. *Etnohistoria del Perú antiguo*. Lima: Universidad Nacional Mayor de San Marcos, 1967.

Valeri, Valerio. "Constitutive history: genealogy and narrative in the legitimation of Hawaiian kingship." In *Culture through time, anthropological approaches*, edited by Emiko Ohnuki Tierney, 154-192. Stanford, CA: Stanford University Press, 1990a.

———. "Diarchy and history in Hawaii and Tonga." In *Culture and History in the Pacific*, edited by Jukka Siikala, TAFAS 27, 45-79. Helsinski: The Finish anthropological society, 1990b.

Van Den Berg, Hans. *La tierra no da así no más, los ritos agrícolas en la religión de los Aymara-Cristianos*. Amsterdam: CEDLA, 1989.

Villanueva Urteaga, Horacio, ed. *Cuzco 1689, economía y sociedad en el sur andino*. Cuzco: Centro de Estudios Andinos Bartolomé de las Casas, 1982.

Villarías Robles, Juan José R. *El sistema económico del imperio inca. Historia crítica de una controversia*. Madrid: CSIC, Centro de estudios Históricos, 1998.

Vivanco Guerra, Alejandro. "Una herranza en el valle de Chancay: San Juan de Viscas en 1963 de los cuadernos de campos de Alejandro Vicanco." *Anthropologica* 19 (2001): 335-345.

Wachtel, Nathan. "Structuralisme et histoire: à propos de l'organisation sociale de Cuzco." *Annales E.S.C.* 21, n°1 (1966): 71-94.

Webster, Steven S. "Parentesco y afinidad en una comunidad indígena Quechua." In *Parentesco y matrimonio en los Andes*, edited by Enrique Mayer and Ralph Bolton, 183-245. Lima: Pontificia Universidad Católica del Perú, 1980.

———. "Interpretation of an Andean social and economic formation." *Man* 16, n°4 (1981): 616-633.

Yaya, Isabel. "The importance of initiatory ordeals: kinship and politics in an Inca narrative." *Ethnohistory* 55, n°1 (2008): 51-85.

Zerubavel, Eviatar. *Hidden rhythms: schedules and calendars in social life*. Chicago: University of Chicago Press, 1981.

Ziólkowski, Mariusz S. "Las fiestas del calendario metropolitano." *Ethnologia Polona* 13 (1987): 183-217, 14 (1988): 221-258.

———. "Knots and oddities: The quipu-calendar or supposed Cuzco luni-sidereal calendar" in *Time and calendars in the inca empire*, edited by Mariusz Ziólkowski and Robert Sadowski, 197-208. Oxford: British Archaeological Reports, 1989.

———. *La guerra de los Wawqi, los objectivos y los mecanismos de la rivalidad dentro de la élite inka, siglos XV-XVI*. Quito: Abya Yala, 1996.

Ziólkowski, Mariusz S., and Sadowski, Robert M. "Los problemas de la reconstrucción de los calendarios prehispánicos andinos." *Estudios Latinoamericanos* 9 (1982-84): 45-87.

———. *Time and calendars in the Inca empire*. Oxford: B.A.R International series 479, 1989.

———. *La arqueoastronomía en la investigación de las culturas andinas*. Quito: Instituto Otavaleño de Antropología, Banco Central del Ecuador, 1992.

Zorn, Elayne. "Dangerous encounters: ritual battles in Andean Bolivia." In *Combat, ritual and performance, Anthropology of the martial arts*, edited by David E. Jones, 120-152. Westport, London: Praeger, 2002.

Zuidema, R. Tom. *The ceque system of Cuzco, the social organization of the capital of the Inca*. Leiden: E. J. Brill, 1964.

———. "Kinship and ancestorcult three peruvian communities." *Bulletin de l'Institut Français d'Etudes Andines* 2, n°1 (1973): 16-33.

———. "La imágen del Sol y la Huaca de Susurpuquio en el sistema astronómico de los Incas en el Cuzco." *Journal de la Société des Américanistes* 63-64 (1976): 199-230.

———. "Shaft-tombs and the Inca empire." *Journal of the Steward Anthropological Society* 9, n°1-2 (1977a): 133-178.

———. "The Inca Calendar." In *Native American Astronomy*, edited by Anthony F. Aveni, 219-259. Austin: University of Texas Press, 1977b.

———. "Parentesco Inca." In *Parentesco y matrimonio en los Andes*, edited by Enrique Mayer and Ralph Bolton, 57-113. Lima: Pontificia Universidad Católica del Perú, 1980.

———. "Inca observations of the solar and lunar passages through zenith and anti-zenith at Cuzco." In *Archaeoastronomy in the Americas*, edited by Ray R. Williamson, 319-342. Maryland: Ballena Press/Center for Archaeoatronomy, 1981.

———. "The sidereal lunar calendar of the Incas." In *Archaeoastronomy in the New World*, edited by Anthony F. Aveni, 59-107. Cambridge: Cambridge University Press, 1982a.

———. "Bureaucracy and systematic knowledge in Andean civilization." In *The Inca and Aztec states 1400-1800: Anthropology and History*, edited by George A. Collier, Renato I. Rosaldo, and John D. Wirth, 419-458. New York: Academic Press, 1982b.

———. "Myth and history in ancient Peru." In *The logic of culture, advances in structural theory and method*, edited by Ino Rossi, 150-175. South Hadley: J. F. Begin Publishers, 1982c.

———. "Llama sacrifices and computation: the roots of the Inca calendar in Huari-Tiahuanaco culture." In *Acts of Congress on Ethnoastronomy*. Washington, DC, 1983a.

———. "Hierarchy and space in Incaic organization." *Ethnohistory* 30, n°2 (1983b): 49-75.

———. "The lion in the city: royal symbols of transition in Cuzco." In *Animal myths and metaphors in South America*, edited by Gary Urton, 183-246. Salt Lake City: University of Utah Press, 1985.

———. *La civilisation inca au Cusco*. Collège de France Essais et Conférences. Paris: PUF, 1986a.

———. "Inka dynasty and irrigation: another look at Andean concepts of history." In *Anthropological history of Andean polities*, edited by John V. Murra, Nathan Wachtel, and Jacques Revel, 177-200. Cambridge/Paris: Cambridge University Press/Maison des Sciences de l'Homme, 1986b.

———. "The place of the *chamay wariqsa* in the rituals of Cuzco." *Amerindia* 11 (1986c): 58-63.

———. "El juego de los Ayllus y el Amaru." [1967] In *Reyes y Guerreros. Ensayos de cultura andina*, 256-272. Lima: Fomciencas, 1989a.

———. "Lugares sagrados e irrigación: tradición histórica, mitos y rituales en el Cusco." [1978] In *Reyes y Guerreros. Ensayos de cultura andina*, 455-487. Lima: Fomciencas, 1989b.

———. "The moieties of Cuzco." In *The attraction of opposites, thought and society in the dualistic mode*, edited by David Maybury-Lewis and Uri Almagor, 255-275. Ann Arbor: University of Michigan Press, 1989c.

———. "El Ushnu." [1979] In *Reyes y Guerreros. Ensayos de cultura andina*, 402-454. Lima: Fomciencas, 1989e.

———. "Batallas rituales en el Cuzco colonial." In *Cultures et sociétés, Andes et Méso-Amériques: melanges en hommage à Pierre Duviols*, edited by Pierre Duviols and Raquel Thiercelin, 811-834. Aix en Provence: Université de Provence, 1991.

———. "Inca cosmos in Andean context, from the perspective of the Capac Raymi Camay Quilla feast celebrating the december solstice in Cuzco" In *Andean cosmologies through time, persistence and emergence*, edited by Robert V. H Dover, Katharine E. Seibold, John H. McDowell, 17-45. Bloomington: Indiana University Press, 1992.

———. "Fête-Dieu et fête de l'Inca, châtiment et sacrifice humain comme rite de communion." In *Le Corps de Dieu en fêtes*, edited by Antoinette Molinié, 175-216. Paris: Les Editions du Cerf, 1996.

———. "La política matrimonial incaica según Juan de Betanzos: un ejemplo implicando a los reyes Inca Roca y Yahuar Huacac." In *Arqueología, antropología e Historia en los Andes: Homenaje a María Rostworowski*, edited by Varón Gabai Rafael, Flores Espinoza Javier, 289-300. Lima: Instituto de Estudios Peruanos, 1997a.

———. "Cosmovisión inca y astronomía en el Cuzco: nuevo año agrícola y sucesión real." In *Pensar América : cosmovisión mesoamericana y andina*, edited by Garrido Aranda A., 249-270. Córdoba: Caja Sur Publicaciones, Ayuntamiento de Montilla, 1997b.

———. "La organización religiosa del sistema de panacas y memoria en el Cusco incaico." In *Incas e Indios cristianos. Elites indígenas e identidades cristianas en los Andes coloniales*, edited by Jean-Jacques Decoster, Travaux de l'Institut Français d'Etudes Andines 149, 19-37. Cuzco: Centro Bartolomé de Las Casas, Institut Français d'Etudes Andines, Asociación Kuraka, 2002.

———. *El calendario inca. Tiempo y espacio en la organización ritual del Cuzco*. Lima: Fondo Editorial del Congreso del Perú, Fondo Editorial de la Pontificia Universidad Católica del Perú, 2011.

Zuidema, R. Tom, and Urton, Gary. "La constelación de la Llama en los Andes peruanos." *Allpanchis Phuturinqa* 9 (1976): 59-119.

INDEX

www.ingramcontent.com/pod-product-compliance
Lightning Source LLC
LaVergne TN
LVHW020651300926
849161LV00004B/904
* 9 7 8 9 0 0 4 2 3 3 8 5 0 *